Methods in Molecular Biology

For further volumes:
http://www.springer.com/series/7651

For over 35 years, biological scientists have come to rely on the research protocols and methodologies in the critically acclaimed *Methods in Molecular Biology* series. The series was the first to introduce the step-by-step protocols approach that has become the standard in all biomedical protocol publishing. Each protocol is provided in readily-reproducible step-by-step fashion, opening with an introductory overview, a list of the materials and reagents needed to complete the experiment, and followed by a detailed procedure that is supported with a helpful notes section offering tips and tricks of the trade as well as troubleshooting advice. These hallmark features were introduced by series editor Dr. John Walker and constitute the key ingredient in each and every volume of the *Methods in Molecular Biology* series. Tested and trusted, comprehensive and reliable, all protocols from the series are indexed in PubMed.

Essential Genes and Genomes

Methods and Protocols

Edited by

Ren Zhang

Center for Molecular Medicine and Genetics, Wayne State University School of Medicine, Detroit, MI, USA

Humana Press

Editor
Ren Zhang
Center for Molecular Medicine and
Genetics
Wayne State University School of Medicine
Detroit, MI, USA

ISSN 1064-3745 ISSN 1940-6029 (electronic)
Methods in Molecular Biology
ISBN 978-1-0716-1719-9 ISBN 978-1-0716-1720-5 (eBook)
https://doi.org/10.1007/978-1-0716-1720-5

This Humana imprint is published by the registered company Springer Science+Business Media, LLC, part of Springer Nature.
The registered company address is: 1 New York Plaza, New York, NY 10004, U.S.A.

Preface

It has been my pleasure to serve as the editor of the book *Essential Genes and Genomes: Methods and Protocols*. Essential genes refer to protein-coding genes and other genomic elements that are required by an organism to survive under specific conditions. Back in 2003, I was discussing with Prof. Chun-Ting Zhang about whether it was a good idea to establish a database of essential genes (DEG), which were, at the time, a novel concept. It was a time when genome-scale essentiality screening was still not available, and thus there were only a low number of essential-gene entries, most of which were determined by single-gene knockout studies. But believing this was a valuable project, we established the DEG. Since then, the number of determined essential genes has increased by more than 100-fold, and the research on gene essentiality screenings in bacteria, archaea, and eukaryotes has advanced significantly.

The progress in essential-gene research has been largely driven by its fundamental and practical implications. The determination of the minimal-gene-set for a living cell addresses the question of what cellular functions are necessary to sustain life. Genome-wide essentiality screenings have elucidated the molecular underpinnings of many biological processes in both prokaryotes and eukaryotes. Interest in essentiality screenings has also been boosted by synthetic biology, which aims to make an artificial self-sustainable living cell. The minimal-gene-set of a bacterium is considered a chassis for further addition of other parts with desirable traits. In terms of practical implications, bacterial essential genes are attractive antibiotic targets, as inhibiting these genes can suppress bacterial survival. Gene essentiality screenings across human cell types can reveal genes that encode factors for regulating tissue-specific cellular processes, and such screenings in cancer cells can disclose factors that determine cancer phenotypes, thus revealing important targets for cancer therapies.

Another driving force for the research on essential genes has been the development of biotechnology. Traditionally, gene essentiality was assessed by deleting the gene (known as knockout) in the genome using gene targeting, a Nobel Prize-winning technology that is based on homologous recombination. Gene targeting technology has been indispensable for determining gene functions in various organisms, such as mice. In terms of genome-wide gene essentiality screenings, however, gene deletion one at a time can be time-consuming and labor-intensive. The recent invention of two technologies, transposon insertion sequencing (Tn-seq) and CRISPR/Cas9, has greatly accelerated the research on the gene essentiality screenings in various species. Tn-seq, invented in 2009, is a method to perform saturated transposon mutagenesis in a genome, followed by parallel sequencing to determine the transposon integration sites, and therefore genes contributing to the phenotypic changes can be identified.

The 2020 Nobel Prize in Chemistry has been awarded to two scientists who pioneered the CRISPR/Cas9 technology, a powerful genome-editing tool that can change the DNA of microorganisms, plants, and animals with high precision. The invention of CRISPR/Cas9 technology has revolutionized biological research in many fields, with gene essentiality screenings being no exception, because Cas9-mediated gene editing is scalable, enabling the examination of gene functions at the systems level. Therefore, as of 2020, according to DEG 15, the current version, most essential genes in prokaryotes and eukaryotes were determined using Tn-seq and CRISPR/Cas9, respectively.

This book provides comprehensive, state-of-the-art information on gene essentiality screenings in a wide variety of organisms. The book opens with chapters that describe CRISPR/Cas9-based approaches, followed by chapters that focus on Tn-seq-based approaches for gene essentiality screenings. The book also contains chapters describing methods that differ from these two categories. The final chapters describe bioinformatic approaches for gene essentiality studies.

This book was completed at the end of 2020. History will tell that the year 2020 is by no means ordinary. Human society has been profoundly impacted by the COVID-19 pandemic. The scale of the changes that people worldwide have experienced, such as shutdowns, economic downturns, social distancing, and the now widespread shift to working from home, has been unprecedented in human history. Many of the authors who contributed to this volume had to delay their submissions as a result of caring for COVID-19 patients, working on COVID-19 manuscripts, or even taking care of kids due to school closures. Therefore, I am grateful to the authors not only for contributing their expertise that made the book possible but also for their patience and understanding in awaiting the completion of this volume during this difficult time. I would also like to thank Dr. John Walker for his support throughout the process in editing the book.

Detroit, MI, USA ***Ren Zhang***

Contents

Contributors

CARMEN AMARO • *Department of Microbiology and Ecology, University of Valencia, Valencia, Spain*

MUSTAPHA AOUIDA • *Division of Biological and Biomedical Sciences, College of Health and Life Sciences, Hamad Bin Khalifa University, Qatar Foundation, Doha, Qatar*

MICHAEL AREGGER • *Center for Cancer Research, National Cancer Institute, Frederick, MD, USA; Donnelly Centre for Cellular and Biomolecular Research, University of Toronto, Toronto, ON, Canada*

JESSE BAKKE • *Department of Foundational Sciences, College of Medicine, Central Michigan University, Mount Pleasant, MI, USA*

FRANCESCA BOLDRIN • *Department of Molecular Medicine, University of Padova, Padova, Italy*

FRANCESCA BOTTACINI • *School of Microbiology and APC Microbiome Ireland, University College Cork, Cork, Ireland*

AMY CAIN • *Department of Molecular Sciences, Macquarie University, Sydney, NSW, Australia*

JASON R. CANTOR • *Morgridge Institute for Research, Madison, WI, USA; Department of Biochemistry, University of Wisconsin-Madison, Madison, WI, USA; Department of Biomedical Engineering, University of Wisconsin-Madison, Madison, WI, USA; University of Wisconsin Carbone Cancer Center, Madison, WI, USA*

MIGUEL CARDA-DIÉGUEZ • *Department of Microbiology and Ecology, University of Valencia, Valencia, Spain*

KATHERINE S. K. CHAN • *Donnelly Centre for Cellular and Biomolecular Research, University of Toronto, Toronto, ON, Canada*

JEFFREY SHIH-CHIEH CHU • *Wuhan Frasergen Bioinformatics, Wuhan East Lake High-tech Zone, Wuhan, China*

ANNE E. CLATWORTHY • *Department of Molecular Biology and Center for Computational and Integrative Biology, Massachusetts General Hospital, Boston, MA, USA; Department of Genetics, Harvard Medical School, Boston, MA, USA; Broad Institute of MIT and Harvard, Cambridge, MA, USA*

PATRICK D. CURTIS • *Department of Biology, University of Mississippi, University, MS, USA*

DIRK DAELEMANS • *KU Leuven Department of Microbiology, Immunology and Transplantation, Laboratory of Virology and Chemotherapy, Rega Institute for Medical Research, Leuven, Belgium*

LUKE DEHART • *Department of Foundational Sciences, College of Medicine, Central Michigan University, Mount Pleasant, MI, USA*

TIMOTHY J. DONOHUE • *Wisconsin Energy Institute and Great Lakes Bioenergy Research Center, University of Wisconsin-Madison, Madison, WI, USA; Department of Bacteriology, University of Wisconsin-Madison, Madison, WI, USA*

KYLE M. FLICKINGER • *Morgridge Institute for Research, Madison, WI, USA; Department of Biochemistry, University of Wisconsin-Madison, Madison, WI, USA*

TATSUO FUKAGAWA • *Graduate School of Frontier Biosciences, Osaka University, Suita, Osaka, Japan*

FENG GAO • *Department of Physics, School of Science, Tianjin University, Tianjin, China; Frontiers Science Center for Synthetic Biology and Key Laboratory of Systems Bioengineering (Ministry of Education), Tianjin University, Tianjin, China; SynBio Research Platform, Collaborative Innovation Center of Chemical Science and Engineering (Tianjin), Tianjin, China*

FENG-BIAO GUO • *School of Pharmaceutical Sciences, Wuhan University, Wuhan, China*

KIMBERLY S. HUGGLER • *Morgridge Institute for Research, Madison, WI, USA; Department of Biochemistry, University of Wisconsin-Madison, Madison, WI, USA*

DEBORAH T. HUNG • *Department of Molecular Biology and Center for Computational and Integrative Biology, Massachusetts General Hospital, Boston, MA, USA; Department of Genetics, Harvard Medical School, Boston, MA, USA; Broad Institute of MIT and Harvard, Cambridge, MA, USA*

THOMAS R. IOERGER • *Department of Computer Science, Texas A&M University, College Station, TX, USA*

BERT KWANTEN • *KU Leuven Department of Microbiology, Immunology and Transplantation, Laboratory of Virology and Chemotherapy, Rega Institute for Medical Research, Leuven, Belgium*

PIYUSH BEHARI LAL • *Wisconsin Energy Institute and Great Lakes Bioenergy Research Center, University of Wisconsin-Madison, Madison, WI, USA*

XIHAO LIAO • *MOE Key Laboratory for Industrial Biocatalysis, Institute of Biochemical Engineering, Department of Chemical Engineering, Tsinghua University, Beijing, China*

GUI-MING LIN • *State Key Laboratory of Freshwater Ecology and Biotechnology, Institute of Hydrobiology, Chinese Academy of Sciences, Wuhan, Hubei Province, China*

YUQIU LU • *Department of Nephrology, Tongji Hospital, Tongji University, Shanghai, China*

HAO LUO • *Department of Physics, School of Science, Tianjin University, Tianjin, China*

BARBARA MAIR • *Boehringer Ingelheim, Vienna, Austria; Donnelly Centre for Cellular and Biomolecular Research, University of Toronto, Toronto, ON, Canada*

RICCARDO MANGANELLI • *Department of Molecular Medicine, University of Padova, Padova, Italy*

ALEX J. MCCARTHY • *Department of Infectious Disease, MRC Centre for Molecular Bacteriology & Infection, Imperial College London, London, UK*

JASON MOFFAT • *Donnelly Centre for Cellular and Biomolecular Research, University of Toronto, Toronto, ON, Canada; Department of Molecular Genetics, University of Toronto, Toronto, ON, Canada; Institute of Biomaterials and Biomedical Engineering, University of Toronto, Toronto, ON, Canada*

KEVIN S. MYERS • *Wisconsin Energy Institute and Great Lakes Bioenergy Research Center, University of Wisconsin-Madison, Madison, WI, USA*

JASPER EDGAR NEGGERS • *KU Leuven Department of Microbiology, Immunology and Transplantation, Laboratory of Virology and Chemotherapy, Rega Institute for Medical Research, Leuven, Belgium; Promakhos Therapeutics, Pagliuca Harvard Life Lab, Allston, Massachusetts, United States of America*

KOHEI NISHIMURA • *Graduate School of Frontier Biosciences, Osaka University, Suita, Osaka, Japan; Division of Biological Science, Graduate School of Science, Nagoya University, Nagoya, Japan*

TIAN-CAI NIU • *State Key Laboratory of Freshwater Ecology and Biotechnology, Institute of Hydrobiology, Chinese Academy of Sciences, Wuhan, Hubei Province, China*

Daniel R. Noguera • *Wisconsin Energy Institute and Great Lakes Bioenergy Research Center, University of Wisconsin-Madison, Madison, WI, USA; Department of Civil & Environmental Engineering, University of Wisconsin-Madison, Madison, WI, USA*

Bradley E. Poulsen • *Department of Molecular Biology and Center for Computational and Integrative Biology, Massachusetts General Hospital, Boston, MA, USA; Department of Genetics, Harvard Medical School, Boston, MA, USA; Broad Institute of MIT and Harvard, Cambridge, MA, USA*

Dindial Ramotar • *Division of Biological and Biomedical Sciences, College of Health and Life Sciences, Hamad Bin Khalifa University, Qatar Foundation, Doha, Qatar*

Nicholas J. Rossiter • *Morgridge Institute for Research, Madison, WI, USA*

Lorena Ruiz • *Department of Microbiology and Biochemistry of Dairy Products, Dairy Research Institute of Asturias (IPLA)-Spanish National Research Council (CSIC), Villaviciosa, Asturias, Spain; Instituto de Investigación Sanitaria del Principado de Asturias (ISPA), Oviedo, Asturias, Spain*

Lucie Semenec • *Department of Molecular Sciences, Macquarie University, Sydney, NSW, Australia*

Rebecca S. Shapiro • *Department of Molecular and Cellular Biology, University of Guelph, Guelph, ON, Canada*

Gayatri Sharma • *Department of Biology, University of Mississippi, University, MS, USA*

Shaolin Shi • *National Clinical Research Center for Kidney Diseases, Jinling Hospital, Nanjing University School of Medicine, Nanjing, China*

Robert C. Shields • *Department of Oral Biology, College of Dentistry, University of Florida, Gainesville, FL, USA; Department of Biological Sciences, College of Sciences and Mathematics, Arkansas State University, Jonesboro, AR, USA*

Peter W. Taylor • *School of Pharmacy, University College London, London, UK*

Amy H. Y. Tong • *Donnelly Centre for Cellular and Biomolecular Research, University of Toronto, Toronto, ON, Canada*

Douwe van Sinderen • *School of Microbiology and APC Microbiome Ireland, University College Cork, Cork, Ireland*

Yaligara Veeranagouda • *GLAVAHCS, Los Angeles, CA, USA; David Geffen School of Medicine, University of California, Los Angeles, Los Angeles, CA, USA*

Alejandro R. Walker • *Department of Oral Biology, College of Dentistry, University of Florida, Gainesville, FL, USA*

Dan Wang • *Department of Physics, School of Science, Tianjin University, Tianjin, China*

Wen Wei • *School of Life Science and Technology, University of Electronic Science and Technology of China, Chengdu, China*

Qing-Feng Wen • *School of Life Science and Technology, University of Electronic Science and Technology of China, Chengdu, China*

Lauren Wensing • *Department of Molecular and Cellular Biology, University of Guelph, Guelph, ON, Canada*

Hannah M. Wexler • *GLAVAHCS, Los Angeles, CA, USA; David Geffen School of Medicine, University of California, Los Angeles, Los Angeles, CA, USA*

Hao Wu • *Department of Physics, School of Science, Tianjin University, Tianjin, China*

Xin-Hui Xing • *MOE Key Laboratory for Industrial Biocatalysis, Institute of Biochemical Engineering, Department of Chemical Engineering, Tsinghua University, Beijing, China; Center for Synthetic and Systems Biology, Tsinghua University, Beijing, China*

Tong Yang • *Department of Physics, School of Science, Tianjin University, Tianjin, China*

ZHI-KAI YANG • *The Fifth Affiliated Hospital of Guangzhou Medical University, Guangzhou, China*
OLIVER P. YOCKEY • *Department of Foundational Sciences, College of Medicine, Central Michigan University, Mount Pleasant, MI, USA*
SHICHENG YU • *Guangzhou Institutes of Biomedicine and Health, Chinese Academy of Sciences, Guangzhou, China; Guangzhou Regenerative Medicine and Health Guangdong Laboratory, Bioland Laboratory, Guangzhou, China; Key Laboratory of Combinatorial Biosynthesis and Drug Discovery, School of Pharmaceutical Sciences, Wuhan University, Wuhan, China*
CHENG-CAI ZHANG • *State Key Laboratory of Freshwater Ecology and Biotechnology, Institute of Hydrobiology, Chinese Academy of Sciences, Wuhan, Hubei Province, China*
CHONG ZHANG • *MOE Key Laboratory for Industrial Biocatalysis, Institute of Biochemical Engineering, Department of Chemical Engineering, Tsinghua University, Beijing, China; Center for Synthetic and Systems Biology, Tsinghua University, Beijing, China*
JU-YUAN ZHANG • *State Key Laboratory of Freshwater Ecology and Biotechnology, Institute of Hydrobiology, Chinese Academy of Sciences, Wuhan, Hubei Province, China*
CHAORAN ZHENG • *Key Laboratory of Combinatorial Biosynthesis and Drug Discovery, School of Pharmaceutical Sciences, Wuhan University, Wuhan, China*

Chapter 1

A Method to Map Gene Essentiality of Human Pluripotent Stem Cells by Genome-Scale CRISPR Screens with Inducible Cas9

Barbara Mair, Michael Aregger, Amy H. Y. Tong, Katherine S. K. Chan, and Jason Moffat

Abstract

Human pluripotent stem cells (hPSCs) have the capacity for self-renewal and differentiation into most cell types and, in contrast to widely used cell lines, are karyotypically normal and non-transformed. Hence, hPSCs are considered the gold-standard system for modelling diseases, especially in the field of regenerative medicine. Despite widespread research use of hPSCs and induced pluripotent stem cells (iPSCs), the systematic understanding of pluripotency and lineage differentiation mechanisms are still incomplete. Before tackling the complexities of lineage differentiation with genetic screens, it is critical to catalogue the general genetic requirements for cell fitness and proliferation in the pluripotent state and assess their plasticity under commonly used culture conditions.

We describe a method to map essential genetic determinants of hPSC fitness and pluripotency, herein defined as cell reproduction, by genome-scale loss-of-function CRISPR screens in an inducible *S. pyogenes* Cas9 H1 hPSC line. To address questions of context-dependent gene essentiality, we include protocols for screening hPSCs cultured on feeder cells and laminin, two commonly used growth substrates. This method establishes parameters for genome-wide screens in hPSCs, making human stem cells amenable for functional genomics approaches to facilitate investigation of hPSC biology.

Key words Human pluripotent stem cells, Genome-wide CRISPR screen, Functional genomics, Essential genes, Growth substrate

1 Introduction

Human pluripotent stem cells (hPSCs) are of great interest as models to study diseases and developmental processes, as well as potential tools for regenerative medicine, due to their intrinsic potential to differentiate into most cell types found in the human body [1, 2]. In contrast to other widely used cell line models, they are capable of self-renewal despite being karyotypically normal and non-transformed. To allow widespread adoption of hPSC as a model system, it is important to understand the genetic

Ren Zhang (ed.), *Essential Genes and Genomes: Methods and Protocols*, Methods in Molecular Biology, vol. 2377, https://doi.org/10.1007/978-1-0716-1720-5_1,

determinants that govern their fitness, self-renewal and pluripotency. In order to systematically investigate these processes, genome-scale functional genomics approaches are typically employed. However, the challenging and costly hPSC culture requirements have delayed the implementation of such technologies in the field.

As a first step for systematically studying hPSC biology in depth, it is necessary to catalogue the general genetic requirements for cell fitness and proliferation in the pluripotent "ground state" under commonly used culture conditions. By definition, such essential genes orchestrate the core processes in a cell, and by now, extensive essentiality catalogues have become available for a number of model organisms, including yeast, bacteria, mouse, fruit fly, zebrafish and plants [3–6]. Until the advent of CRISPR technology, reliably and comprehensively cataloguing human essential genes was challenging due to the size and complexity of the human genome, but has now become routine for model cell lines [7, 8].

These experiments across hundreds of cell lines from different origins have also demonstrated that gene essentiality is not absolute, but rather context-dependent: whereas some genes are constitutively essential (i.e. core essential genes), others only impact cellular fitness in certain conditions or genetic backgrounds [9–15]. This suggests that identifying genetic dependencies in a certain cell line or tissue holds the key to deciphering fundamental processes specific to this context.

Here, we present a method employing CRISPR/Cas9-mediated loss-of-function screening to investigate core and context-dependent gene essentiality in hPSCs [16–19]. We describe the generation and characterization of karyotypically and phenotypically normal, genome editing proficient hPSC lines stably expressing Cas9 under a doxycycline-inducible promoter. We also provide a detailed protocol for genome-scale CRISPR screening in these cells on two different growth substrates. This includes guide RNA (gRNA) library amplification, lentivirus production, hPSC infection, propagation and sampling, genomic DNA isolation, sequencing library preparation, sample sequencing and data analysis. This method allows identification of sets of essential genes based on a reduction in proliferation, which encompasses phenotypes affecting cellular fitness and viability, but also differentiation due to slowed proliferation or self-renewal.

This method is also suitable for identifying genes required for other phenotypes of interest, including drug sensitivity, expression of a reporter gene or protein, or function of a specific pathway.

2 Materials

2.1 CRISPR gRNA Library

2.1.1 gRNA Library Selection and Amplification

1. pLCKO2::TKOv3 or other CRISPR gRNA library (e.g. www.addgene.org/crispr/libraries/),
2. gRNA plasmid library,
3. TE buffer: 10 mM Tris–HCl pH 7.5 or 8.0 containing 1 mM EDTA.
4. electrocompetent bacterial cells (e.g. Lucigen Endura).
5. recovery medium (or SOC medium).
6. LB agar plates containing carbenicillin (10 cm and 15 cm).
7. Liquid LB medium containing carbenicillin.
8. Maxi- or mega plasmid purification kit.
9. Sterile bacterial culture tubes and bottles.
10. Cell spreaders.
11. Electroporator with corresponding cuvettes.
12. Centrifuge with sterile buckets.
13. Shaking incubator.
14. Plate incubator.

2.1.2 gRNA Library Lentivirus Production

1. Class IIA biosafety cabinet.
2. Humidified incubator at 37 °C and 5% CO_2.
3. Standard tissue culture equipment, lentivirus-work-approved (*see* **Note 1**).
4. Mycoplasma-free, low passage HEK293T cells.
5. Low-antibiotic medium: DMEM (Dulbecco's Modification of Eagle's Medium) with 10% FBS (can be heat inactivated) and 0.1 penicillin/streptomycin.
6. High-BSA medium: DMEM with 1.1 g/l BSA (from 20 g/100 ml stock dissolved in DMEM, filter sterilized) and 1% penicillin/streptomycin.
7. Opti-MEM serum-free medium.
8. Transfection-quality psPAX2 (lentivirus packaging) and pMD2.G (lentivirus envelope) plasmids.
9. gRNA library plasmid pool.
10. X-tremeGENE 9 DNA transfection reagent (or transfection reagent of choice).
11. Sterile tubes (screw-cap).

2.2 Screening Cell Line

2.2.1 Cell Line Selection

1. biosafety cabinet, incubator and tissue culture equipment as above (also *see* **Note 1**).
   [cell line, growth substrate, media, etc., of choice, here:]
2. H1 human embryonic stem cell line (WA01).
3. Irradiated mouse embryonic fibroblasts.
4. Human embryonic stem cell (hESC) medium [20]: DMEM/F12, 20% knockout serum replacement (KOSR), 1% penicillin/streptomycin, 2 mM L-glutamine, 1% non-essential amino acids (NEAA), 55μM β-mercaptoethanol, 15 ng/ml human basic fibroblast growth factor (hbFGF, added immediately prior to use).
5. TrypLE Express.
6. ROCK inhibitor (ROCKi) Y-27632.
7. hESC freezing medium: 50% hESC medium with 40% FBS and 10% DMSO.
8. Recombinant human laminin 521.
9. PBS (phosphate-buffered saline) with Ca^{2+} and Mg^{2+}.
10. mTeSR1 complete medium.
11. ReLeSR enzyme-free dissociation reagent.
12. mFreSR medium.
13. Reagents for STR profiling (e.g. GenePrint 10 System).
14. Regents for karyotyping (*see*, e.g., www.cuh.nhs.uk).
15. PBS.
16. FACS buffer: PBS containing 2% FBS or BSA.
17. PBS containing 4% PFA.
18. PBS containing 0.1% TritonX-100.
19. 7-AAD viability dye.
20. Flow cytometer.
21. Centrifuges for different tube sizes.
22. Fluorescently labelled antibodies targeting KDR, CD235a, CD34, CD43, CD117, CD184, CD73, CD41, SSEA4, SSEA5, TRA-1-60, OCT3/4 and SOX2.
23. Mycoplasma testing, e.g. MycoAlert Detection Kit or e-Myco Plus Mycoplasma PCR Detection Kit.

2.2.2 Generation of Stable, Inducible Cas9-Expressing Cell Line

1. Biosafety cabinet, incubator and tissue culture equipment as above (also *see* **Note 1**).
2. Cell counting device and/or viability assay.
3. Quality control assays and reagents as described in sections above and below.
4. Sanger sequencing facilities.

5. Standard PCR reagents and thermocycler.
6. Primers for AAVS1 genomic locus.
7. Electroporation-grade plasmids stocks of AAVS1 T2 CRISPR in pX330 (Addgene plasmid #72833) and all-in-one targeting construct (AAVS1-CAG-TetOn3G-Cas9-P2A-BFP2, Addgene plasmid #125519).
8. Optional: AAVS1-targeting TALENs, GeneArt Genomic Cleavage Selection Kit.
9. Genomic DNA extraction kit.
10. Mammalian cell electroporator and corresponding cuvettes and reagents.
11. Primers for assessing correct targeting and backbone integrations:

Integration (left arm)	FW	GCTGCCGTCTCTCTCCTGAGTC
Integration (left arm)	RV	GCTTTAACAGAGAGAAGTTCGTGGCAC
Integration (right arm, nested PCR 1)	FW	CAGGCCGATGCTGTACTTCTTGTC
Integration (right arm, nested PCR 1)	RV	CAAAGCTGCCCATCTGGAGGAG
Integration (right arm, nested PCR 2)	FW	TCACTGATAGGGAGTAAACTCGAGGGAG
Integration (right arm, nested PCR 2)	RV	CTCCTCTCTCAAACCCAGGAGC
Backbone PCR (left arm)	FW	CTGCTCTTTGAGCCTGCAGACAC
Backbone PCR (left arm)	RV	GCTTTAACAGAGAGAAGTTCGTGGCAC
Backbone PCR (right arm)	FW	CAGGCCGATGCTGTACTTCTTGTC

(continued)

Backbone PCR (right arm)	RV	GATATGATCCTGATGCAGCTAGATTAACCC

12. Doxycycline.
13. G418.
14. PBS.
15. PBS containing 4% PFA.
16. PBS containing 0.5% TritonX-100.
17. PBS with 2% BSA and 0.05% Tween-20.
18. PBS containing 0.05% Tween-20.
19. DAPI (4′,6-diamidino-2-phenylindole).
20. Cas9 antibody (e.g. from Diagenode).
21. Fluorescently labelled secondary antibody.
22. Immunofluorescence microscope.

2.2.3 Prescreen Optimization, Cell Line Characterization and Quality Control Procedures

1. Biosafety cabinet, incubator and tissue culture equipment as above (also *see* **Note 1**).
2. Cell lines, growth substrates, culture media and reagents as above.
3. Cell counting device, viability assay.
4. Doxycycline.
5. Flow cytometer.
6. PBS.
7. FACS buffer: PBS containing 2% FBS or BSA.
8. Puromycin.
9. Transfection-grade lentiviral gRNA vectors (same vector backbone as selected gRNA library) targeting essential genes such as PSMD1 and PSMB2 and non-essential control regions (e.g. Addgene plasmids #74179, #74180, #74181, #74184, #74185, #74190 or sequences in Refs. 10, 19 for pLCKO).
10. Reagents for virus production and transduction as described above and below.

2.3 CRISPR Screen

2.3.1 Determination of Lentivirus Multiplicity of Infection (MOI)

1. Biosafety cabinet, incubator and tissue culture equipment as above (also *see* **Note 1**).
2. Cell lines, growth substrates, culture media and reagents as above.
3. Aliquot of the same gRNA library virus preparation (without prior thaws) that will be used in the screen.
4. Tissue culture plates with growth substrate or feeder cells.

5. PBS.
6. Puromycin.
7. Cell counting device.

2.3.2 gRNA Library Transduction, Cell Passaging, Sampling and Quality Control

1. Biosafety cabinet, incubator and tissue culture equipment as above (also *see* **Note 1**).
2. Cell lines, growth substrates, culture media and reagents as above.
3. gRNA library virus preparation (without prior thaws).
4. Tissue culture plates with growth substrate or feeder cells.
5. Quality control assays and reagents as described in sections above.
6. Doxycycline.
7. PBS.
8. Puromycin.
9. Cell counting device.
10. −20 or −80 °C storage space.
11. 1.5-ml screw cap tubes.

2.3.3 Genomic DNA Isolation

1. Dedicated workstation.
2. Water, 10% bleach, 70% ethanol for cleaning.
3. Wizard Genomic DNA Purification Kit (Promega) containing Nuclei Lysis Solution, Protein Precipitation Solution and DNA Rehydration Solution (or TE buffer: 10 mM Tris–HCl pH 7.5 or 8.0, 1 mM EDTA).
4. RNaseA, (20 mg/ml).
5. UltraPure water and pure 95% ethanol to prepare 70% ethanol for washing DNA.
6. 1.5-ml DNA LoBind tubes.
7. 50-ml tubes.
8. Swinging bucket rotor centrifuge.
9. Qubit 2.0 Fluorometer.
10. Qubit dsDNA BR Assay Kit with BR reagent and BR Buffer (or TE buffer as above).
11. 0.5-ml tubes.

2.3.4 Sequencing Library Preparation

1. Dedicated PCR1 and PCR2 work stations.
2. PCR hood.
3. Water (tap and purified), 10% bleach, 70% ethanol, 0.1 N HCl for cleaning.
4. UltraPure agarose.

5. SYBR Safe DNA stain.
6. UltraPure water.
7. 2× NEBNext Ultra II Q5 Master Mix.
8. Thermocycler.
9. Gel electrophoresis equipment.
10. Blue light transilluminator.
11. 0.2-ml PCR strip tubes.
12. Razor blades.
13. Gel extraction kit.
14. 1.5-ml DNA LoBind tubes.
15. PCR1 primers (*see* **Note 2**):

PCR1	FW	GAGGGCCTATTTCCCATGATTC
PCR1 pLCKO	RV	CAAACCCAGGGCTGCCTTGGAA

PCR2 primers with Illumina TruSeq adapters (i5 and i7 indices denoted by NNNNNNNN):

PCR2 pLCKO (U6)	FW	AATGATACGGCGACCACCGAGATCTACACNNNNNNNNACACTCTTTCCCTACACGACGCTCTTCCGATCTTTGTGGAAAGGACGAGGTACCG
PCR2 pLCKO (U6)	RV	CAAGCAGAAGACGGCATACGAGATNNNNNNNNGTGACTGGAGTTCAGACGTGTGCTCTTCCGATCTACTTGCTATTTCTAGCTCTAAAAC

16. Qubit 2.0 Fluorometer.
17. Qubit dsDNA HS Assay Kit with HS reagent and HS Buffer (or TE buffer as above).

2.3.5 Next-Generation Sequencing

1. Illumina HiSeq2500 or NextSeq550 (single-read 50−/75-cycle chemistry).
2. Standard primers for dual indexing.
3. For HiSeq 2500: HiSeq SR Cluster Kit v4 cBot, HiSeq SBS Kit v4; NextSeq 550: NextSeq 500/550 High Output Kit v2.5.
4. Sequencing library quality control equipment and reagents as needed (e.g. BioAnalyzer, TapeStation).

2.3.6 Data Analysis

1. MD5 hashing algorithm.
2. BOWTIE.
3. BAGEL or other pooled CRISPR screen analysis tool.
4. gRNA library design file.
5. Essential and non-essential reference gene sets (e.g. from github.com/hart-lab/bagel).
6. Quality control software as needed (e.g. Scikit-learn and Scipy Python libraries).

3 Methods

Many steps in this protocol are similar to a generic pooled lentiviral CRISPR-Cas9 screening protocol for mammalian cells that was recently published by our group [21, 22]. For convenience, we reiterate all required steps here, adapting the protocol specifically for hPSCs and updating some technical details.

3.1 CRISPR gRNA Library

3.1.1 gRNA Library Selection

1. For essential gene identification in a system of interest, we recommend using a genome-scale gRNA library. The following protocol will focus on the pooled lentiviral pLCKO2::TKOv3 library, a sequence-optimized human gRNA library containing 71,090 gRNAs targeting 18,053 human protein-coding genes with four gRNAs each [19, 21, 23]. The library also contains 142 gRNA sequences targeting EGFP, LacZ and luciferase as controls. In contrast to pLCV2::TKOv3 [21, 23], pLCKO2::TKOv3 is a two-component system, requiring Cas9 expression in the screening cell line. For hPSCs, this is beneficial due to the possibility of obtaining higher library virus titres.
2. In principle, the following protocol can also be used with other CRISPR libraries. We indicate modifications required for use of the pLCKO2::TKOv3 precursor libraries (pLCV2::TKOv3 and TKOv1) [10, 21, 23]. Further modifications may be required for different libraries.
3. All TKO libraries, as well as many other high-quality genome-scale and focused CRISPR gRNA libraries, can be obtained through Addgene (www.addgene.org/crispr/libraries/). We recommend their tutorials and library-associated publications, as well as suggestions herein to choose an appropriate library for the desired purpose.

3.1.2 gRNA Library Amplification

1. Dilute the gRNA plasmid library to 50 ng/μl in TE buffer (*see* **Note 3**).
2. Electroporate the library into electrocompetent bacteria according to the manufacturer's instructions. Set up 4 reactions

of 2μl diluted gRNA library to 25μl of bacteria. After electroporation, add 975μl recovery medium (or SOC medium) and transfer cells to a culture tube with an additional 1 ml of recovery medium. Recover cells in a shaking incubator at 200 rpm for 1 h at 37 °C.

3. Pool all 8 ml of recovered bacteria and mix well.
4. Prepare a dilution plate to estimate library representation: add 10μl recovered bacteria to 990μl of recovery medium, mix well and plate 20μl onto a prewarmed 10-cm LB agar plate containing carbenicillin. This represents a 40,000-fold dilution.
5. Plate the full library by spreading 400μl of recovered cells on each of 20 prewarmed 15-cm LB agar plates containing carbenicillin.
6. Incubate the dilution and library plates at 30 °C for 14–16 h.
7. Count the number of colonies on the dilution plate. Multiply this number by the dilution factor 40,000 to obtain the total number of colonies plated. Proceed if this number exceeds a library coverage of at least 200-fold (optimally 500–1000). For the TKOv3 library, this corresponds to a minimal colony number of 1.4×10^7 (=200 colonies per gRNA for 71,090 gRNAs). Sufficient coverage is crucial for maintaining full library representation.

   If the obtained number of colonies is lower than 200-fold library coverage, optimize the transformation protocol or increase the number of electroporation reactions.
8. Harvest colonies by adding 7 ml of LB medium containing carbenicillin per 15-cm plate and scraping bacteria off with a cell spreader. Transfer the harvested cells into a sterile bottle using a 10-ml pipette, pooling cells from all 20 plates. Rinse every scraped plate with an additional 5 ml of LB medium containing carbenicillin and add this to the bottle with the cells. Stir well for 1 h at room temperature.
9. Transfer cells to centrifuge bottles and harvest at 7000 × *g*. Remove the supernatant thoroughly. Weigh the wet cell pellet.
10. Purify the library plasmid pool using a maxi- or mega-scale plasmid purification kit. Perform multiple maxi or mega preps according to column capacity.

3.1.3 gRNA Library Lentivirus Production

1. All cell incubation steps are carried out in a humidified incubator at 37 °C and 5% CO_2 (also *see* **Note 1**).
2. Seed mycoplasma-free HEK293T packaging cells at $8x10^6$ cells per 15-cm plate in 20 ml low-antibiotic medium. For 500 ml virus prepare ~30 plates.
3. Incubate cells until the following afternoon. The cells should be 70–80% confluent at the time of transfection.

4. Prepare a ~1:1:1 molar mixture of the three transfection plasmids in Opti-MEM (for pLCKO2::TKOv3 per plate: 100µl Opti-MEM, 7µg psPAX2, 4µg pMD2.G and 5µg gRNA library plasmid pool).
5. Per plate, prepare 800µl Opti-MEM with 48µl X-tremeGENE transfection reagent and incubate at room temperature for 5 min.
6. Combine the plasmid mix and the X-tremeGene mix to obtain a 3:1 ratio of transfection reagent:DNA complex. Mix by gently flicking the tube. Incubate the mix for 30 min at room temperature (*see* **Note 4**).
7. Add the transfection mix to the packaging cells in a dropwise manner, distributing the mix across the plate.
8. 18–24 h later, remove medium containing the transfection reagent and replace with high-BSA medium for viral harvests (18–20 ml-/15-cm plate).
9. 24 h later, harvest the lentivirus-containing supernatant and centrifuge at 500 × *g* for 5 min to pellet any accidentally collected packaging cells. Aliquot supernatant into sterile screw cap storage tubes (*see* **Note 5**).

3.2 Screening Cell Line

3.2.1 Cell Line Selection

1. In order to establish gene essentiality and substrate dependency in hPSCs, many cell line options have become available (*see*, e.g., hpscreg.eu for a comprehensive registry). Cell lines, growth substrates and media conditions should be carefully considered from a biological perspective, taking available infrastructure and cost into account. While the procedures in the following sections should be widely transferable to other cell lines and culture conditions, we will focus on the commonly used H1 human embryonic stem cell line (WA01).
2. In general, we recommend careful optimization of culture conditions. This includes the type of culture medium and growth substrates, conditions and reagents for passaging, freezing and thawing cells, optimal cell densities for each size of culture vessels (here 24-well to 10-cm plates) and the frequency of cell passaging and media changes. Furthermore, individual procedures such as transfection and viral transduction have to be optimized for each cell line, condition and application. We summarize our optimal culture conditions on mouse embryonic fibroblast (MEF) feeder cells and laminin below, and transfection and transduction procedures in the relevant following sections. Ideally, the cell lines should be single-cell adapted.
3. We culture H1 hPSCs on irradiated mouse embryonic fibroblasts in human embryonic stem cell (hESC) medium [20] (*see* Subheading 2). Cells are passaged every 4–5 days using TrypLE

Express and plated in presence of 10μM ROCK inhibitor Y-27632 (ROCKi) for 24 h. Cells are frozen in hESC freezing medium.

4. For feeder-free conditions, tissue culture plates are coated with recombinant human laminin at a concentration of 0.5μg/cm^2 (diluted in PBS with Ca^{2+} and Mg^{2+}). Cells are cultured in mTeSR1 complete medium and passaged every 4–5 days using ReLeSR enzyme-free dissociation reagent in the absence of ROCKi (only used for thawing and transduction). Cells are cryopreserved in mFreSR medium.
5. Optimizations should always be performed such that "best practice" hPSC procedures for keeping cells pluripotent are maintained, e.g. constant morphological checks, daily media changes (occasional "double feeds" are acceptable) and minimal passaging in single-cell suspension. Furthermore, we recommend regularly authenticating cell lines by short tandem repeat (STR) profiling and checking karyotypic integrity, especially after genome editing and clonal expansion steps. STR profiling can be performed by PCR of 9 regions plus Amelogenin, e.g. using the GenePrint 10 System (Promega). Karyotype is determined by G-banding analysis of fixed cells, e.g. by the Medical Genetics Service Cytogenetics Laboratory (Cambridge, UK). *See* www.cuh.nhs.uk for details.
6. Further, we recommend frequent checks for expression of pluripotency and absence of differentiation markers by flow cytometry following standard protocols:

   In brief, dissociate cells and wash once in PBS or medium and once in FACS buffer. For cell-surface antigens, antibody staining is performed in FACS buffer for 30 min at 4 °C in the dark. For intracellular antigens or if more convenient, cells can be first fixed with 4% PFA in PBS for 10 min at room temperature or on ice, followed by permeabilization (0.1% TritonX-100 in PBS for 5 min at room temperature). Wash cells with FACS buffer and stain with antibodies for 30 min at 4 °C on ice in the dark. Stained cells are again washed with FACS buffer. 7-AAD viability dye can be added to live cells before quantification on a flow cytometer. We recommend probing for the following antigens: KDR, CD235a, CD34, CD43, CD117, CD184, CD73, CD41, SSEA4, SSEA5, TRA-1-60, OCT3/4 and SOX2 (*see* Ref. 19 for antibody suppliers and dilutions).
7. Finally, cells should always be free of mycoplasma. We recommend testing cells frequently, e.g. using the MycoAlert Detection Kit (Lonza) or PCR-based mycoplasma detection kits (e.g. e-Myco Plus Mycoplasma PCR Detection Kit).

3.2.2 Generation of Stable, Inducible Cas9-Expressing Cell Line

1. Prepare electroporation-grade plasmids stocks of the AAVS1 cleavage plasmid (AAVS1 T2 CRISPR in pX330) and the all-in-one targeting construct (AAVS1-CAG-TetOn3G-Cas9-P2A-BFP2) (*see* **Notes 6–8**).
2. Establish sensitivity of cells to G418 to ensure optimal selection after targeting using a dose–response curve. Aim for a concentration slightly higher than required to kill 100% of cells in ~5 days. For H1 hESCs, we used 100μg/ml.
3. We recommend sequencing the AAVS1 genomic locus in the chosen cell line to confirm absence of single nucleotide polymorphisms in the homology arms or at the CRISPR target site.
4. Electroporate hPSCs with all-in one targeting construct and AAVS1 T2 CRISPR according to manufacturer's guidelines. Optimization of cell numbers, recovery conditions (e.g. ROCKi) and the electroporation protocol may be required. For H1 hESCs, we used 1100–1200 V for 10 ms.
5. Change the medium 24-h postelectroporation. Add G418 at the predetermined concentration and select resistant clones for 5–7 days (*see* **Note 9**).
6. Expand and cryobank clones.
7. Confirm correct integration of targeting construct by isolating genomic DNA, PCR amplification and Sanger sequencing of left-arm and right-arm products according to standard protocols. *See* Subheading 2 for primer sequences.
8. Upon correct targeting of the AAVS1 locus, there should be no integration of the all-in-one plasmid backbone. Perform a PCR and Sanger sequencing on genomic DNA for random integration of targeting construct using two primer pairs that generate a PCR product only when the plasmid backbone is present (*see* Subheading 2). Genomic DNA isolation, PCR and Sanger sequencing can be performed according to standard protocols.
9. Perform thorough quality controls on correctly targeted clones, including karyotyping, STR profiling, mycoplasma tests, Cas9 induction, etc., as outlined above and below (*see* **Note 10**).
10. In addition, we recommend checking Cas9 expression by immunofluorescence staining and imaging to assess protein localization (nuclear localization required) and expression heterogeneity:

    Retreat cells with G418 for 5–12 days. Treat with doxycycline (see below) for 48–72 h. Fix cells in PBS with 4% PFA for 20 min at room temperature and permeabilize using 0.5% TritonX-100 in PBS for 20 min at room temperature. Block with 2% BSA and 0.05% Tween-20 in PBS for 30 min and stain with DAPI and/or Cas9 antibody for 2 h at room temperature.

Wash cells three times in PBS with 0.05% Tween-20 and incubate for 1 h at room temperature in the dark with secondary antibody. Wash cells three times in PBS and acquire images.

3.2.3 Prescreen Optimization, Cell Line Characterization and Quality Control Procedures

1. Cell plating density optimization:

   We recommend culturing cells on 10-cm plates (or 15 cm, if necessary) for the screen to accommodate high cell numbers. Measure the approximate doubling time of your cells under the chosen conditions and ensure that cells are passaged before full confluency, cell death in colony centres or spontaneous differentiation occur (*see* **Note 11**).

2. Cas9 induction and expression optimization:

   For the generated inducible Cas9 cells, it is crucial to optimize concentration, duration and timing of doxycycline treatment for optimal induction. This is cell line dependent. For H1 hPSCs, we use 1.5μg/ml doxycycline for 48–72 h. We induce Cas9 48 h prior to gRNA transduction, but for other setups (e.g. during a differentiation protocol), a different timing may be suitable.

   We recommend checking BFP-Cas9 expression by flow cytometry. Dissociate cells and wash once in PBS or medium and once in FACS buffer prior to measurement on a flow cytometer.

3. Puromycin sensitivity:

   To ensure optimal selection for transduced cells, puromycin sensitivity has to be determined with a dose–response curve. We recommend a concentration range between 0μg/ml and 5μg/ml. Aim for a concentration slightly higher than required to kill 100% of untransduced cells in 48 h.

4. Screenability:

   To make sure that the selected cell line is suitable for screening, CRISPR editing efficiency has to be assessed. We recommend transducing cells with lentiviral gRNA vectors in the same vector backbone as the selected gRNA library (e.g. Addgene plasmids #74179, #74180, #74181, #74184, #74185, #74190 or sequences in Refs. 10, 19 for pLCKO). The test gRNAs should target essential genes such as PSMD1 and PSMB2 and non-essential regions as controls. See the previous and following sections for virus production and transduction procedures.

   After 48 h of puromycin selection, plate cells at low density to allow for proliferation over >5 days. Determine cell viability by cell counting or viability dye. When editing is efficient, a strong proliferation defect should be visible upon targeting essential genes (*see* **Note 12**).

3.3 CRISPR Screen

3.3.1 Determination of Lentivirus Multiplicity of Infection (MOI)

1. Before starting a screen, it is crucial to determine the Multiplicity of Infection (MOI) of the gRNA library virus. It is desirable to achieve a MOI of ~0.3 to minimize the probability of multiple gRNA integrations in the same cell.
2. MOI determination should always be carried out under the same cell culture conditions as planned for the primary screen (type, size and volume of tissue culture vessels, media and growth substrate conditions, cell plating density, ROCKi treatment, time line, etc.). Use an aliquot of the same virus preparation (without prior thaws) that will be used in the screen. MOI results obtained from different formats do not always scale reliably. Also *see* **Note 1** for virus handling.
3. Design a dilution series of virus and prepare the required number of plates with growth substrate or feeder cells. Typically, we test 3 concentrations of virus (between 5 and 100μl pLCKO2::TKOv3 per 10-cm dish) and include 2 controls (uninfected selected and infected unselected) (*see* **Note 13**).
4. For hPSCs grown on laminin, plate cells at the predetermined optimal density in the presence of 10μM ROCKi and incubate for 24 h. We recommend dissociating the hPSC colonies to a near-single-cell suspension and adding ROCKi at this step to ensure higher efficiency and a more uniform rate of infection.
5. Thaw a fresh aliquot of pooled CRISPR gRNA library lentivirus.
6. For hPSCs grown on laminin, change the medium to a total of 10 ml per plate with 10μM ROCKi and the desired amount of virus. Mix well by tilting the plates. Extended ROCKi treatment helps mitigate potential virus toxicities at this step (*see* **Note 14**).
7. For hPSCs grown on feeder cells, dissociate the cells to a to a near-single-cell suspension and incubate the cells in suspension with the desired amounts of virus in 10 ml medium per plate with 10μM ROCKi for 30 min. Mix well by inverting the infection vessel. Plate on prepared plates with feeder cells. Infection in suspension allows virus entry before plating on feeder cells (which will also get infected).
8. Prepare two plates with the highest amount of virus (one as infected unselected control) and one plate without virus (uninfected selected control).
9. 24 h later, gently remove medium with virus (and from uninfected control plate), wash cells once with PBS and add medium with the predetermined concentration of puromycin. Do not add puromycin to the infected unselected control plate. ROCKi is not required.
10. Change medium with puromycin (except on unselected control plate) 24 h later.

11. 48 h after infection, cells on the uninfected selected control plate should be dead. Remove medium from all plates, collect remaining cells and count.
12. Determine the virus volume that gives 30–40% survival with puromycin selection compared to the infected unselected control. This is the volume required for MOI of ~0.3–0.4 and will be used in the screen.

3.3.2 gRNA Library Transduction, Cell Passaging, Sampling and Quality Control

1. Prepare plates with growth substrates or feeder cells as needed for cell expansion, infection and every passage. Typically, 20–30 plates are required for infection and 4–10 plates per replicate per passage ($1–3 \times 10^6$ cells/plate).
2. Taking the proliferation rate of cells into account, the total number of cells infected should be such that at MOI ~0.3, ~1.20×10^8 cells are present after puromycin selection at T0. We recommend calculating generously, allowing for MOI and proliferation fluctuations.
3. Infection itself should be performed at least at 200-fold library coverage, i.e. for pLCKO2::TKOv3: 71,090 gRNAs * 200-fold/0.3 MOI = ~5×10^7 cells. Also *see* **Note 1** for virus handling.
4. Thaw and expand cells from low passage, mycoplasma-free, quality-controlled stock such that 8–9×10^7 cells will be available on infection day.
5. 48 h prior to infection, add the predetermined doxycycline concentration to the medium to induce Cas9 expression. Collect samples for pluripotency checks, baseline Cas9 expression and mycoplasma test.
6. For hPSCs grown on laminin, plate cells onto required number of plates for infection 24 h prior to infection. Use medium with 10μM ROCKi and doxycycline. Include uninfected selected and infected unselected control plates as during MOI determination. For precise screen MOI determination, uninduced (i.e. not treated with doxycycline) infected unselected and selected control plates can be included. Collect samples for Cas9 expression and pluripotency checks.
7. Thaw fresh aliquots of pooled CRISPR gRNA library lentivirus.
8. For hPSCs grown on laminin, change the medium to a total of 10 ml per plate with 10μM ROCKi and the required amount of virus. Mix well by tilting the plates.
9. For hPSCs grown on feeder cells, dissociate the cells to a to a near-single-cell suspension and incubate the cells in suspension with the required amounts of virus in 10 ml medium per plate with 10μM ROCKi for 30 min. Mix well by inverting the

infection vessel. Plate on prepared plates with feeder cells. Include infected unselected and uninfected selected control plates as before; for precise screen MOI determination, uninduced (i.e. not treated with doxycycline) infected unselected and selected control plates can be included as well. Collect samples for Cas9 expression and pluripotency checks.

10. 24 h later, gently remove medium, wash cells once with PBS and add medium with puromycin (except for the unselected control plates).
11. Change medium with puromycin (except on unselected control plates) 24 h later.
12. 48 h after infection, cells on the uninfected selected control plates should be dead. Remove medium from all plates and wash well with PBS to remove all dead cells, which can skew cell counts and contaminate genomic DNA extraction. Thoroughly collect and pool remaining cells (keeping control cells separate) and count.
13. Calculate screen MOI using control plates as reference. Induced unselected cell numbers may be slightly lower than uninduced unselected cells (=100% viability).
14. Collect 3 samples of at least 200-fold library coverage each. These are the T0 reference samples, one of which will be subjected to library preparation and sequencing (the other two serve as backup). For pLCKO2::TKOv3, we typically collect 3×10^7 cells per sample to buffer against loss during genomic DNA extractions.
15. Spin samples at $250 \times g$ for 5 min, wash once with PBS, thoroughly remove supernatant and freeze cell pellets "dry" at −20 or −80 °C.
16. From the cell pool, plate 3 replicates groups of cells onto the required number of plates. For dropout screens, at least a 200-fold coverage of the library should be maintained throughout the screen for each replicate. For pLCKO2::TKOv3, this equals 1.5×10^7 cells per replicate across the required number of plates. Plate the same number of cells on each plate, and the same cell number for each replicate, e.g. plate 5 plates at 3×10^6 cells for Replicate A and repeat for Replicates B and C (*see* **Note 15**).
17. Optionally, doxycycline can be added to cells again for 24 h.
18. From the remaining cells in the pool, collect samples for mycoplasma test and pluripotency checks.
19. Change medium every 24 h as required for standard hPSC culture and passage cells every 4–5 days when they reach typical confluence. A transient decrease in proliferation speed after

infection may occur. Passaging continues until the screen end point. We recommend at least 12 doublings.

20. At every passage, collect cells, pooling cells from all plates of one replicate, i.e. one pool each for Replicate A, B and C. Count cells and collect 3 samples of at least 200-fold library coverage each, one of which will be subjected to library preparation and sequencing (the other two serve as backup). For pLCKO2::TKOv3, we typically collect 2×10^7 cells per sample at every passage after T0 to buffer against loss during genomic DNA extractions.
21. Spin samples at 250 × *g* for 5 min, wash once with PBS, thoroughly remove supernatant and freeze cell pellets "dry" at −20 or −80 °C. We label these samples by time point and replicate (i.e. T4_A, T4_B and T8_A).
22. Also collect samples for pluripotency checks at every passage and for mycoplasma testing at the end point.
23. Replate cells at 200-fold coverage for each replicate as described at T0.

3.3.3 Genomic DNA Isolation

1. Clean the genomic DNA station with 10% bleach, followed by water and finally 70% ethanol. Prepare aliquots of required amounts of Wizard Genomic DNA Purification Kit solutions. Prewarm TE buffer (Rehydration Solution) to 65 °C and prepare 70% ethanol (for samples) using UltraPure water.
2. Thaw cell pellets at room temperature for 5–10 min (*see* **Note 16**).
3. Resuspend cells in 1 ml PBS and transfer to a 50-ml centrifuge tube. Rinse the original tube with 400μl PBS and add to the same 50-ml tube.
4. Add 5 ml Nuclei Lysis Solution to the resuspended cells. Mix by pipetting with 10-ml pipette (five times).
5. Add 32μl RNase A (20 mg/ml stock; final concentration of 100μg/ml) to the nuclear lysate and mix by inverting the tube five times. Incubate at 37 °C for 15 min and then let cool down to room temperature (~10 min).
6. Add 1670μl Protein Precipitation Solution to the nuclear lysate and vortex vigorously for 20 s.
7. Centrifuge at 4500 × *g* for 10 min at room temperature (*see* **Note 17**).
8. Using a 10-ml pipette, transfer the supernatant to a 50-ml centrifuge tube containing 5 ml isopropanol. Be very careful not to transfer any precipitate (*see* **Note 18**).
9. Gently mix the solution ten times by inversion, until the white thread-like strands of DNA form a visible mass.

10. Centrifuge at 4500 × *g* for 5 min at room temperature. The DNA will be visible as a small white pellet.
11. Carefully remove the supernatant, avoid dislodging the pellet. Add 5 ml 70% ethanol to the DNA. Gently rotate the tube to wash the DNA pellet and the sides of the centrifuge tube.
12. Centrifuge at 4500 × *g* for 5 min at room temperature.
13. Carefully and thoroughly remove all ethanol, avoid dislodging the pellet. Air-dry the genomic DNA at room temperature for 10–30 min until pellet appears translucent. Do not overdry (before visible cracks appear). Avoid contamination by dust particles.
14. Add 400μl TE buffer (DNA Rehydration Solution) and let DNA dissolve by incubating at 65 °C for 1–4 h. Mix DNA by gently flicking the tube every 15 min until the pellet is completely dissolved. Suspension may be slightly viscous but should not contain any clumps of DNA (*see* **Note 19**).
15. Centrifuge at 4500 × *g* for 1 min at room temperature and transfer genomic DNA to a 1.5-ml low-binding tube.
16. Quantitate DNA using a Qubit fluorometer (*see* **Notes 20** and **21**), and store at −20 °C.

3.3.4 Sequencing Library Preparation

1. The following protocol is optimized for the TKOv3 library in the modified pLCKO backbone [19]. For TKOv3 in pLCV2, *see* Ref. 21; for TKOv1, *see* Ref. 10.
2. Always include the T0 sample to determine library representation in the screen and allow gRNA fold change calculation.
3. Clean the PCR1 and PCR2 work stations with 10% bleach, followed by water and finally 70% ethanol.
4. Prepare a 1% agarose gel containing SYBR Safe DNA stain for checking PCR1 amplified products. PCR1 enriches the gRNA cassette from the genome.
5. In the PCR hood: set up one diagnostic reaction (do not add genomic DNA yet) and one no-template negative control.

   At the work station: add genomic DNA to diagnostic reaction. For no-template control, use water instead of DNA.

   Reaction setup:

Reagent	Volume (μl)
2× NEBNext ultra II Q5 master mix	25
10μM PCR1 forward primer	2.5
10μM PCR1 reverse primer	2.5
Genomic DNA (3.5μg)	

(continued)

Reagent	Volume (μl)
Water	Up to 50μl

Amplify diagnostic PCR1 reaction in a thermocycler using the following program:

Step	Temperature	Time	
1	98 °C	30 s	
2	98 °C	10 s	25 cycles (steps 2–4)
3	65 °C	30 s	
4	72 °C	15 s	
5	72 °C	2 min	
6	10 °C	Hold	

6. Check PCR amplification by running 2μl of the PCR product on the prepared agarose gel. Expected product size is ~600 bp.
7. If the correct PCR product is detected, proceed to set up 14 reactions to reach the desired coverage (*see* **Note 22**):
   In the PCR hood: Prepare a 15x master mix according to the reaction setup above. Do not add genomic DNA yet.
   At the work station: Add genomic DNA to PCR master mix. Distribute master mix into 14 0.2-ml PCR strip tubes (50μl/tube).
8. Amplify PCR1 reactions in a thermocycler using the program outlined above.
9. Proceed to the PCR2 work station and pool all 14 + 1 (diagnostic) PCR1 reactions in a 1.5-ml microfuge tube.
10. Soak agarose gel equipment (gel tray, comb and gel tank) for purifying amplified PCR2 products with 0.1 N HCl for 10 min prior to casting gel. Rinse 4 times with tap water and once with purified water.
11. Prepare a 2% agarose gel containing SYBR Safe DNA stain for purifying PCR2 amplified products. Use large combs and allow for ample space between samples. PCR2 amplifies the gRNA cassette and adds Illumina TruSeq adapters with i5 and i7 indices (*see* **Note 23**).
12. In the PCR hood: set up PCR2 reactions (do not add template PCR1 product yet). Prepare no-template controls as desired.
    At the work station: add 5μl of pooled PCR 1 product. For no-template controls, use water instead.
    Reaction setup:

Reagent	Volume (μl)
2× NEBNext ultra II Q5 master mix	25
10μM PCR2 reverse (i7) primer	2.5
10μM PCR2 forward (i5) primer	2.5
Pooled PCR1 product	5
Water	15

Amplify PCR2 reactions in a thermocycler using the following program:

Step	Temperature	Time	
1	98 °C	30 s	
2 3 4	98 °C 55 °C 65 °C	10 s 30 s 15 s	10 cycles (steps 2–4)
5	65 °C	5 min	
6	10 °C	Hold	

13. Run the complete PCR2 reaction on the prepared 2% agarose gel at low voltage (1.5–2 h run). Visualize the PCR product on a blue light transilluminator, the expected product size is ~200 bp. Multiple bands may be visible; it is important to separate the bands well.
14. Excise the 200 bp band with a new razor blade, and transfer the gel slice to a 1.5-ml tube.
15. Purify DNA from agarose gel slice using a gel extraction kit. We recommend repeating each binding step twice and including intermittent washes with binding buffer if binding volumes are large. We also recommend a double elution to maximize yield.
16. Store gel-purified DNA in a 1.5-ml low-binding tube.
17. Quantify DNA on Qubit fluorometer (*see* **Notes 21** and **24**).

3.3.5 Next-Generation Sequencing

1. Due to differences in sample submission requirements and quality controls applied by next-generation sequencing facilities and providers, we only provide general recommendations in this section.
2. Sequencing libraries can now be pooled. The degree of multiplexing depends on the desired read depth per sample. For dropout screens, we recommended to read each sample at 200-fold library coverage and 500-fold coverage for T0 samples to ensure robustness during data analysis due to the

absence of independent replicates at T0. For strong positive selection screens aiming for identification of enriched gRNAs, 50-fold coverage may be sufficient.

3. We recommend sequencing on Illumina HiSeq2500 or NextSeq550 with standard primers for dual indexing. For HiSeq2500, we use standard Single-Read (SR) 50-cycle chemistry with dual index. For NextSeq550, we use standard Single-Read (SR) 75-cycle chemistry with dual index.
4. We recommend 22 dark cycles (base additions without imaging) for the constant region at the end of the U6 promoter. The actual 26-bp read begins after the dark cycles and contains two index reads (i7 first, followed by i5). The recommended sequencing strategy is as follows:

   Dark Cycle: 22 bp; Read 1: 26 bp; Index Read 1: 8 bp; Index Read 2: 8 bp.

   Also *see* **Note 25**.

3.3.6 Data Analysis

1. Here, we describe data analysis using the BAGEL algorithm [24], which is optimized for essential gene determination as calculated by gRNA drop out from a given population. Depending on the screen setup and biological question (e.g. enrichment screens, genetic interactions), there may be more suitable analysis tools, of which a wide variety have become available.
2. Demultiplex sequencing data and verify integrity using the MD5 hashing algorithm.
3. Trim the last 6 bp of the sample reads to extract the 20 bp gRNA sequence. Save untrimmed read number for quality control.
4. Perform a quality control alignment to the gRNA library design file using BOWTIE (allowing for max. 2 mismatches, ignoring qualities and discarding any read mapping to more than one sequence in the library). We exclude sequencing samples with an aligned read ratio below 80% from further analysis.
5. Combine all available reads from different sequencing runs for each sample (if applicable) and align using BOWTIE as above. Tally gRNA counts.
6. Normalize read counts to ten million reads per sample by dividing each read count by the sum of all read counts in the sample and then multiplying by ten million.
7. Calculate fold change to the reference sample (usually T0).
8. Use BAGEL to generate Bayes Factor scores [24]. FDR is defined as 1—precision.
9. We recommend assessing screen performance using reference essential and non-essential genes:

Calculate precision and recall using BF scores, discarding scores for genes that are not included in training sets (union of essential and non-essential gene sets from github.com/hart-lab/bagel, 684 essential and 926 non-EGs). The essential set can be used as the true positive list, e.g. for the precision_recall_curve function of the Scikit-learn library for Python.

Calculate fold changes for essential and non-essential reference genes (mean fold change of all guides across replicates). We recommend performing a Gaussian kernel density estimation of separately for essential and non-essential training (e.g. using gaussian_kde from Scipy Python library) and plotting essential and non-essential data separately for visual quality control.

4 Notes

1. When working with active lentivirus (production, MOI determination and primary infection steps), safety procedures and registration protocols according to institution's guidelines must be followed. We suggest wearing double gloves and treating all waste with bleach before removing it from the hood. Do not use vacuum pumps, remove liquid waste manually and bleach for 30 min before discarding.
2. The PCR1 reverse primer and the PCR1 annealing temperature differ between pLCKO- and pLCV2-based gRNA libraries. Here, we describe pLCKO2::TKOv3. For pLCV2-TKOv3, *see* Ref. 21.
3. For all steps of this protocol, it is crucial to work in a clean environment to avoid cross-contamination with other plasmid DNA or gDNA. Cleaning the working surface, pipettes, tube racks, etc., with 10% bleach is recommended. Wipe all bleached material with water to remove bleach. Aliquot required reagents to reduce contamination. Use filter tips and avoid creating aerosols when ejecting tips.
4. For best results, prepare individual mixes for each plate or small batches for few plates and stagger incubation times.
5. Virus can be stored at 4 °C if used within hours or days, but should be stored at −80 °C long-term. Freeze/thaw cycles should be kept to a minimum to prevent loss of activity.
6. For the generation of stable cell lines expressing Cas9 after random (instead of targeted) integration, *see*, e.g., Ref. 10. Working with a pooled population of Cas9 expressing cells (i.e. Cas9 cell line or Cas9 delivered via one-component library) works well for most cell lines. However, selecting a

single-cell clone is recommended for targeted Cas9 integration (and for selection of a clone with high editing efficiency for improved screen performance) but may introduce a clonal effect.

7. Correct insertion of the inducible Cas9-BFP2 cassette into the AAVS1 locus yields expression of a promoterless P2A-Neomycin resistance gene fused to a splice acceptor site, resulting in cell survival upon G418 treatment.
8. AAVS1 T2 CRISPR should not cleave the homology arms of the target construct. However, well established AAVS1-targeting TALENs [25] may be used as an alternative strategy. The functionality of the chosen target locus cleavage strategy can be tested using the GeneArt Genomic Cleavage Selection Kit.
9. Individual clones can be isolated directly from larger wells or through limiting dilution. The efficiency and genomic stability may be cell line dependent.
10. If desired for future applications, lineage differentiation assays can be included in the quality control procedures.
11. For MOI determination, non-selected cells have to keep proliferating during the assay duration of 72 h; i.e., choose the plating density accordingly.
12. In addition to the proliferation-based assays, surveyor nuclease assays or Western blotting can be performed to determine editing on DNA or protein level, respectively.
13. One-component systems encoding Cas9 on the gRNA plasmid will produce much lower viral titres and higher volumes or virus concentration are required.
14. Polybrene (up to 8μg/ml) or protamine sulphate (up to 5μg/ml) can be added to improve infection efficiency, especially with one-component systems. In our hands, both were too toxic to hPSCs and were therefore omitted from the protocol. If used, the concentration needs to be optimized to minimize toxicity, effects on pluripotency, etc.
15. Positive selections screens can be performed at lower representation.
16. This protocol is optimized for cell pellets containing 2–5 × 10^7.
17. We recommend using a swinging-bucket rotor centrifuge for best results during genomic DNA isolation.
18. If the supernatant still contains precipitate, aliquot to 1.5-ml centrifuge tubes and spin in a table-top centrifuge at full speed for 5–10 min at room temperature.

19. If DNA does not dissolve completely, leave tube at 4 °C overnight and incubate at 65 °C for additional time the next day.
20. Use the dsDNA broad range ("BR") kit at this step. Dilute genomic DNA 1:10 in TE buffer and measure 2μl per sample. Use 2μl standard 2 (100 ng/μl) as positive control.
21. The Qubit is very sensitive to pipetting errors. Pipette buffer, reagent, standards and samples (genomic DNA) as accurately as possible. Minimize exposure of BR/HS reagent to light and store it in a tube wrapped in aluminium.
22. Assuming a diploid human genome is ~6.5 pg and contains one guide-RNA, 50μg of genomic DNA yields ~100-fold coverage of the TKOv3 library. In our experience, this is sufficient in most cases; for higher coverage, the number of PCR reactions can be increased.
23. Alternatively, sequencing libraries can be generated in a one-step PCR amplification (*see*, e.g., Ref. 10).
24. Use the dsDNA high sensitivity ("HS") kit at this step.
25. If there is no option to run dark cycles, i5 forward primers have to be modified to include stagger regions of different length. These separate the constant sequence regions between samples to maintain the sequence diversity across the flow-cell at each imaging cycle.
    (a) Stagger regions are introduced into i5 forward primer. We recommend varying the length of the stagger from 0 to 5 bases: (N = i5 barcode, S = staggered region):
    (b) AATGATACGGCGACCACCGAGATCTACAC [NNNNNNNN] ACACTCTTTCCCTACAC GACGCTCTTCCGATCT [S] TTGTGGAAAGGAC GAAACACCG.
    (c) The i7 reverse primers do not require modification.
    (d) When using stagger regions, the sequencing strategy has to be adjusted by increasing Read 1 to 26 bp + the length of longest stagger region.
    (e) Staggered primers can also be used in combination with dark cycles to increase sequence diversity when reading beyond the guide sequence into the constant tracrRNA sequence. The sequence diversity can be further increased by using primer pairs in reverse orientation (i.e. the i5 forward primer binds to the tracrRNA region and the i7 reverse primer binds to the U6 promoter).

Acknowledgements

We thank members of the Moffat laboratory and collaborators for helpful additions to the protocol. This work was supported by the Canadian Institutes for Health Research (MOP-142375 to JM), Medicine-by-Design First Research Excellence Fund (CITPA-2016-11 to JM) and Genome Canada (OGI-157 to JM). BM holds a postdoctoral fellowship from the Canadian Institutes of Health Research. MA was supported by a Swiss National Science Foundation postdoctoral fellowship. JM is a Canada Research Chair in Functional Genomics.

References

1. Romito A, Cobellis G (2016) Pluripotent stem cells: current understanding and future directions. Stem Cells Int 2016:9451492. https://doi.org/10.1155/2016/9451492
2. Trounson A, DeWitt ND (2016) Pluripotent stem cells progressing to the clinic. Nat Rev Mol Cell Biol 17:194–200. https://doi.org/10.1038/nrm.2016.10
3. Chen W-H, Lu G, Chen X et al (2017) OGEE v2: an update of the online gene essentiality database with special focus on differentially essential genes in human cancer cell lines. Nucleic Acids Res 45:D940–D944. https://doi.org/10.1093/nar/gkw1013
4. Chen W-H, Minguez P, Lercher MJ, Bork P (2012) OGEE: an online gene essentiality database. Nucleic Acids Res 40:D901–D906. https://doi.org/10.1093/nar/gkr986
5. Zhang R, Ou H-Y, Zhang C-T (2004) DEG: a database of essential genes. Nucleic Acids Res 32:271D–272D. https://doi.org/10.1093/nar/gkh024
6. Luo H, Lin Y, Gao F et al (2014) DEG 10, an update of the database of essential genes that includes both protein-coding genes and non-coding genomic elements: table 1. Nucleic Acids Res 42:D574–D580. https://doi.org/10.1093/nar/gkt1131
7. Meyers RM, Bryan JG, McFarland JM et al (2017) Computational correction of copy number effect improves specificity of CRISPR-Cas9 essentiality screens in cancer cells. Nat Genet 49:1779–1784. https://doi.org/10.1038/ng.3984
8. Behan FM, Iorio F, Picco G et al (2019) Prioritization of cancer therapeutic targets using CRISPR-Cas9 screens. Nature 568:511–516. https://doi.org/10.1038/s41586-019-1103-9
9. Hart T, Brown KR, Sircoulomb F et al (2014) Measuring error rates in genomic perturbation screens: gold standards for human functional genomics. Mol Syst Biol 10:733
10. Hart T, Chandrashekhar M, Aregger M et al (2015) High-resolution CRISPR screens reveal fitness genes and genotype-specific cancer liabilities. Cell 163:1515–1526. https://doi.org/10.1016/j.cell.2015.11.015
11. Blomen VA, Májek P, Jae LT et al (2015) Gene essentiality and synthetic lethality in haploid human cells. Science 350:1092–1096. https://doi.org/10.1126/science.aac7557
12. Wang T, Birsoy K, Hughes NW et al (2015) Identification and characterization of essential genes in the human genome. Science 350:1096–1101. https://doi.org/10.1126/science.aac7041
13. Larrimore KE, Rancati G (2019) The conditional nature of gene essentiality. Curr Opin Genet Dev 58–59:55–61. https://doi.org/10.1016/J.GDE.2019.07.015
14. Rancati G, Moffat J, Typas A, Pavelka N (2017) Emerging and evolving concepts in gene essentiality. Nat Rev Genet 19:34–49. https://doi.org/10.1038/nrg.2017.74
15. Bartha I, di Iulio J, Venter JC, Telenti A (2018) Human gene essentiality. Nat Rev Genet 19:51–62. https://doi.org/10.1038/nrg.2017.75
16. Ihry RJ, Worringer KA, Salick MR et al (2018) p53 inhibits CRISPR–Cas9 engineering in human pluripotent stem cells. Nat Med 24:939–946. https://doi.org/10.1038/s41591-018-0050-6
17. Ihry RJ, Salick MR, Ho DJ et al (2019) Genome-scale CRISPR screens identify human pluripotency-specific genes. Cell Rep 27:616–630.e6. https://doi.org/10.1016/J.CELREP.2019.03.043

18. Yilmaz A, Peretz M, Aharony A et al (2018) Defining essential genes for human pluripotent stem cells by CRISPR–Cas9 screening in haploid cells. Nat Cell Biol 20:610–619. https://doi.org/10.1038/s41556-018-0088-1
19. Mair B, Tomic J, Masud SN et al (2019) Essential gene profiles for human pluripotent stem cells identify uncharacterized genes and substrate dependencies. Cell Rep 27:599–615.e12. https://doi.org/10.1016/j.celrep.2019.02.041
20. Ditadi A, Sturgeon CM (2016) Directed differentiation of definitive hemogenic endothelium and hematopoietic progenitors from human pluripotent stem cells. Methods 101:65–72. https://doi.org/10.1016/j.ymeth.2015.10.001
21. Aregger M, Chandrashekhar M, Tong AHY et al (2019) Pooled lentiviral CRISPR-Cas9 screens for functional genomics in mammalian cells. Methods Mol Biol 1869:169–188. https://doi.org/10.1007/978-1-4939-8805-1_15
22. Chan K, Tong AHY, Brown KR et al (2019) Pooled CRISPR-based genetic screens in mammalian cells. J Vis Exp. https://doi.org/10.3791/59780
23. Hart T, Tong AHY, Chan K et al (2017) Evaluation and Design of Genome-Wide CRISPR/SpCas9 knockout screens. G3 7:2719–2727. https://doi.org/10.1534/g3.117.041277
24. Hart T, Moffat J (2016) BAGEL: a computational framework for identifying essential genes from pooled library screens. BMC Bioinformatics 17:164. https://doi.org/10.1186/s12859-016-1015-8
25. González F, Zhu Z, Shi Z-D et al (2014) An iCRISPR platform for rapid, multiplexable, and inducible genome editing in human pluripotent stem cells. Cell Stem Cell 15:215–226. https://doi.org/10.1016/j.stem.2014.05.018

Chapter 2

CRISPR/Cas9 Screening to Identify Conditionally Essential Genes in Human Cell Lines

Kimberly S. Huggler, Nicholas J. Rossiter, Kyle M. Flickinger, and Jason R. Cantor

Abstract

Forward genetic screens across hundreds of cancer cell lines have started to define the genetic dependencies of proliferating human cells. However, most such screens have been performed in vitro with little consideration into how medium composition might affect gene essentiality. This protocol describes a method to use CRISPR/Cas9-based loss-of-function screens to ask how gene essentiality in human cell lines varies with medium composition. First, a single-guide RNA (sgRNA) library is packaged into lentivirus, and an optimal infection titer is determined for the target cells. Following selection, genomic DNA (gDNA) is extracted from an aliquot of the transduced cells. The remaining transduced cells are then screened in at least two distinct cell culture media. At the conclusion of the screening period, gDNA is collected from each cell population. Next, high-throughput sequencing is used to determine sgRNA barcode abundances from the initial and each of the final populations. Finally, an analytical pipeline is used to identify medium-essential candidate genes from these screen results.

Key words CRISPR, Genetic screen, Conditional gene essentiality, HPLM, Physiologic medium

1 Introduction

The CRISPR/Cas9 system is a revolutionary DNA-editing technology that can be exploited to inactivate any of the nearly 20,000 genes in the human genome. Forward genetic screens based on CRISPR have been used in recent years to characterize protein function, map gene interaction networks, and identify mediators of either drug or toxin resistance [1–3]. In addition, there has been substantial interest in leveraging CRISPR-based loss-of-function screens to identify genes that are required for cell growth and survival, as such "essential" genes could potentially suggest targetable liabilities in human cancers [4]. Nonetheless, it is now well-appreciated that genetic contributions to cell fitness are quantitative

Kimberly S. Huggler and Nicholas J. Rossiter contributed equally to this work.

Ren Zhang (ed.), *Essential Genes and Genomes: Methods and Protocols*, Methods in Molecular Biology, vol. 2377, https://doi.org/10.1007/978-1-0716-1720-5_2,

and can also be context-dependent [5]. Consistent with this, CRISPR screen results across hundreds of human cancer cell lines have not only defined a common set of core essential genes but also cell-essential genes that depend on either genotype or lineage [6–8].

Environmental factors influence cell physiology, and interestingly, recent in vitro studies have demonstrated that gene essentiality can in fact vary with either oxygen tension or cell culture in 2D monolayers versus 3D spheroids [9–12]. However, there has been little investigation into how the composition of culture media might influence gene essentiality in human cells. Moreover, most forward genetic screens based on either RNAi or CRISPR have been performed in vitro using cells cultured in media with little relevance to biochemical conditions in human blood [13]. Previously, we developed human plasma-like medium (HPLM), a physiologic medium designed to more closely reflect the metabolic composition of human blood [14]. To test the hypothesis that human cells harbor medium-essential genes, we recently performed CRISPR/Cas9 screens of human blood cancer cell lines in HPLM versus conventional media. Collectively, our screen results revealed that medium composition has a profound impact on gene essentiality [15]. Here, we outline a general strategy for performing CRISPR-based conditional gene essentiality profiling in human cell lines (Fig. 1). Additional considerations related to (1) sgRNA library transformation, (2) validation of candidate hit genes, and (3) elucidation of gene-nutrient interactions will not be discussed in this chapter, and instead, we refer the reader to [15, 16].

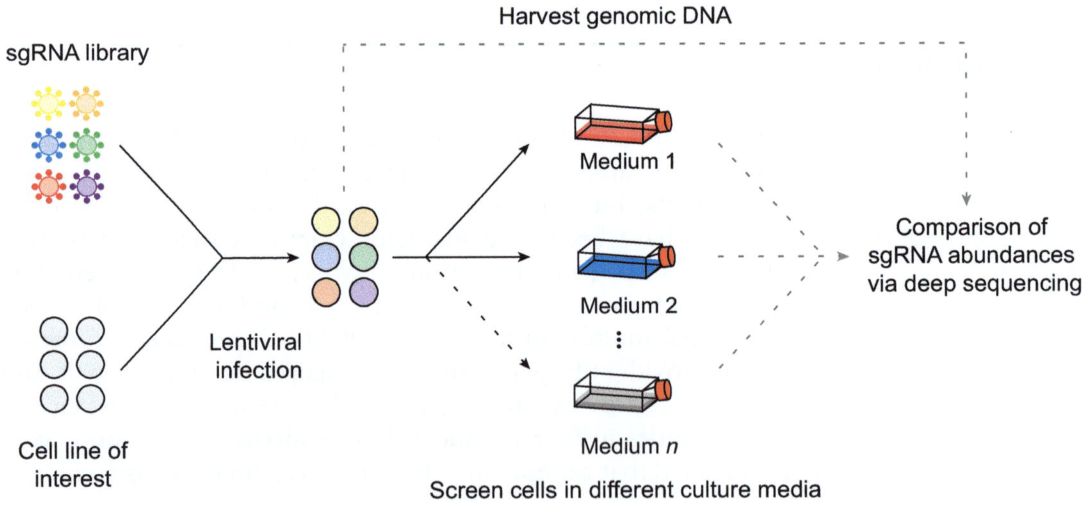

Fig. 1 Schematic depicting parallel CRISPR screens of a human cell line in multiple cell culture media. Following lentiviral transduction, an initial population of cells stably expressing a sgRNA library is split into two or more conditions and then passaged for ~13–14 population doublings. Final sgRNA abundances from cells screened in each condition are then compared versus the initial cell population by high-throughput sequencing

2 Materials

Please consult the appropriate Material Data Safety Sheets and your institution's Environmental Health and Safety office for proper handling of equipment and lentiviruses used in this protocol.

2.1 Viral Packaging

Dulbecco's Modified Eagle Medium (DMEM), high glucose, GlutaMAX Supplement.

Fetal Bovine Serum (FBS), Heat Inactivated.

Penicillin (10,000 U/mL)–Streptomycin (10 mg/mL).

DMEM25^{+S}: DMEM with 10% FBS and 0.5% penicillin–streptomycin.

DMEM25^{+2S}: DMEM with 20% FBS and 0.5% penicillin–streptomycin.

X-tremeGENE 9 DNA Transfection Reagent (Sigma Aldrich #6365779001).

Bottle-top vacuum filter, cellulose acetate membrane, pore size 0.22 μm.

Tissue culture (TC)-treated culture dish, 6 cm.

Luer-Lok Syringe, 5 mL.

25-mm Polyethersulfone (PES) Syringe Filter, 0.45 μm.

Polypropylene centrifuge tube, sterile, 15 mL.

Cryogenic vial, 2 mL.

pCMV_VSV-G envelope plasmid (Addgene #8454).

pCMV_deltaVPR8.2 packaging plasmid (Addgene #8455).

Lentiviral sgRNA library (acquire by methods described elsewhere or by sourcing from Addgene, www.addgene.org).

Human embryonic kidney (HEK) 293T cell line (ATCC CRL-3216).

2.2 Titer Test

Polybrene Infection/Transfection Reagent.

Puromycin.

RPMI 1640.

RPMI 1640, no glucose.

Glucose, powder.

RPMI11^{+S}: RPMI 1640 with 10% FBS and 0.5% penicillin–streptomycin.

RPMI^{+S}: RPMI 1640 (no glucose) with 5 mM glucose, 10% FBS, and 0.5% penicillin-streptomycin.

T-75 cell culture-treated flask, with vent cap.

Tissue culture (TC)-treated six-well plate.

2.3 Screen Infection

T-75 cell culture-treated flask, with vent cap.

T-175 cell culture-treated flask, with vent cap.

Tissue culture (TC)-treated six-well plate.

Polypropylene centrifuge tube, sterile, 15 mL.

2.4 Screen Cell Culture and Library Preparation

T-225 cell culture-treated flask, with vent cap.

QIAamp DNA Blood Maxi Kit (QIAGEN # 51194).

Phosphate-buffered saline (PBS).

SnakeSkin dialysis tubing, 3.5 K MWCO, 35 mm (Thermo Fisher # PI88244).

Pyrex Low Form Griffin Beaker, 4 L.

RPMI^{+dS}: RPMI 1640 (no glucose) with 5 mM glucose, 10% dialyzed FBS, and 0.5% penicillin–streptomycin.

RPMI 1640 Vitamins Solution, 100× (Sigma Aldrich # R7256).

Human Plasma-Like Medium (HPLM): We refer the reader elsewhere for a complete list of defined components and corresponding product numbers [14].

HPLM^{+dS}: HPLM with 10% dialyzed FBS, using RPMI 1640 100× Vitamins Solution, and 0.5% penicillin–streptomycin, as prepared in [15].

Single-guide RNA (sgRNA) barcode polymerase chain reaction (PCR) primers:

Forward: AATGATACGGCGACCACCGAGATCTACACGAA TACTGCCATTTGTCTCAAGATCTA.

Reverse:

CAAGCAGAAGACGGCATACGAGATC nnnnnn TTTCTTGGG TAGTTTGCAGTTTT.

(The sequence "nnnnnn" denotes a user-specified sample barcode sequence.)

Primers for Illumina HiSeq:
Sequencing:

CGGTGCCACTTTTTCAAGTTGATAACGGACTAGCCT-TATTTTAACTTGCTATTTCTAGCTCTAAAAC.

Indexing:

TTTCAAGTTACGGTAAGCATATGATAGTCCATTTTAAAA-CATAATTTTAAAACTGCAAACTACCCAAGAAA.

TaKaRa Ex Taq DNA polymerase (TaKaRa # RR001A).

SYBR Safe DNA Gel Stain.

Agarose gel (1.0%).

QIAquick PCR Purification Kit (QIAGEN # 28104).

3 Methods

The following protocol assumes that the user has generated or acquired a suitable sgRNA library and begins at the viral packaging step.

3.1 Viral Packaging

Day 1

1. Add the following components to make DMEM25^{+S}.

DMEM (high glucose, GlutaMAX Supplement)	447.5 mL
FBS, Heat Inactivated	50 mL
Penicillin (10,000 U/mL)–Streptomycin (10 mg/mL)	2.5 mL

2. Add the following components to make DMEM25^{+2S}.

DMEM (high glucose, GlutaMAX Supplement)	397.5 mL
FBS, Heat Inactivated	100 mL
Penicillin (10,000 U/mL)–Streptomycin (10 mg/mL)	2.5 mL

3. Filter each medium through a 0.22-μm bottle-top vacuum filter in a tissue culture hood.
4. Seed 1,200,000 HEK-293T cells in a 6-cm dish containing 4 mL DMEM25^{+S}. Incubate cells overnight at 37 °C in a tissue culture incubator.

Day 2

5. Assemble the following transfection mixture.

DMEM (high glucose, GlutaMAX Supplement), prewarmed	120 μL
Lentiviral sgRNA library	1 μg
pCMV_VSV-G	150 ng
pCMV_deltaVPR8.2	900 ng
X-tremeGENE 9 DNA Transfection Reagent (added last)	6 μL

6. Mix well by pipetting and incubate for 15–20 min at room temperature.
7. Add dropwise to HEK-293T cells and incubate for 18 h at 37 °C in a tissue culture incubator.

Day 3

8. Replace medium in 6-cm dish with 4 mL DMEM25^{+2S}, and incubate overnight at 37 °C in a tissue culture incubator.

Day 4

9. Transfer viral supernatant to 5-mL Luer-Lock syringe and filter into a 15-mL polypropylene centrifuge tube through a 0.45-μm PES syringe filter.
10. Aliquot and store harvested virus at −80 °C (*see* **Note 1**).

3.2 Titer Test

The following steps are intended for suspension cell lines but can be adapted for adherent cell lines by incorporating appropriate steps for cell detachment and by replacing T-25 flasks with relevant tissue culture-treated plates.

Day 1

1. Create a 2 mg/mL stock solution of polybrene in water.
2. Create a 500 mM stock solution of glucose in water.
3. Add the following components to make RPMI11^{+S}.

RPMI 1640	447.5 mL
FBS, Heat Inactivated	50 mL
Penicillin (10,000 U/mL)–Streptomycin (10 mg/mL)	2.5 mL

4. Add the following components to make RPMI^{+S}.

RPMI 1640, no glucose	442.5 mL
FBS, Heat Inactivated	50 mL
500 mM glucose	5 mL
Penicillin (10,000 U/mL)–Streptomycin (10 mg/mL)	2.5 mL

5. Set up the following infections in a six-well tissue culture-treated plate (*see* **Note 2**).

Target cells	5,000,000
Polybrene (2 mg/mL)	8 μL
Filtered virus	0, 100, 200, 400, or 1000 μL
RPMI^{+S} or RPMI11^{+S}	to 2 mL

6. Centrifuge plate at 600 × *g* for 45 min in a prewarmed centrifuge and then incubate cells overnight at 37 °C in a tissue culture incubator.

Day 2

7. Pellet cells and aspirate virus-containing medium.
8. Resuspend cells into a T-75 flask with 20 mL fresh medium and then incubate cells overnight at 37 °C in a tissue culture incubator.

Day 3

9. From each infected culture, pellet 3,000,000 cells and resuspend into a T-75 flask with 20 mL fresh medium containing appropriate selection dose of puromycin (*see* **Note 3**).
10. From the uninfected culture, pellet 3,000,000 cells and resuspend into a T-75 flask with 20 mL fresh medium containing appropriate selection dose of puromycin. From the uninfected culture, pellet an additional 3,000,000 cells and instead resuspend into a T-75 flask with 20 mL fresh medium lacking antibiotic.

Day 6

11. Measure the viability of puromycin-treated cultures versus the uninfected, unselected culture.
12. Confirm that puromycin treatment of uninfected cells caused >95% cell death.
13. Identify the viral dose required for ~40–60% cell survival and then discard all flasks.

3.3 Screen Infection

The following steps assume that the user has generated, aliquoted, and frozen a single batch of sgRNA library lentivirus that was initially used for testing titer in the preceding section.

Day 1

1. Calculate the total cells, volume of virus, and number of wells in six-well tissue culture-treated plates required to achieve 1000-fold coverage of the library (e.g., for a 20,000-sgRNA library, one needs 20,000,000 infected cells and thus 40,000,000 total cells based on choosing a viral dose that permits ~50% cell survival).
2. Assemble a large-scale cell-virus infection mixture in a six-well tissue culture-treated plate according to the following amounts per well:

Target cells	5,000,000
Polybrene	(2 mg/mL) 8 μL
Viral dose	giving ~40–60% cell survival
RPMI^{+S} or RPMI11^{+S}	to 2 mL

3. Dispense 2 mL aliquots of the mixture into six-well plates.
4. As a negative control, identically seed a single well using a similar mixture that instead lacks virus.
5. Centrifuge the plates at 600 × *g* for 45 min in a prewarmed centrifuge and then incubate cells for 18 h at 37 °C in a tissue culture incubator.

Day 2

6. Pellet a single infected well and the uninfected well, aspirate the media, and resuspend each into a T-75 flask with 20 mL fresh medium and then incubate cells overnight at 37 °C in a tissue culture incubator.
7. Pellet the remaining infected cells in pools of six total wells each (~12 mL total volume per 15-mL polypropylene tube).
8. Resuspend each pool into a T-175 flask with 120 mL fresh medium and then incubate cells overnight at 37 °C in a tissue culture incubator.

Day 3

9. From the infected T-75 culture, pellet 3,000,000 cells and resuspend into a T-75 flask with 20 mL fresh medium containing an appropriate selection dose of puromycin.
10. From the uninfected T-75 culture, pellet 3,000,000 cells and resuspend into a T-75 flask with 20 mL fresh medium containing an appropriate selection dose of puromycin. Pellet an additional 3,000,000 cells and instead resuspend into a T-75 flask with 20 mL fresh medium lacking antibiotic.
11. Pellet all remaining infected cells (from T-175 flasks) and aspirate virus-containing media. Resuspend and pool the cells in fresh medium and then measure the resulting density.
12. Pellet all the cells and then resuspend to a density of 150,000 cells/mL with fresh medium containing an appropriate dose of puromycin and seed the cells in T-175 flasks.
13. Incubate all cultures for 3 days at 37 °C in a tissue culture incubator.

Day 6

14. Measure the viability of puromycin-treated T-75 cultures versus the uninfected, unselected T-75 culture.
15. Confirm that puromycin treatment of uninfected cells caused >95% cell death.
16. If cell survival is ~40–60% in the infected population, proceed to Subheading 3.4.

3.4 Screen Cell Culture and Library Preparation

The following steps describe screens performed in HPLM^{+dS} versus RPMI^{+dS} but can be identically used to execute parallel screens across any two (or more) culture media (*see* **Note 4**). **Steps A–D** relate to manual FBS dialysis and should be completed at least 24 h prior to **step 1**.

A. Transfer FBS to SnakeSkin dialysis tubing, 3.5 K MWCO, 35 mm according to manufacturer's instructions.

B. Place tubing into a 4 L beaker that contains PBS at a volume 20-fold greater versus that of FBS within the dialysis tubing. Mix by stirring at 4 °C.

C. Perform a complete PBS exchange every 9–12 h for 48 h.

D. Store dialyzed FBS at −20 °C until needed to prepare complete media.

Day 1

1. Pellet infected cells from T-175 flasks and pool by resuspension in the initial infection medium (e.g., $RPMI^{+S}$ or $RPMI11^{+S}$).
2. Pellet and freeze cells from the resuspension required to achieve at least 500-fold coverage of the library for eventual DNA extraction (e.g., for a 20,000-sgRNA library, pellet at least 10,000,000 total cells) (*see* **Note 5**).
3. For each screening medium of interest (e.g., $RPMI^{+dS}$ and $HPLM^{+dS}$), pellet and reseed cells from the resuspension in **step 1** to a density of 150,000 cells/mL into a total number of T-225 flasks required to maintain 1000-fold coverage of the library.
4. Continue to passage each screen every 48 h, maintaining 1000-fold coverage of the library at each reseeding (*see* **Note 6**).
5. After ~13–14 population doublings, pellet and freeze cells required to achieve at least 500-fold coverage of the library (e.g., for a 20,000-sgRNA library, pellet at least 10,000,000 total cells) (*see* **Note 5**).
6. Extract genomic DNA (gDNA) from the initial and final cell pellets using the QIAamp DNA Blood Maxi kit according to the manufacturer's instructions.
7. Calculate the total number of polymerase chain reactions (PCRs) required. Use a 250-fold coverage of the library as input for sgRNA amplification, assuming an input of 3 μg genomic DNA per 50 μL reaction and an estimated mass of 6.6 pg per diploid human genome (e.g., for a 20,000-sgRNA library, requiring 5,000,000 genome equivalents, one needs 33 μg for diploid human cells—i.e., 11 reactions, each with 3 μg of genomic DNA).
8. Assemble a total mixture based on the following per-sample PCR reaction and dispense into PCR strip tubes in 50 μL aliquots:

10× TaKaRa Ex Taq DNA polymerase buffer	5 μL
dNTPs (2.5 mM)	4 μL
Forward sgRNA PCR primer (10 uM)	2 μL

(continued)

Sample-specific barcoded reverse PCR primer (10 uM)	2 μL
Genomic DNA	3 μg
H_2O	to 50 μL

9. Amplify reactions in a thermocycler using the following program:

1 cycle	95 °C	5 min
30 cycles	95 °C	10 s
	61 °C	15 s
	72 °C	45 s
1 cycle	72 °C	5 min
1 cycle	4 °C	hold

10. Pool the reactions and run a small aliquot (~5–10 μL) on a 1% agarose gel with SYBR Safe DNA gel stain and visualize the PCR product bands using a gel imager (*see* **Note 7**).
11. Purify up to 500 μL of the remaining pooled product using the QIAquick PCR Purification Kit according to manufacturer's instructions, ultimately eluting the DNA into 50 μL of water.
12. Submit cleaned PCR products for high-throughput sequencing on an Illumina HiSeq using the custom sequencing and indexing primers described in Subheading 2.4. A single end run with a 6-bp indexing read should be performed.

3.5 Data Analysis

Several robust pipelines exist for the analysis of CRISPR screen data that can increase statistical power by correction of copy number effect, integration of multiple replicates, model selection, and estimation of individual sgRNA efficacies [6, 17, 18]. Below, we detail a straightforward and easily interpretable protocol to identify medium-essential candidate genes. Determining the most suitable method for a given experiment is left to the reader.

The following steps describe a method for calculating differential gene scores from raw sequencing data. First, depletion scores for each sgRNA are determined and averaged to generate gene scores. After assessing data quality, the gene scores are scaled to reduce potential bias when comparing datasets between screen conditions (*see* **Note 8**). Finally, differential scores are calculated to identify medium-essential candidate genes. In addition to the raw sequencing data obtained from each screen, this method requires the use of a library that contains nontargeting sgRNAs and others that instead target a predefined set of core essential genes (CEGs) [7]. These genes and sgRNAs will serve as controls both for evaluating data quality and normalizing gene scores.

1. Obtain read counts by aligning raw sequencing data to the sgRNA sequences that comprise the user library, allowing only for exact matches. The ratio of mapped to total reads should ideally exceed 60 percent (*see* **Note 9**).
2. Add a pseudocount of 1 to each sgRNA count to prevent potentially undefined values in **step 5a**.
3. Calculate the fractional abundance of each sgRNA by dividing its read count by the total number of reads in the sample.
4. Remove sgRNAs with low reads within the initial population from further downstream analysis (*see* **Note 10**).
5. For each screen:
   (a) Calculate sgRNA depletion scores as the $\log_2$ fold change in fractional abundance for each sgRNA between respective final and initial cell populations.
   (b) Determine gene-level scores by averaging the depletion scores of all sgRNAs targeting each gene. Remove genes for which less than 70% of the targeting sgRNAs from the initial library remain after the filtering in **step 4** (*see* **Note 11**).
   (c) To ensure adequate separation of screen controls, calculate the strictly standardized mean difference (SSMD) between gene-level CEG scores and nontargeting sgRNA depletion scores [19] (*see* **Note 12**).
   (d) Scale gene scores according to the following equation:

   $$\text{scaled gene score} = \frac{\text{gene score} - \text{NT}}{\text{NT} - \text{CEG}}$$

   where NT denotes the median score of all postfiltering nontargeting sgRNAs in the library and CEG denotes the median score of all postfiltering core essential genes targeted by the library (Fig. 2a).
6. For each gene, determine the differential gene score between any two screen datasets and then calculate its "differential dependency" as the standard score relative to the entire set of targeted genes.
7. To identify conditionally essential candidate genes, set a differential dependency cutoff (e.g., standard score *(Z-score)* greater than 2) (Fig. 2b) (*see* **Note 13**).

4 Notes

1. Viral supernatants can be stored at −80 °C for long-term storage, but freeze-thawing will likely cause a reduction in transduction efficiency. Nonetheless, if a large batch of

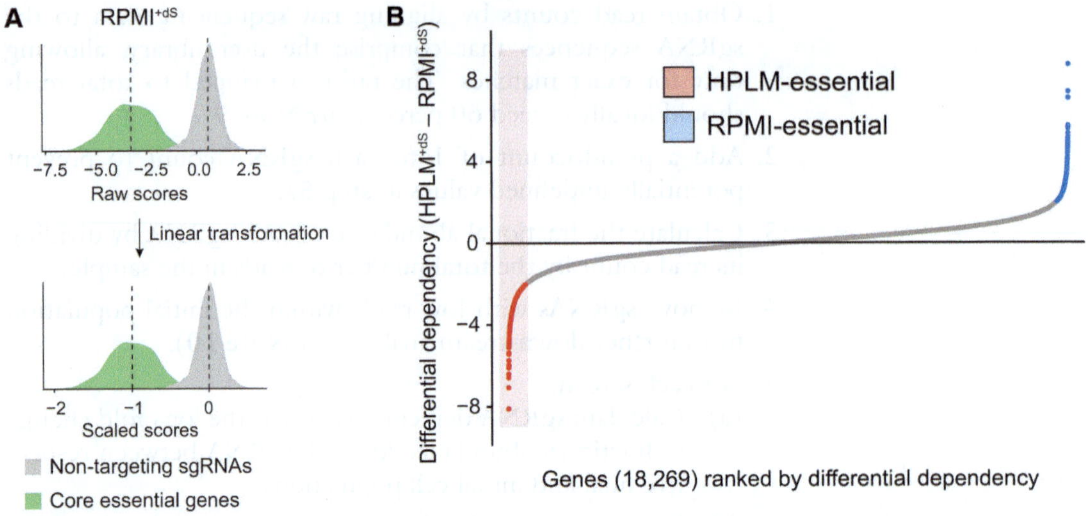

Fig. 2 Normalization and analysis of CRISPR screen results. (**a**) Linear transformation of scores for a screen performed in RPMI^{+dS}. Following normalization, the medians (dashed lines) of the CEG and NT distributions equal −1 and 0, respectively. (**b**) Genes ranked by differential dependency for genome-wide screens of the K562 cell line in HPLM^{+dS} versus RPMI^{+dS}. Highlighted regions depict medium-essential genes based on a differential dependency cutoff of 2. (Panels are adapted from [15])

harvested virus is generated by pooling several identically prepared 6-cm dishes in parallel, then one can perform the titer test and ensuing screen infection using frozen aliquots derived from the same stock.

2. First, some cell lines might not tolerate spin infection and overnight incubation at this density. Adjust cell numbers accordingly for the cell line of interest. Second, the culture medium used for the titer test should be the same one used for screen infection. Here, we denote each of two possible RPMI-based media containing either conventional (11.1 mM) or physiologic (5 mM) glucose.
3. The optimal dose of puromycin (or alternative antibiotic if necessary) should be determined by performing a puromycin kill curve. The minimum concentration of puromycin that causes death in >95% of uninfected cells after 72 h should be used as a selection dose.
4. The total number of parallel screening conditions should be considered in calculations required in Subheading 3.3, **step 1**.
5. It is recommended to freeze at least two such pellets from the initial infected population.
6. At each passage, all cells in a given condition should be pelleted and pooled by resuspension in fresh screening medium. After measuring the resuspension density, pellet and reseed cells to a density of 150,000 cells/mL across the same total number of

T-225 flasks. Further, as screens approach termination (population doubling ≥ 10), it is advisable to freeze at least two pellets—at 500-fold library coverage—at each passage in case of a DNA extraction failure or other downstream bottleneck.

7. Using the suggested primers and sgRNA library cloned into the pLentiCRISPR backbone, a PCR product band should be observed at ~300 bp.
8. Screens performed in different media conditions can introduce discrepancies in global selection that affect the dynamic range of gene scores [17], but the calculation of differential scores assumes equivalent gene score distributions between conditions. To reduce potential bias, gene scores are instead normalized based on the assumption that sets of core essential genes and nontargeting sgRNAs exhibit the same selection across different screens. To achieve this, a linear transformation is applied such that the medians of the nontargeting sgRNAs and core essential genes for each condition equal 0 and −1, respectively.
9. Mapping efficiencies markedly below this threshold can indicate issues with either gDNA extraction or PCR amplification and may produce low-quality screen datasets.
10. This threshold will vary based on library size and sequencing coverage. For screens maintained at 1000-fold coverage of a library that contained ~200,000 sgRNAs, a cutoff of 50 counts has been reported [2, 15].
11. For example, for a library that contains 10 sgRNAs per gene, the user could consider including in downstream analysis only those genes targeted by a minimum of 7 sgRNAs after filtering for low read counts in the initial population.
12. Multiple measures can be used to evaluate screen data quality. SSMD and area under a receiver operating characteristic curve (AUROC) are metrics that can be calculated to ensure sufficient separation of positive and negative controls in the screen. Notably, others propose that SSMD magnitudes greater than 2 indicate an excellent separation of such controls [19].
13. Significance testing of genes (e.g., by using Welch's *t*-test, Kolmogorov–Smirnov test, Mann–Whitney *U* test) can be performed by comparing sgRNA-level score distributions versus those for the nontargeting controls or, similarly, between screen datasets.

Acknowledgements

This work was supported by grant and fellowship support from the NIH (T32HG002760 to K.S.H. and K22CA225864 to J.R.C.). J.R.C. is an inventor on a patent application for HPLM

(PCT/US2017/061377) assigned to the Whitehead Institute. Kimberly S. Huggler and Nicholas J. Rossiter contributed equally to this manuscript.

References

1. Kory N, Wyant GA, Prakash G et al (2018) SFXN1 is a mitochondrial serine transporter required for one-carbon metabolism. Science 362:eaat9528
2. Wang T, Yu H, Hughes NW et al (2017) Gene essentiality profiling reveals gene networks and synthetic lethal interactions with oncogenic Ras. Cell 168:890–903.e15
3. Gilbert LA, Horlbeck MA, Adamson B et al (2014) Genome-scale CRISPR-mediated control of gene repression and activation. Cell 159:647–661
4. Wang T, Birsoy K, Hughes NW et al (2015) Identification and characterization of essential genes in the human genome. Science 350:1096–1101
5. Larrimore KE, Rancati G (2019) The conditional nature of gene essentiality. Curr Opin Genet Dev 58:55–61
6. Meyers RM, Bryan JG, McFarland JM et al (2017) Computational correction of copy number effect improves specificity of CRISPR–Cas9 essentiality screens in cancer cells. Nat Genet 49:1779
7. Hart T, Tong AHY, Chan K et al (2017) Evaluation and Design of Genome-Wide CRISPR/SpCas9 knockout screens. G3 genes genomes. Genetics 7:2719–2727
8. Behan FM, Iorio F, Picco G et al (2019) Prioritization of cancer therapeutic targets using CRISPR–Cas9 screens. Nature 568:511–516
9. Faubert B, Solmonson A, DeBerardinis RJ (2020) Metabolic reprogramming and cancer progression. Science 368:eaaw5473
10. Kaymak I, Williams KS, Cantor JR et al (2020) Immunometabolic interplay in the tumor microenvironment. Cancer Cell 39:28–37
11. Jain IH, Calvo SE, Markhard AL et al (2020) Genetic screen for cell fitness in high or low oxygen highlights mitochondrial and lipid metabolism. Cell 181(3):716–727.e11
12. Han K, Pierce SE, Li A et al (2020) CRISPR screens in cancer spheroids identify 3D growth-specific vulnerabilities. Nature 580:136–141
13. Cantor JR (2019) The rise of physiologic media. Trends Cell Biol 29:854–861
14. Cantor JR, Abu-Remaileh M, Kanarek N et al (2017) Physiologic medium rewires cellular metabolism and reveals uric acid as an endogenous inhibitor of UMP synthase. Cell 169:258–272.e17
15. Rossiter NJ, Huggler KS, Adelmann CH et al (2021) CRISPR screens in physiologic medium reveal conditionally essential genes in human cells. Cell Metab. [online ahead of print]
16. Joung J, Konermann S, Gootenberg JS et al (2017) Genome-scale CRISPR-Cas9 knockout and transcriptional activation screening. Nat Protoc 12:828–863
17. Wang B, Wang M, Zhang W et al (2019) Integrative analysis of pooled CRISPR genetic screens using MAGeCKFlute. Nat Protoc 14:756–780
18. Hart T, Moffat J (2016) BAGEL: a computational framework for identifying essential genes from pooled library screens. BMC Bioinformatics 17:164
19. Zhang XD (2008) Novel analytic criteria and effective plate designs for quality control in genome-scale RNAi screens. J Biomol Screen 13:363–377

Chapter 3

Target Identification of Small Molecules Using Large-Scale CRISPR–Cas Mutagenesis Scanning of Essential Genes

Bert Kwanten, Jasper Edgar Neggers, and Dirk Daelemans

Abstract

Target deconvolution of new bioactive agents identified from phenotypic screens remains a challenging task. The discovery of mutations that confer resistance to such agents is regarded as the gold standard proof of target identification. Here, we describe a method that exploits the error-prone repair of CRISPR-induced DNA double-strand breaks to enhance mutagenesis and increase the incidence of drug resistance mutations in essential genes. As each DNA double-strand break is introduced at a targeted genomic site predefined by the presence of a protospacer adjacent motif (PAM) and a particular CRISPR single guide RNA (sgRNA), the genetic location of drug resistance mutations can be easily uncovered through targeted sequencing of CRISPR sgRNAs. Moreover, the method allows for the identification of not only the drug target gene, but also the drug-binding domain within the target gene.

Key words CRISPR–Cas, Cas9, Cas12a, CRISPRres, sgRNA tiling library, Genetic screen, Drug target identification, Resistance mutation, Target deconvolution

1 Introduction

Identification of the molecular target and mechanism of action of newly discovered bioactive agents remains challenging. Yet, solid knowledge on this matter is a prerequisite to further their development into a clinical drug candidate. Such knowledge provides mechanistic insight, helps to predict efficacy and potential side effects, and supports biomarker development and hit-to-lead optimization. A phenotypic screening assay often harbors considerable biological complexity with multiple underlying mechanisms and targets, which makes pinpointing the molecular target of a hit from such a screen especially challenging. There is no single, off-the-shelf methodology for target deconvolution, and the process often requires a combination of different approaches. However, the discovery of mutations conferring drug resistance within a cellular context is viewed as the gold standard proof for target engagement [1, 2].

Ren Zhang (ed.), *Essential Genes and Genomes: Methods and Protocols*, Methods in Molecular Biology, vol. 2377,
https://doi.org/10.1007/978-1-0716-1720-5_3,

One classic method for target deconvolution is based on the spontaneous generation of resistance mutations. In this process, cancer cells are cultured under selective pressure with stepwise increasing drug concentrations. Although in many cases drug resistance will eventually occur, this may take months to years. In addition, the observed resistance is not always a result of mutations directly in the target protein of the bioactive agent, but rather caused by other indirect mechanisms, such as the upregulation of multidrug efflux pumps or antiapoptotic signaling cascades. Moreover, this approach requires full transcriptome sequencing of several clones to reveal the resistance mutations [3, 4] as the human's large genome size and mutational heterogeneity of cancer cell lines obscure the identification of resistance-conferring mutations in individual clones.

Here, we describe the protocol for a new genetic target deconvolution approach we named CRISPRres (**CRISPR**-induced **res**istance in **es**sential genes) [5]. CRISPRres was developed to speed up the drug resistance selection process and to rapidly identify resistance mutations without the need for full transcriptome sequencing. More specifically, the method relies on the introduction of double-strand breaks in the genomic DNA by targeted CRISPR endonucleases and the subsequent error-prone DNA repair by the cellular machinery. The concept is similar to the use of ultraviolet (UV) radiation to induce DNA double-strand breaks to enhance mutagenesis [6], with the key difference that UV-mediated double-strand breaks occur randomly across the genome, while CRISPR-mediated double-strand breaks are targeted to a specific locus.

In human cells, genomic DNA double-strand breaks are mainly repaired by the cellular nonhomologous end-joining (NHEJ) and microhomology-mediated end-joining (MMEJ) machinery. These DNA repair processes are error-prone and generally introduce a variety of local mutations at the break site, mainly consisting of small nucleotide insertions and deletions (indels). In protein-coding sequences, these indels lead to either out-of-frame mutations, which result in a frameshift leading to a loss of protein function (knockout) due to the occurrence of premature stop codons and nonsense-mediated decay, or in-frame mutations—on which CRISPRres relies—that often resolve in a mutated, but functional protein. Cells can become resistant to a drug when they acquire mutations inside the agent's target protein that prevent drug target engagement but maintain protein function. By deriving functional in-frame drug resistance mutations, CRISPRres can be used to directly identify the target gene of an anticancer agent. As such, CRISPRres is complementary to typical CRISPR knockout/loss-of-function screens, which usually do not identify the direct target an anticancer drug is engaging with but rather (nonessential) genes involved in the mechanism of action of the drug.

For a novel anticancer compound with an unknown molecular target, CRISPRres can be applied to a large set of candidate genes by generating a lentiviral tiling library containing all possible sgRNAs targeting the coding sequence of the candidate genes. Cells are subsequently transduced with this library in such a way that each cell contains one sgRNA. Hence, each cell will carry CRISPR-induced mutations at only one specific genomic locus as determined by the sgRNA targeting sequence. Treatment of the mutated cell population with an anticancer compound will then kill the majority of cells, except those rare cells that contain a functional mutation in the protein domain that is targeted by the compound and which confers drug resistance or that contain a resistance mutation in an allosteric site. Of note, cells with frameshift mutations in essential target genes will deplete from the cell pool and will not be identified, lowering the noise signal.

Each drug-resistant cell carries the sgRNA sequence, which will directly annotate the genomic locus at which the resistance mutations were introduced. This avoids the need for transcriptome sequencing of multiple clones to retrieve the resistance-conferring mutations. Indeed, targeted amplicon sequencing of the sgRNA cassette is sufficient to identify both the target gene and the locus containing the resistance mutations. When followed by focused sequencing of the genomic locus targeted by the enriched sgRNA (s), the specific resistance-conferring mutations can also be identified.

The protocol described here uses the canonical CRISPR *Sp*Cas9 endonuclease, but it can easily be adapted to other nucleases with different PAM-specificities such as *As*Cas12a [7]. Alternatively, while this procedure relies on in-frame mutations generated by DNA repair of CRISPR-induced DNA double-strand breaks, others have used CRISPR mutants coupled to base editing enzymes to generate resistance mutations [8–10].

This method is part of patent application WO 2018/032063 "Drug-Target Identification by Rapid Selection of Drug Resistance Mutations"; In brief, the protocol is as follows:

1. Cancer cells expressing the *Sp*Cas9 endonuclease are lentivirally transduced with a sgRNA tiling library targeting the protein-coding sequences of a predefined set of genes. Although the protocol is described for cells stably expressing *Sp*Cas9, it can also be applied to cells that do not. However, in that case, the lentivector should express both the endonuclease and the CRISPR guide RNA.
2. Transduced cells are selected for with an antibiotic and subsequently treated with the investigational anticancer agent at a concentration that kills all nonresistant cells.

3. Cancer cells containing mutations that confer resistance to the anticancer drug will survive treatment and are grown out over a period of 1–3 weeks, in which the cell culture medium and the investigational anticancer agent are regularly refreshed.
4. The genomic DNA of the resistant cells is extracted, and the sgRNA coding cassette present in the genomic DNA is sequenced by targeted next-generation sequencing. The sequence of the sgRNAs in the resistant colonies will directly identify the target gene that contains the resistance-conferring mutations as well as define the location of the mutations within that gene.
5. [optional] The target locus is sequenced by targeted next-generation sequencing, revealing the exact mutations that are responsible for drug resistance.

2 Materials

All solutions are made using nuclease-free, sterile-filtered water.

2.1 sgRNA Library Cloning

1. Oligonucleotide pool containing sgRNA target sequences: dissolved at 5 ng/μL in water (Twist Bioscience).
2. Forward primer: 5′-GTACTTCACGTGAGTCATAGC-3′, dissolved at 10 μM in water.
3. Reverse primer: 5′-GCTGCATTCATAGTAGCAGTC-3′, dissolved at 10 μM in water.
4. Phusion high-fidelity PCR master mix with HF buffer (New England Biolabs).
5. Thermocycler.
6. QIAquick nucleotide removal kit (Qiagen).
7. Instrument for DNA quantification (e.g., UV spectrophotometer).
8. FastDigest BsmBI (Esp3I) (Thermo Fisher Scientific).
9. 10× FastDigest buffer (Thermo Fisher Scientific).
10. Dithiothreitol (DTT): dissolved at 20 mM in water (freshly prepared).
11. XCell SureLock Mini-Cell electrophoresis system (Invitrogen).
12. Novex TBE gel, 6% (Invitrogen).
13. Novex TBE running buffer (Invitrogen).
14. Novex Hi-Density sample buffer (Invitrogen).
15. 50 bp DNA ladder for DNA electrophoresis

16. DNA stain and imaging system of choice for gel electrophoresis (We use Midori Green Direct and a FAS V gel imaging system (Nippon Genetics)).
17. plentiGuide-Puro plasmid DNA (Addgene #52963),
18. FastAP thermosensitive fast alkaline phosphatase (Thermo Fisher Scientific).
19. Agarose for extraction of nucleic acids (Seakem GTG agarose, Lonza).
20. 1 kb DNA ladder for DNA electrophoresis
21. DNA gel loading dye for DNA gel electrophoresis.
22. TAE buffer (Tris-acetate-EDTA).
23. Agarose gel tank system.
24. PureLink quick gel extraction kit (Invitrogen).
25. T4 DNA ligase (New England Biolabs).
26. T4 DNA ligase buffer (New England Biolabs).
27. One Shot Stbl3 chemically competent *E. coli* (including SOC medium) (Invitrogen).
28. LB agar plates with 100 μg/mL ampicillin (both 10-cm and 15-cm-diameter Petri dishes).
29. Endura electrocompetent cells (including 2.4 mL recovery medium per reaction) (Lucigen).
30. Gene Pulser Xcell (Bio-Rad).
31. Gene Pulser electroporation cuvettes (0.1 cm gap) (Bio-Rad).
32. Phosphate-buffered saline (PBS).
33. L-shaped cell spreaders.
34. PureLink HiPure plasmid maxiprep kit (Invitrogen).

2.2 Lentivector Production

1. plentiGuide-Puro sgRNA plasmid library (transfer plasmid).
2. pMD2.G plasmid (Addgene #12259).
3. psPAX2 plasmid (Addgene #12260).
4. HEK293T cells (*see* **Note 1**) cultured in DMEM supplemented with 10% heat-inactivated fetal calf serum and 20 μg/mL gentamicin. Cells are incubated at 37 °C and 5% CO_2.
5. X-tremeGENE 9 DNA transfection reagent (Roche).
6. Opti-MEM (Gibco).
7. Bovine serum albumin for cell culture purposes.

2.3 Lentivector Titration

1. Cell line of choice selected for screening. This cell line should stably express *Sp*Cas9 (*see* Subheading 3.5) or, alternatively, one could use a single vector system with both sgRNA and endonuclease on the same plasmid.

2. Appropriate growth medium (e.g., use IMDM supplemented with 10% fetal calf serum and 20 μg/mL gentamicin for HAP1 and K562).
3. VenorGeM OneStep mycoplasma detection kit (Minerva Biolabs).
4. Trypsin-EDTA 0.05% (for adherent cells, e.g., HAP1).
5. Cell count instrument with trypan blue exclusion method.
6. Trypan blue dye.
7. Polybrene: 8 mg/mL in water. Store aliquots at −20 °C and do not refreeze after thawing.
8. One or more aliquots of lentiviral library (thaw on ice and do not refreeze).
9. Puromycin dihydrochloride: 1 mg/mL in water. Store aliquots at −20 °C.
10. Centrifuge with plate adapters and heating function.

2.4 Stable SpCas9 Cell Line Generation

1. plentiCRISPRv2-Blast plasmid DNA (Addgene #83480).
2. Blasticidin: 10 mg/mL in water.

2.5 CRISPR–SpCas9 Compound Screen

1. Bioactive compound of choice: typically dissolved in DMSO.

2.6 DNA Library Preparation

1. QIAamp DNA Mini Kit (Qiagen).
2. NEBnext Ultra II Q5 master mix (New England Biolabs).
3. QIAquick PCR purification kit (Qiagen).
4. Primers for PCR 1 and PCR 2 (Table 7).

2.7 Validation of Individual Hits

1. For each presumed hit to be validated, two oligos should be designed such that the forward and reverse strands are complementary and pairing results in overhangs for cloning into plentiGuide-Puro (Addgene #52963). Also include a few safe-targeting sgRNAs to serve as control [11].

   Forward oligo: 5′-CACCGN_{20}-3′, dissolved at 100 μM in water.

   Reverse oligo: 5′-AAACN_{20}C-3′, dissolved at 100 μM in water.

   N_{20} represents the 20 bp sgRNA target sequence.
2. T4 polynucleotide kinase (PNK) (New England Biolabs).

3 Methods

3.1 In Silico sgRNA Tiling Library Design

1. A list of candidate target genes for the compound under investigation is used as input on the Ensembl BioMart platform [12, 13].

2. Start a new query within the "Ensembl Genes" database and continue by selecting the "Human genes" dataset.
3. Set filters to "Gene" followed by "Input external references ID list" and select "gene names." Here, you can paste the list of genes.
4. Now continue with selecting the following attributes for export: "Gene name," "Chromosome scaffold/name," "Genomic coding start," and "Genomic coding end."
5. Next, export the results as a XLS file. Remove duplicate rows and rows with a blank value for "Genomic coding start" and "Genomic coding end," since these originate from exons that are not translated.
6. This file can now be used as input file for the RStudio script "sgRNAs_from_ensembl(SpCas9).R" (*see* "Electronic Supplementary Material" [5]) to generate a comprehensive list of all sgRNAs targeting the genes of interest. When running this script, make sure the genome assembly code (e.g., GRCh38.p13) is identical to the one used in Ensembl. Remove duplicate sgRNAs from the output list as well as sequences with a BsmBI restriction site (CGTCTC) and T nucleotide repeats of five or more (e.g., TTTTT).
7. To the 20 bp sgRNA target sequence, add adaptor sequences at both ends. The complete sequence to be synthesized should look as follows: 5′-GTACTTCACGTGAGTCATA GCCGTCTCTCACCGN_{20}GTTTGGAGACGGACTGCTA CTATGAATGCAGC-3′. These adaptor oligonucleotides contain a generic priming site for PCR amplification as well as a BsmBI cut site required for further cloning into plentiGuide-Puro (*see* **Note 2**).
8. Synthesis of large batches of DNA oligonucleotides requires specialized chip synthesis technology, which is offered by several companies. We recommend Twist Bioscience Corp. (San Francisco, CA).

3.2 sgRNA Library Cloning

The protocol for the sgRNA library cloning was adapted from [14, 15].

1. Perform PCR on the oligo pool to amplify the DNA and generate double-stranded DNA oligos. To obtain a sufficient yield of DNA, set up five parallel reactions (Tables 1 and 2) (*see* **Note 3**).
2. Pool the PCR products together and purify using a single column of the QIAquick nucleotide removal kit (Qiagen). Elute in 30 μL and measure the DNA concentration (spectrophotometer or fluorometric measurement; there is no need for more advanced methods [e.g., Bioanalyzer]).

Table 1
PCR reaction mixture for amplification of oligo pool DNA

	V (μL)
Oligo pool DNA (5 ng/μL)	1
Forward primer (10 μM)	1.25
Reverse primer (10 μM)	1.25
Phusion master mix (New England Biolabs)	12.5
Water	9
Total	**25**

Table 2
Thermocycler settings for PCR on oligo pool DNA

		T (°C)	*t* (s)	Cycles
1	Initial denaturation	98	60	
2	Denaturation	98	5	Steps 2–4: 10 cycles
3	Annealing	61	10	
4	Elongation	72	5	
5	Final elongation	72	300	
6	Pause	10	∞	

Table 3
Reaction mixture for restriction digestion of PCR product with BsmBI

	V (μL)
PCR product	0.2 μg
10× FastDigest buffer (Thermo Fisher Scientific)	3
FastDigest BsmBI (Thermo Fisher Scientific)	1
DTT 20 mM (freshly prepared)	1.5
Water	Up to 30
Total	**30**

3. Cut the PCR product with the FastDigest BsmBI restriction enzyme (Thermo Fisher Scientific) to generate sticky ends, which will enable ligation cloning (Table 3). Again, we recommend setting up five parallel reactions. However, always keep

some undigested PCR product to use as a control when checking for digestion efficiency (*see* **step 6**).

4. Incubate in a warm water bath at 37 °C for 15 min. Inactivate at 65 °C for 10 min.
5. Pool the products together and purify using a single column of the QIAquick nucleotide removal kit. Elute in 30 μL and measure the DNA concentration (spectrophotometer or fluorometric measurement; there is no need for more advanced methods [e.g., Bioanalyzer]).
6. To check for complete digestion, perform DNA PAGE with a sample of undigested and digested PCR product. For each sample, mix 50 ng of DNA with 2 μL of Novex Hi-Density sample buffer (Invitrogen), 0.3 μL Midori Green Direct (Nippon Genetics), and water up to 10 μL. Set aside all other digested product.
7. Mix 1 μL 50 bp DNA ladder with 1 μL of Midori Green Direct, 2 μL of Hi-Density sample buffer and 6 μL of water.
8. Install the gel cassette (Novex TBE gel, 6%) into the electrophoresis tank and fill the chambers with 1 × TBE running buffer.
9. Load the DNA ladder and samples on the gel and run at 200 V for 20 min.
10. Remove the gel from the cassette and rinse with water.
11. When using a dye other than preadded Midori Green (e.g., SYBR Green 1 or ethidium bromide), incubate the gel in the appropriate staining solution or use your own optimized protocol for detection of the DNA bands.
12. Image the gel with the FAS V gel-reading instrument (or equivalent).
13. The digested DNA fragments are 25, 28, and 32 bp in length. Most importantly, there should be no intact 85 bp PCR product band in the digested sample.
14. Cut open and dephosphorylate the plentiGuide-Puro backbone plasmid with BsmBI (Table 4). Set up five parallel reactions.
15. Incubate in a warm water bath at 37 °C for 30 min. Inactivate at 65 °C for 10 min.
16. Prepare a 1% agarose gel with TAE buffer. Use agarose suitable for extraction of nucleic acids (SeaKem GTG Agarose, Lonza).
17. Place the gel into an electrophoresis tank and fill with TAE buffer.

Table 4
Reaction mixture for restriction digestion and dephosphorylation of plentiGuide-Puro with BsmBI

	V (μL)
plentiGuide-puro DNA prep	5 μg
10× FastDigest buffer	6
FastDigest BsmBI	3
FastAP (Thermo Fisher Scientific)	3
DTT 20 mM (freshly prepared)	3
Water	Up to 60
Total	**60**

18. Pool the digested plasmid DNA together and add loading dye (final concentration 1×) and Midori Green (0.6 μL per 10 μL). Carefully and slowly, load the sample onto the gel. Try to get as much volume in one well as possible without causing it to overflow. Mix 10 μL 1 kb DNA ladder with 2 μL Midori Green and load as well.
19. Run the electrophoresis at 120 V for 1.5 h.
20. Take a clean, empty 15-mL tube (or other recipient) and record its weight.
21. After imaging, you should see two clear bands at 8298 bp (main product) and 1885 bp (stuffer). The presence of a larger smear or bands above >9000 bp indicates incomplete digestion. Carefully excise the 8298 bp band not to include any excess agarose gel. Transfer the excised gel fragment to the preweighted 15-mL tube. Neglect the 1885 bp stuffer band.
22. Extract the 8298 bp DNA from the excised gel fragment following the protocol of the PureLink quick gel extraction kit (Invitrogen). Weigh the 15-mL tube containing the excised gel and calculate the weight of the excised gel fragment. During gel extraction, use one purification column per 400 mg agarose gel.
23. Pool together the gel-purified DNA and measure its concentration.
24. Next, ligate the digested PCR product into the gel-extracted plentiGuide-Puro plasmid. Maintain a 5:1 molar ratio of insert: vector (Table 5). Set up five parallel reactions.
25. Incubate at room temperature for 10 min.
26. Pool all five reactions together and purify the DNA with one column of the QIAquick nucleotide removal kit. Elute the

Table 5
Ligation reaction of digested PCR-amplified sgRNA pool into plentiGuide-Puro backbone

	V (μL)
Digested PCR product	25.5 ng
Digested, gel-purified plentiGuide-puro	500 ng
T4 DNA ligase 400 units/μL (New England Biolabs)	1
T4 DNA ligase buffer (New England Biolabs)	2
Water	Up to 20
Total	**20**

DNA in 30 μL of pure water rather than elution buffer since the latter might have a negative impact on transformation efficiency.

27. Before continuing with electroporation, we recommend doing a test transformation in Stbl3 chemically competent cells (Invitrogen). Transform 2 μL of ligation mixture into one vial of competent cells and follow the manufacturer's protocol.
28. Plate out a serial dilution of the transformed bacteria on 10-cm agar plates (+ampicillin).
29. The next day, count colonies and calculate back to the original transformation volume (300 μL). The total number of colony-forming units should easily exceed 2000.
30. Next, continue with the electroporation (*see* **Note 4**). Thaw the Endura electrocompetent cells (Lucigen) on ice and resuspend by gently flicking the tube.
31. Add 2 μL ligation mixture to one reaction volume of cells. Do not mix by pipetting up and down, but by gently flicking the tube.
32. Transfer the mixture of cells and DNA to the bottom of a chilled 0.1-cm electroporation cuvette (Bio-Rad). Make sure to avoid air bubbles. Place the cuvette in the ShockPod cuvette chamber of the Gene Pulser Xcell (Bio-Rad). Use a new cuvette for each reaction.
33. Pulse once using the settings suggested by the manufacturer (Lucigen).
34. Remove the cuvette from the chamber and immediately add 1 mL of recovery medium to the cuvette. Quickly but gently, resuspend the cells and transfer to a round-bottom tube already containing 1.4 mL recovery medium.
35. Incubate in a shaking incubator at 32 °C for 1 h.
36. Pool all the reactions together and mix.

37. Spread out 800 μL transformation mixture per 15-cm selective plate with an L-shaped cell spreader (3 agar plates per electroporation reaction).
38. In addition, plate out a serial dilution on 10-cm agar plates.
39. Incubate agar plates at 32 °C for up to 18 h.
40. The next day, count the colonies from the serial dilution plates and calculate the total number of colony-forming units (cfu). This should easily exceed 10^6 colonies per electroporation reaction. The coverage is calculated by dividing the total number of cfu by the number of unique sgRNA sequences in the library.
41. Weigh an empty centrifugation bottle.
42. To recover the bacterial colonies, add 8 mL PBS to an agar plate and use a cell spreader to resuspend the cells. Transfer the cell suspension to the centrifugation bottle.
43. Repeat with 8 mL of fresh PBS to remove as much bacteria as possible.
44. After scraping of all agar plates, spin the cell suspension at $4000 \times g$ for 15 min.
45. Decant the supernatant and weigh the bottle to determine the bacterial pellet's mass.
46. Continue with extracting the plasmid DNA using the PureLink HiPure plasmid maxiprep kit (Invitrogen). Use one column per 2 g bacteria and follow the provided protocol for low-copy plasmids.
47. Pool the eluted DNA and measure the final plasmid DNA concentration.

3.3 Lentivector Production

Working with lentivector must be done in an appropriate biosafety cabinet in a BSL-2 laboratory. Always wear double gloves, a (disposable) laboratory coat and use bleach to decontaminate materials in the biosafety cabinet for 30 min before handling waste.

1. The day before transfection, seed 9×10^6 HEK293T cells per 150-cm^2 culture flask in 20 mL DMEM (supplemented with 10% heat-inactivated fetal calf serum and 20 μg/mL gentamicin). Prepare 10 flasks (or more) to obtain a sufficient stock of lentivector.
2. After a 24-h incubation period at 37 °C, 5% CO_2, flasks should be 70–80% confluent. Prepare a 1:1:1 molar ratio mixture of the three plasmids to be transfected (Table 6).
3. Allow the X-tremGENE reagent (Roche) and opti-MEM (Gibco) to equilibrate to room temperature. Next, using a separate tube for each flask of HEK293T cells, dilute 48 μL X-tremGENE transfection reagent in 800 μL opti-MEM. Mix

Table 6
DNA mixture for transfection of HEK293T packaging cells

	Size (bp)	*m* (μg)
plentiGuide-puro sgRNA plasmid pool	8323	5.37
pMD2.G (Addgene #12259)	5822	3.75
psPAX2 (Addgene #12260)	10,668	6.88
Total		**16**

by gently flicking the tube. Incubate at room temperature for 5 min.

4. Now add the mixture of DNA (16 μg) to each X-tremGENE transfection tube and mix by gently flicking it. Incubate at room temperature for 30 min.
5. Carefully transfer the transfection complex to the HEK293T cells by adding dropwise to the flask. Tilt the flask back and forth to mix the reagent with the culture medium.
6. After 18-h incubation, replace the medium with 20 mL DMEM supplemented with 1.1 g/100 mL bovine serum albumin (no calf serum).
7. After another ~24 h of incubation, harvest lentivector by collecting media in 50-mL conical tubes.
8. Spin at 200 × *g* for 5 min to pellet any cell debris. Collect the supernatant and pool all media together in one new flask.
9. Aliquot the lentivector in 2- or 5-mL cryogenic vials and store at −80 °C.

3.4 Lentivector Titration

Perform titration on the same cell line that you will use for screening (e.g., HAP1 or K562) (*see* **Note 5**). This cell line should stably express *Sp*Cas9.

1. Prewarm the centrifuge to 37 °C.
2. Thaw an aliquot of lentivector and polybrene on ice.
3. Trypsinize and count cells with the trypan blue dye exclusion method.
4. Prepare 10.5 mL of a 1.33×10^6 cells/mL suspension in supplemented cell culture medium (IMDM with 10% FBS, 20 μg/mL gentamicin) with 10, 67 μg/mL polybrene (*see* **Note 6**).
5. Plate out the cell suspension at 1.5 mL/well in six wells of a 12-well culture plate such that each well will contain 2×10^6 cells. Fill the remaining six wells with 2 mL PBS.

6. Now add a designated volume of lentivector to each well (0, 10, 25, 50, 100 and 200 μL). Next, add additional supplemented cell culture medium to a final volume of 2 mL in each well. The end concentration of polybrene is 8 μg/mL. Mix well by pipetting up and down.
7. Spin the 12-well plate at 750 × *g* for 2 h at 37 °C (spinfection) and place the cells back into the incubator overnight (37 °C, 5% CO_2).
8. The next day, infection should be complete. Aspirate medium, wash with PBS and trypsinize cells. Resuspend cells in each well separately with a total volume of 2.2 mL supplemented cell culture medium.
9. For each condition, prepare two 75-cm^2 flasks with 19 mL of supplemented cell culture medium (*see* **Note 7**). Add puromycin to only one of the two flasks for each condition at a final concentration of 1 μg/mL (*see* **Note 8**).
10. For each condition, transfer 1 mL of cell suspension to a flask without and 1 mL to a flask with puromycin. Return flasks to the incubator (37 °C, 5% CO_2).
11. After 48–72 h, all cells in the control flask with puromycin but without any lentivector should be dead.
12. Aspirate media, wash with PBS, trypsinize, resuspend in FBS-supplemented cell culture medium and count cells. Keep the volume the same across conditions.
13. For each lentivector volume of the dilution series, calculate the lentivector transduction efficiency as follows:

$$Transduction\ Efficiency\ (\%) = \frac{Cell\ Count_{+Puromycin}}{Cell\ Count_{-Puromycin}} \times 100$$

14. From the transduction efficiency, we can obtain the multiplicity of infection (MOI), i.e., the ratio of infecting viral particles per cell, as follows:

$$MOI = -\ln\left(1 - \frac{Transduction\ Efficiency}{100}\right)$$

15. Plot all calculated MOI values versus lentivector volumes and fit a linear regression curve. From this curve, obtain the required volume for a MOI of 0.25 (*see* **Note 9**).

3.5 Stable SpCas9 Cell Line Generation

There are many different strategies to establish a *Sp*Cas9 stable cell line. One of the more straightforward methods uses lentiviral transduction of the *Sp*Cas9 coding sequence [16].

1. Starting with the plentiCRISPRv2-Blasti plasmid, produce lentivector (*see* Subheading 3.3) and titrate lentivector (*see* Subheading 3.4) as described above.
2. Instead of puromycin, select with 10 μg/mL blasticidin (*see* **Note 8**).
3. Since this transfer plasmid is larger than plentiGuide-Puro (due to the presence of the complete coding sequence for the *Sp*Cas9 endonuclease), a lower viral titer is expected. Therefore, titrate volumes from 0 to 750 μL.
4. Select the condition with a MOI <0.25 since these cells will most likely carry only one copy of the insert.

Following transduction and blasticidin selection, in most cases, establishing a clonal cell line with uniform *Sp*Cas9 expression is required. *Sp*Cas9 expression and activity can be assessed in different ways such as mutation detection by surveyor nuclease assay [17], sgRNA target location sequencing, or the EGFP reporter assay (Addgene #107145).

3.6 CRISPR–SpCas9 Compound Screen

1. Test all used cell lines for mycoplasma contamination using the VenorGeM OneStep mycoplasma detection kit (Minerva Biolabs).
2. Prewarm the centrifuge to 37 °C.
3. Thaw one or more aliquots of lentivector and polybrene on ice.
4. Calculate the required number of cells for transduction as below. We recommend starting at a 5000-fold coverage (*see* **Note 10**).

$$Required\ Number\ of\ Cells = \frac{Library\ Size \times Coverage}{MOI}$$

5. Round up to the nearest multiple of 2×10^6. Now prepare the required volume of a 1×10^6 cells/mL cell suspension in supplemented cell culture medium (10% FBS, 20 μg/mL gentamicin) with 8 μg/mL polybrene and lentivector (the volume obtained from the viral titration multiplied by the number of wells to be used). Mix well.
6. Plate out in 12-well plates at 2 mL per well.
7. In addition, prepare six wells with 2 mL 1×10^6 cells/mL cell suspension with 8 μg/mL polybrene but no lentivector. This condition serves as a control for puromycin selection.
8. Spin the 12-well plates at 750 × *g* for 2 h at 37 °C (spinfection) and place the cells back in the incubator overnight (37 °C, 5% CO_2).

9. The next day, aspirate medium, wash with PBS, trypsinize and resuspend cells in fresh, supplemented medium (10% FBS, 20 μg/mL gentamicin).
10. Transfer all transduced cells to a sterile container and count. Likewise, collect and count cells from the wells without lentivector.
11. Plate out cells in 150-cm^2 flasks at 10×10^6 cells per flask in 30 mL medium with 1 μg/mL puromycin. Furthermore, plate out one flask with 10×10^6 nontransduced cells in 30 mL medium with 1 μg/mL puromycin (*see* **Note 11**).
12. Plate out transduced cells in two 75-cm^2 flasks at 1×10^6 cells per flask in 20 mL medium. To one of two flasks, add puromycin at a final concentration of 1 μg/mL. These two flasks serve as a control to check whether the foreseen MOI matches the actual one.
13. After 48–72 h of puromycin selection, check whether (just about) all cells in the flask with nontransduced cells and puromycin are killed.
14. Count cells in both 75-cm^2 flasks and determine the transduction efficiency to verify the true MOI as described above (*see* Subheading 3.4).
15. Culture the transduced cells for another 2–3 days to assure that all genome-editing events are completed.
16. After this incubation period, seed puromycin selected cells in new 150-cm^2 flasks at 10×10^6 cells per flask. In these flasks, start treatment with the investigational compound (*see* **Note 12**). Make sure one or multiple control flasks with nontransduced cells are in place to monitor treatment progress.
17. Keep cells under selection for a period of 2–3 weeks and refresh the cell culture medium (10% FBS, 20 μg/mL gentamicin) with investigational compound at least once a week (more if the compound is unstable or metabolized). When refreshing cell culture medium for adherent cells, be gentle not to disturb or detach growing colonies. During the first refreshment, many dead cells will likely be removed.
18. If successful, compound-resistant colonies should start to appear after 7–10 days in the transduced population but not in the control flasks.
19. To harvest resistant cells, carefully aspirate the medium, wash the cells gently with PBS and detach the cells by incubation in trypsin-EDTA 0.05% for 5 min (or until detachment). Add fresh, supplemented cell culture medium (10% FBS, 20 μg/mL gentamicin) and collect cells in 15- or 50-mL tubes. Use low volumes if only few resistant colonies are present. Harvest all collected cells by centrifugation at $200 \times g$ for 5 min. Very

carefully remove the supernatant and resuspend resistant cells in fresh, supplemented cell culture medium. The resuspension volume can be adjusted to reflect the size of the cell pellet. Pool all resistant cells and count the number of cells (trypan blue exclusion). At a minimum, if possible, set aside 5×10^6 resistant cells for genomic DNA (gDNA) extraction. Spin down these cells at $200 \times g$ for 5 min, remove supernatant, wash with 10 mL PBS and spin again. Remove supernatant and store this cell pellet at −20 °C, awaiting DNA extraction (*see* **Note 13**). Remaining cells can either be frozen down as a viable cell culture, cultured further, or be used for gDNA extraction at the user's discretion.

20. We recommend repeating the entire screen once with a new batch of transduced cells (*see* **Note 14**).

3.7 DNA Library Preparation

1. Thaw cell pellets and extract gDNA with the QIAamp DNA Mini Kit (Qiagen) according to the protocol provided. Measure gDNA concentrations.
2. In a first PCR reaction, the sgRNA regions are enriched (Tables 7, 8, and 9). Perform PCR on the gDNA samples as well as the plasmid library pool (*see* **Note 15**).
3. Purify the product with the QIAquick PCR purification kit (Qiagen) and elute in 50 μL.
4. Check whether the PCR reaction was successful by running 5 μL on agarose gel. PCR 1 should yield a product of 867 bp.
5. Use 5 μL PCR 1 product as input for PCR 2. This reaction will further amplify the sgRNA sequence and attach TruSeq adapters for Illumina sequencing (Tables 7, 10 and 11). For each sample, use a different set of i5 and i7 indices to enable demultiplexing after sequencing.
6. Use gel extraction to purify the PCR 2 product as described above (*see* Subheading 3.2). This time, use a 2% agarose gel instead of 1% since the expected product is only ~300 bp. Measure the DNA concentration.
7. Dilute all samples to the lowest concentration and pool together a designated volume from each sample, considering the anticipated sequencing coverage for each sample (*see* **Note 16**).

3.8 Next-Generation Sequencing of sgRNAs

Most laboratories rely on a third party for Illumina next-generation sequencing. Take into account the following guidelines:

1. Both MiSeq, HiSeq, and NextSeq platforms are suitable for sgRNA sequencing. Platform choice depends on the required number of reads (*see* **Note 16**).

Table 7
PCR 1 and PCR 2 primers for next-generation sequencing library preparation

PCR 1	primer sequences 5′–3′
forward primer	GAGGGCCTATTTCCCATGATTC
Reverse primer	GTTGCGAAAAAGAACGTTCACGG
PCR 2 i5 forward primer sequences 5′–3′	
D501-F	AATGATACGGCGACCACCGAGATCTACACTATAGCCT[a]ACACTCTTTCCCTA CACGACGCTCTTCCGATCT**TTGTGGAAAGGACGAAACACCG**[b]
D502-F stagger 1	AATGATACGGCGACCACCGAGATCTACACATAGAGGCACACTCTTTCCCTA CACGACGCTCTTCCGATCTC**TTGTGGAAAGGACGAAACACCG**
D503-F stagger 2	AATGATACGGCGACCACCGAGATCTACACCCTATCCTACACTCTTTCCCTA CACGACGCTCTTCCGATCTAC**TTGTGGAAAGGACGAAACACCG**
D504-F stagger 3	AATGATACGGCGACCACCGAGATCTACACGGCTCTGAACACTCTTTCCCTA CACGACGCTCTTCCGATCTCAC**TTGTGGAAAGGACGAAACACCG**
D505-F stagger 4	AATGATACGGCGACCACCGAGATCTACACAGGCGAAGACACTCTTTCCCTA CACGACGCTCTTCCGATCTACAC**TTGTGGAAAGGACGAAACACCG**
D506-F stagger 5	AATGATACGGCGACCACCGAGATCTACACTAATCTTAACACTCTTTCCCTA CACGACGCTCTTCCGATCTGACAC**TTGTGGAAAGGACGAAACACCG**
D507-F stagger 6	AATGATACGGCGACCACCGAGATCTACACCAGGACGTACACTCTTTCCCTA CACGACGCTCTTCCGATCTGGACAC**TTGTGGAAAGGACGAAACACCG**
PCR 2 i7 reverse primer sequences 5′–3′	
D701-R	CAAGCAGAAGACGGCATACGAGATCGAGTAATGTGACTGGAGTTCA GACGTGTGCTCTTCCGATCT **TACTGCCATTTGTCTCAAGATCTAGTTAC**
D702-R stagger 1	CAAGCAGAAGACGGCATACGAGATTCTCCGGAGTGACTGGAGTTCA GACGTGTGCTCTTCCGATCTG **TACTGCCATTTGTCTCAAGATCTAGTTAC**
D703-R stagger 2	CAAGCAGAAGACGGCATACGAGATAATGAGCGGTGACTGGAGTTCA GACGTGTGCTCTTCCGATCTAG **TACTGCCATTTGTCTCAAGATCTAGTTAC**
D704-R stagger 3	CAAGCAGAAGACGGCATACGAGATGGAATCTCGTGACTGGAGTTCA GACGTGTGCTCTTCCGATCTGAG **TACTGCCATTTGTCTCAAGATCTAGTTAC**
D705-R stagger 4	CAAGCAGAAGACGGCATACGAGATTTCTGAATGTGACTGGAGTTCA GACGTGTGCTCTTCCGATCTTGAG **TACTGCCATTTGTCTCAAGATCTAGTTAC**
D706-R stagger 5	CAAGCAGAAGACGGCATACGAGATACGAATTCGTGACTGGAGTTCA GACGTGTGCTCTTCCGATCTATGAG **TACTGCCATTTGTCTCAAGATCTAGTTAC**
D707-R stagger 6	CAAGCAGAAGACGGCATACGAGATAGCTTCAGGTGACTGGAGTTCA GACGTGTGCTCTTCCGATCTCATGAG **TACTGCCATTTGTCTCAAGATCTAGTTAC**

[a]i5 and i7 index sequences underlined
[b]PCR 2 priming sites in bold

Table 8
PCR 1 reaction mixture

	V (μL)
Plasmid library DNA/gDNA from selected cells	100 ng/2000 ng
PCR 1 forward primer (10 μM)	2.5
PCR 1 reverse primer (10 μM)	2.5
NEBnext ultra II Q5 master mix (New England Biolabs)	25
Water	Up to 50 μL
Total	**50**

Table 9
Thermocycler settings for PCR 1

		T (°C)	*t* (s)	Cycles
1	Initial denaturation	98	30	
2	Denaturation	98	10	**Steps 2–4**: 25 cycles
3	Annealing	66	30	
4	Elongation	72	15	
5	Final elongation	72	120	
6	Pause	10	∞	

Table 10
PCR 2 reaction mixture

	V (μL)
PCR 1 product	5
PCR 2 forward primer (10 μM)	2.5
PCR 2 reverse primer (10 μM)	2.5
NEBnext ultra II Q5 master mix	25
Water	15
Total	**50**

2. Before sequencing, exact DNA concentrations should be determined with qPCR using p5 and p7 specific primers.
3. Spike-in of 20–30% PhiX is required to increase library variability.

Table 11
Thermocycler settings for PCR 2

		T (°C)	*t* (s)	Cycles
1	Initial denaturation	98	30	
2	Denaturation	98	10	**Steps 2–4**: 10 cycles
3	Annealing	55	30	
4	Elongation	65	15	
5	Final elongation	65	300	
6	Pause	10	∞	

4. Sequencing of sgRNAs requires a single read of at least 50 bp. A longer read length and/or paired-end sequencing are also possible.
5. When combining i5 and i7 indices, dual-index reads are necessary. i7 sequences on the sample sheet should be the reverse complement of primer sequences (Table 7).
6. After obtaining demultiplexed FASTQ files, sgRNA counts can be extracted with EdgeR [18, 19] or comparable software tools. A concise script for this purpose is available from the "Electronic Supplementary Material" [5].
7. Fold-change of enriched sgRNAs is calculated as below.

$$log_2 FC = log_2 \left(\frac{Counts\ Per\ Million_{after\ selection} + 1}{Counts\ Per\ Million_{plasmid\ library} + 1} \right)$$

8. sgRNAs emerging from both replicate screens with a $\log_2$FC greater than 2–4 are considered hits (though the precise cutoff value may vary from screen to screen and library size). These should be further validated individually.

3.9 Validation of Individual Hits

The protocol for cloning of individual sgRNA target sequences into the plentiGuide-Puro backbone is as previously described [16].

1. Dissolve oligos at 100 μM in water and setup the phosphorylation and annealing reaction (Tables 12 and 13).
2. Dilute annealed oligos at 1:200 in water.
3. Ligate the annealed oligo pair into the digested, gel-extracted plentiGuide-Puro vector (Table 14).
4. Transform the ligation product into Stbl3 competent cells and extract plasmid DNA. Use Sanger sequencing to confirm correct integration of the sgRNA sequence. Continue with producing lentivector (one flask), transduction into *Sp*Cas9 stable

Table 12
Reaction mixture for phosphorylation and annealing of complementary oligos

	V (μL)
Forward oligo (100 μM)	1
Reverse oligo (100 μM)	1
10× T4 DNA ligation buffer	1
T4 PNK 10 units/μL (New England Biolabs)	0.5
Water	6.5
Total	**10**

Table 13
Thermocycler settings for phosphorylation and annealing of complementary oligos

		T (°C)	**t**	**Cycles**	**ΔT (°C)**
1	Phosphorylation	37	30 min		
2	Denaturation	95	5 min		
3	Annealing	95	12 s	**Step 3**: 85 cycles	−1
4	Pause	10	∞		

Table 14
Ligation reaction of annealed oligos into plentiGuide-Puro backbone

	V (μL)
Annealed oligos (1:200 dilution)	1
Digested, gel-purified plentiGuide-puro	50 ng
T4 DNA ligase 400 units/μL	1
T4 DNA ligase buffer	2
Water	Up to 20
Total	**20**

cells (one well), and puromycin selection as described above (*see* Subheadings 3.1–3.4). Establish a cell line for each hit and safe-targeting sgRNA.

5. For each sgRNA-expressing cell line, seed 5×10^6 cells in a 75-cm^2 flask and select with compound for resistance over 7–14 days. Select until cells with a safe-targeting control sgRNA are completely dead and allow resistant colonies to form in the flasks with hit sgRNAs.

6. Collect resistant cells and count live cell numbers with the trypan blue method. The ability to induce resistant cells effectively validates the hit sgRNA.
7. Harvest 1×10^6 resistant cells by centrifugation. Wash with 10 mL PBS and spin again. Remove supernatant and store cell pellet at −20 °C.

3.10 Sample Preparation and Next-Generation Sequencing of Target Site Variants

1. Thaw cell pellets, extract gDNA, and prepare samples for next-generation sequencing as described above (*see* Subheading 3.7). PCR 1 primers should be designed such that they amplify the sgRNA target region of interest. For the nested PCR 2 primers, the priming sites should be adjusted (Table 7) to target a region of ~100 bp (primers not included) around the sgRNA cut site.
2. Next-generation sequencing is as described above with an optimal read length of >150 bp and paired-end sequencing (*see* Subheading 3.8).
3. Target site variants are identified and quantified with the "CrispRVariants – Exon analysis.R" script (*see* "Electronic Supplementary Material" [5] adapted from [20]).

4 Notes

1. Make sure to obtain a clonal cell line suitable for lentivector production.
2. It is possible to use different sets of PCR priming sites in order to construct several libraries from one and the same pool of oligonucleotides.
3. When using a different set of primers, adjust the annealing temperature accordingly. The New England Biolabs Website offers a tool to estimate annealing temperatures for each different available polymerase (https://tmcalculator.neb.com/#!/main).
4. Determining the required number of electroporation reactions in advance can be a challenge. This depends on both library size and desired coverage. Coverage is defined as the average number of cells containing a given sgRNA sequence after electroporation. A high-coverage electroporation ensures that (almost) all sgRNA sequences are represented (e.g., a library of 10,000 sgRNA sequences and a desired coverage of 1000-fold means there should be at least 10×10^6 colony-forming units after transformation). The safest option here is to start with a single electroporation, count the number of colonies, and scale up the reactions accordingly.

5. Preferably, select a cell line that is easy to culture, fast-growing, suitable for lentiviral transduction, amenable to puromycin and blasticidin selection, and susceptible to the examined bioactive molecule.
6. Some cell lines require a lower concentration of polybrene or do not tolerate it at all.
7. Make sure to select a flask big enough to support growth during the entire puromycin selection process or alternatively, seed fewer cells. Avoid cell overgrowth in the conditions without puromycin as this will cause overestimation of the multiplicity of infection.
8. Make sure to use a concentration of puromycin/blasticidin high enough to kill all nontransduced cells in a time span of 48–72 h. For most cell lines, concentrations of 1 μg/mL puromycin and 10 μg/mL blasticidin are adequate.
9. At high lentivector volumes and transduction percentages, linearity between MOI and volume might be lost due to experimental error. These values should be excluded from the regression curve.
10. The minimal required coverage for a screen is hard to predict beforehand. Certainly, this depends on the compound under investigation and more specifically, the type of mutations required for resistance—a three-base-pair deletion is easily generated while a specific point mutation will, by chance, occur less frequently. Nevertheless, we recommend performing initial screens with a coverage of ~5000, meaning that for each unique sgRNA in the library, at least ~5000 cells should be infected. Empirically, this number ensures that a decent amount of genetic diversity is generated at each sgRNA target site, although it will most certainly not reach saturation of all possible mutations for each target site.
11. Cell numbers mentioned here are for HAP1 cells and might be different for other cell lines. Make sure you know the doubling time of your cell line and keep in mind that cells might temporarily grow slower the first days after transduction.
12. An eligible drug should have a single known or presumed molecular target. Prior to screening, establish a dose–response curve to determine the drug concentration required to kill all nonresistant cells. This is vital to minimize background signal. When uncertain about the optimal concentration, screen two different concentrations in parallel.
13. When in doubt whether surviving cells are truly resistant, harvest part of the cells and continue treatment of the remaining cells over an additional amount of time.

14. In some cases, enrichment of a particular sgRNA might occur for reasons other than inducing resistance mutations: as a passenger sgRNA in a double transduced cells or in cells that are inherently resistant because of a spontaneous mutation. Such false positives can be efficiently eliminated by repeating a screen and only considering enriched hits that emerge from both replicate screens.
15. The amount of gDNA used as template for PCR 1 is of importance to maintain adequate coverage. In general, a diploid human cell would yield around 6.6 pg of gDNA, though most cancer cells are not perfectly diploid and HAP1 cancer cells are haploid. This means that under optimal conditions, using 2000 ng of gDNA would represent about 300,000 unique genomic diploid genomes/cells. However, for selected cells, only a handful of drug resistance generating sgRNAs are expected to be found. Hence, using 2000 ng would be more than sufficient to cover the depth of the sample.
16. For sequencing of the plasmid pool, a sequencing depth of 200-fold per unique sgRNA in the library should be considered a minimum. For samples from drug-selected cells, as little as 1×10^6 sequencing reads can be more than sufficient to cover the enriched sgRNAs. This also implicates that when combining a plasmid pool sample of the entire library and the enriched sgRNAs from drug-selected samples into a single next-generation sequencing run, the latter might have to be diluted down to achieve the desired ratio of plasmid library: resistant sample(s).

References

1. Nijman SMB (2015) Functional genomics to uncover drug mechanism of action. Nat Chem Biol 11:942–948
2. Kapoor TM, Miller RM (2017) Leveraging Chemotype-specific resistance for drug target identification and chemical biology. Trends Pharmacol Sci 38:1100–1109
3. Kasap C, Elemento O, Kapoor TM (2014) DrugTargetSeqR: a genomics- and CRISPR-Cas9-based method to analyze drug targets. Nat Chem Biol 10:626–628
4. Smurnyy Y, Cai M, Wu H et al (2014) DNA sequencing and CRISPR-Cas9 gene editing for target validation in mammalian cells. Nat Chem Biol 10:623–625
5. Neggers JE, Kwanten B, Dierckx T et al (2018) Target identification of small molecules using large-scale CRISPR-Cas mutagenesis scanning of essential genes. Nat Commun 9:502
6. Greinert R, Volkmer B, Henning S et al (2012) UVA-induced DNA double-strand breaks result from the repair of clustered oxidative DNA damages. Nucleic Acids Res 40:10263–10273
7. Neggers JE, Jacquemyn M, Dierckx T et al (2020) Enhanced AsCas12a enables CRISPR-directed evolution to screen for functional drug resistance mutations in sequences inaccessible to SpCas9. Mol Ther 29(1):208–224
8. Hess GT, Frésard L, Han K et al (2016) Directed evolution using dCas9-targeted somatic hypermutation in mammalian cells. Nat Methods 13:1036–1042
9. Komor A, Kim Y, Packer M et al (2016) Programmable editing of a target base in genomic DNA without double-stranded DNA cleavage. Nature 533:420–424. https://doi.org/10.1038/nature17946

10. Hanna RE, Hegde M, Fagre CR et al (2021) Massively parallel assessment of human variants with base editor screens. Cell 184 (4):1064–1080.e20
11. Morgens DW, Wainberg M, Boyle EA et al (2017) Genome-scale measurement of off-target activity using Cas9 toxicity in high-throughput screens. Nat Commun 8:15178
12. Yates AD, Achuthan P, Akanni W et al (2020) Ensembl 2020. Nucleic Acids Res 48:682–688
13. Kinsella RJ, Kähäri A, Haider S et al (2011) Ensembl BioMarts: a hub for data retrieval across taxonomic space. Database 2011:bar030
14. Hart T, Chandrashekhar M, Aregger M et al (2015) High-resolution CRISPR screens reveal fitness genes and genotype-specific cancer liabilities. Cell 163:1515–1526
15. Shalem O, Sanjana NE, Hartenian E et al (2014) Genome-scale CRISPR-Cas9 knockout screening in human cells. Science 343:84–87
16. Sanjana NE, Shalem O, Zhang F (2014) Improved vectors and genome-wide libraries for CRISPR screening. Nat Methods 11:783–784
17. Ran FA, Hsu PD, Wright J et al (2013) Genome engineering using the CRISPR-Cas9 system. Nat Protoc 8:2281–2308
18. Robinson MD, Mccarthy DJ, Smyth GK (2010) edgeR: a Bioconductor package for differential expression analysis of digital gene expression data. Bioinformatics 26:139–140
19. McCarthy DJ, Chen Y, Smyth GK (2012) Differential expression analysis of multifactor RNA-Seq experiments with respect to biological variation. Nucleic Acids Res 40:4288–4297
20. Lindsay H, Burger A, Biyong B et al (2016) CrispRVariants charts the mutation spectrum of genome engineering experiments. Nat Biotechnol 34:701–702

Chapter 4

Design and Generation of a CRISPR Interference System for Genetic Repression and Essential Gene Analysis in the Fungal Pathogen *Candida albicans*

Lauren Wensing and Rebecca S. Shapiro

Abstract

Studying life-threatening fungal pathogens such as *Candida albicans* is of critical importance, yet progress can be hindered by challenges associated with manipulating these pathogens genetically. CRISPR-based technologies have significantly improved our ability to manipulate the genomes of countless organisms, including fungal pathogens such as *C. albicans*. CRISPR interference (CRISPRi) is a modified variation of CRISPR technology that enables the targeted genetic repression of specific genes of interest and can be used as a technique for studying essential genes. We recently developed tools to enable CRISPRi in *C. albicans* and the repression of essential genes in this fungus. Here, we describe a protocol for CRISPRi in *C. albicans*, including the design of the single-guide RNAs (sgRNAs) for targeting essential genes, the high-efficiency cloning of sgRNAs into *C. albicans*-optimized CRISPRi plasmids, transformation into fungal strains, and testing to monitor the repression capabilities of these constructs. Together, this protocol will illuminate efficient strategies for targeted genetic repression of essential genes in *C. albicans* using a novel CRISPRi platform.

Key words CRISPR, CRISPRi, *Candida albicans*, Fungal pathogens, Fungal genetics, Genetic regulation, Genetic repression

1 Introduction

Developing novel strategies to combat infectious diseases and drug-resistant microbial pathogens is one of the most pressing global health challenges. Pathogenic fungi are a growing public health concern, and invasive fungal infections are an important cause of mortality, particularly among the increasing populations of immunocompromised individuals worldwide [1, 2]. Among fungal pathogens, *Candida* species are a leading cause of fungal infection, with *Candida albicans* serving as the primary cause of all invasive candidiasis infections [3, 4]. Like many systemic fungal infections, invasive candidiasis is associated with very high rates of mortality: ~40–60% depending on the patient group and clinical

Ren Zhang (ed.), *Essential Genes and Genomes: Methods and Protocols*, Methods in Molecular Biology, vol. 2377,
https://doi.org/10.1007/978-1-0716-1720-5_4,

setting [5]. Despite the pressing need to combat these deadly fungal infections, the current arsenal of antifungal drugs is limited to very few classes of effective, nontoxic therapeutic agents [6], and continuing research efforts must focus on the identification of novel therapeutic targets for the treatment of these life-threatening infections.

Improving research on prospective targets for antifungal drug therapy requires effective molecular genetic strategies to investigate the function of key fungal genes, including essential genes, which are likely targets for antifungal therapy. Despite the importance of essential genes as prospective targets for antifungal drugs, these genes can be difficult to identify and study, particularly in diploid fungal organisms such as *C. albicans*, as both alleles of a gene of interest must be targeted by genetic manipulation techniques. To date, several strategies have been developed to identify essential genes in *C. albicans*, including gene disruption strategies in both the diploid [7, 8] and haploid [9] cell states. Additional platforms have been developed for the study of essential genes, including strategies to conditionally regulate essential gene expression via deletion of one gene allele and promoter replacement of the second allele in *C. albicans* diploids [10]. Additional tools to rapidly and efficiently downregulate essential genes in *C. albicans* can further help promote the investigation of these genes encoding prospective drug targets.

CRISPR-based techniques have been rapidly adopted for genetic manipulation across a diversity of biological organisms and have revolutionized the study of genetically intractable microbial organisms [11], including *C. albicans* [12–18], and numerous other fungal pathogens [19]. CRISPR interference (CRISPRi) is a modified CRISPR-based platform that exploits a nuclease-dead version of the Cas9 enzyme (dCas9) targeted to a specific genomic location by a single-guide RNA (sgRNA) to achieve steric hindrance of RNA polymerase, thus blocking transcription and enabling repression of gene expression from a gene of interest (Fig. 1) [20–22]. CRISPRi can be used to efficiently and specifically target essential genes for repression to enable the investigation of these genes, and this has been widely applied across numerous bacterial species, including many critical bacterial pathogens [20, 23–29]. Recently, we and others demonstrated the first application of CRISPRi in a fungal pathogen—*C. albicans* [30, 31]—and further validated the utility of this system for the repression and phenotypic analysis of essential fungal genes [30].

Here, we present a protocol for the application of CRISPRi in *C. albicans* as a tool to study essential genes. This detailed protocol includes an efficient cloning strategy for the generation of targeted CRISPRi plasmids, a fungal transformation protocol, and methodology to validate transcriptional repression by CRISPRi in *C. albicans*.

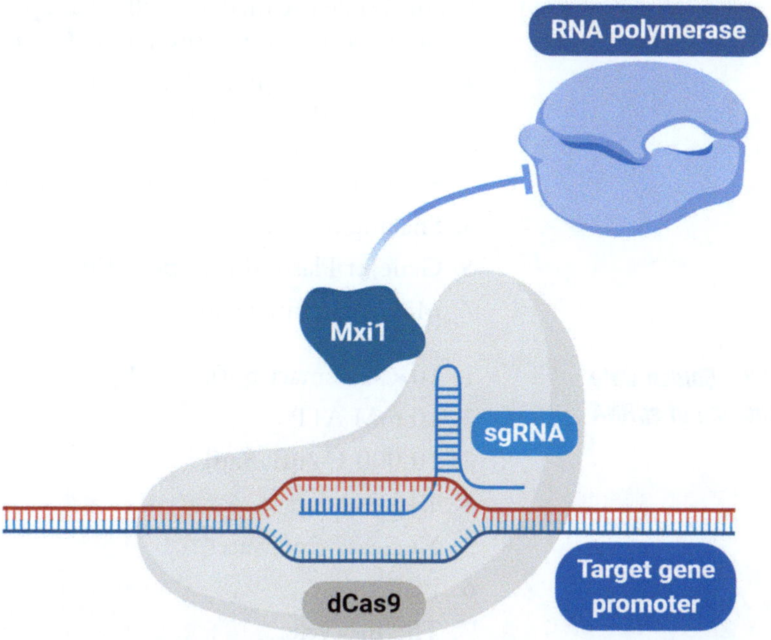

Fig. 1 CRISPRi as a strategy to inhibit gene expression. CRISPRi exploits a nuclease-dead Cas9 (dCas9), fused with transcriptional repressors such as Mxi1. The dCas9-Mxi1 fusion is guided to the promoter region of a gene of interest via a targeted sgRNA. The dCas9-Mxi1-sgRNA complex will bind the promoter region of interest and block transcription from this promoter by preventing RNA polymerase from binding, thus inhibiting gene expression

2 Materials

2.1 Duplexing of Forward and Reverse Oligonucleotides for the sgRNA

1. Forward and reverse oligonucleotides for sgRNA.
2. Nuclease-free duplexing buffer.
3. Heating block.
4. Microcentrifuge tubes.

2.2 Plasmid Extraction

1. *C. albicans* CRISPRi backbone plasmid (aka pRS159, plasmid can be purchased from Addgene, ref. No 122378, https://www.addgene.org/122378/).
2. 100 mg/mL Ampicillin: Combine 0.1 g of ampicillin with 1 mL of sterile water (autoclaved MilliQ water). Filter to sterilize using a 0.2-μm filter syringe. Store at −20 °C for 4 to 6 months.
3. LB-ampicillin plates (LB-AMP): Combine 1 L of water, 25 g of LB broth powder, and 20 g of bacteriological grade agar. Stir and autoclave media. Cool media to ~50 °C on a stir plate and mix in 1 mL of ampicillin stock (100 mg/mL) and continue to

stir. While the media is still warm, pour into plates and let them sit at room temp. Stored at 4 °C for 6 months.

4. LB broth: Combine 1 L of MilliQ water and 25 g of LB broth powder Mix thoroughly using a stir plate until dissolved entirely. Autoclave and allow to cool to room temperature. Store at room temperature for 1 year.
5. Shaking incubator.
6. GeneJet Plasmid Miniprep Kit.
7. Microcentrifuge tubes.

2.3 Golden Gate Cloning of sgRNA

1. 10× Cutsmart buffer.
2. 10 mM ATP.
3. 10,000 U/mL SapI.
4. T4 DNA ligase.
5. Nuclease-free water.
6. PCR tubes.
7. Thermocycler.

2.4 Transformation into Competent E. coli Cells

1. Chemically competent bacterial cells.
2. Water bath.
3. SOC medium.
4. Glass beads.
5. LB-AMP plates.

2.5 PCR Verification

1. 1× TAE buffer: Combine 242.4 g/L of Tris base, 57.1 mL/L Acetic acid, and EDTA 18.6 g/L EDTA (pH should be ~8.2). This makes a 50× reaction. Combine 70 mL of 50× TAE, 3430 water and stir. Store at room temp.
2. Agarose gel: Combine 1 g of electrophoresis grade agar, 100 mL of 1× TAE buffer. Microwave till boiling in increments of 15 to 30 s. Cool to touch at room temp and then add 5 μL of RedSafe and mix thoroughly. Pour in the contents of the flask into an assembled casting tray. Leave the gel at room temperature to set gel (~30 min).
3. 2× Taq mix.
4. Nuclease-free water.
5. Forward primer (forward oligonucleotide for sgRNA) and reverse primer (*ACCCACTGAATTCTACATCGAAC* , sequence found in the plasmid).
6. Thermocycler.
7. Electrophoresis Gel equipment.

8. 1 kb ladder.
9. Gel imager.

2.6 Archiving Plasmids

1. 40% glycerol solution: Combine 596 mL of sterile water with 404 mL of glycerol. Autoclave solution. Store at room temperature.
2. Cryogenic vials.

2.7 Linearization of the Cloned Plasmid

1. PacI enzyme.
2. Microcentrifuge tube.
3. 10× CutSmart buffer.

2.8 Transformation of Cloned Plasmids into C. albicans

1. Culture tubes.
2. YPD broth: Combine 20 g of peptone, 20 g of glucose, 10 g of yeast extract, and 1 L of water. Stir media using a stir plate and magnetic spinner until completely dissolved. Autoclave media and then cool and store at room temperature.
3. Spectrophotometer.
4. Cuvette.
5. Microcentrifuge tube.
6. Tabletop centrifuge (for microcentrifuge tubes).
7. Heating block.
8. Salmon Sperm DNA Solution.
9. Nourseothricin (250 mg/mL concentration): Combine 1 g of Nourseothricin sulfate and 4 mL of water. Filter solution by using a 0.2-μm filter syringe. Store at −20 °C.
10. 50% (wt/vol) PEG: Combine 250 g of PEG 3350 and 250 mL of MilliQ water and vortex thoroughly. Allow mixture to rest for 10 min. Continue to add water up to a total volume of 500 mL. Vortex mixture until the solution appears transparent. Using a 0.2-μm filter, sterilize by filtration. Store solution at 4 °C for up to 18 months.
11. 10× TE buffer: Combine 10 mL of Tris–HCl (1 M, pH 8.0), 88 mL of MilliQ water, and 2 mL of EDTA (0.5 M, pH 8.0). Autoclave solution and cool at room temp and store for up to 3 years at room temperature.
12. 1 M lithium acetate (LiAc) pH 7.4: Combine 10.2 g of LiAc and 100 mL of water. Using a pH meter, adjust to 7.4. Autoclave solution and cool at room temperature and store for 1 year.
13. Water bath.
14. Shaking incubator.
15. Large centrifuge (for culture tubes).

16. YPD-NAT plates: Combine 10 g of yeast extract, 20 g of agar, 20 g of peptone, 20 g of glucose, and 1 L of water. Stir and autoclave media. Cool media to ~50 °C on a stir plate. To the media, add 1 mL of NAT (250 mg/mL). Mix media well. Pour contents of the flasks into Petri dishes, ~20–25 mL into each plate. Allow plates to cool at room temperature. Store plates at 4 °C for 6 months.

2.9 Verification of Integration into C. albicans

1. 1× TAE buffer.
2. 1% Agarose gel.
3. 2× Taq mix.
4. Nuclease-free water.
5. Forward primer (*ACTATTAAAGAACGTGGACTC-CAACGTCA*) and reverse primer (*CAAGTTTG-CACTCTTTTGTCTA*).
6. Thermocycler.
7. Electrophoresis Gel equipment.
8. 1 kb ladder.
9. Gel imager.

2.10 qRT-PCR to Confirm Repression of Gene of Interest

1. Culture tubes.
2. Spectrophotometer.
3. Shaking incubator.
4. RNA extraction Kit for fungal samples.
5. Centrifuge.
6. YPD media.
7. Zymolase solution: 0.0452 g of stock zymolase stock powder (concentration of 22,100 U/g) and 50 μL of sorbitol buffer mix well by vortexing.
8. β-mercaptoethanol.
9. 70% ethanol.
10. DNase 1.
11. DNase 1 reaction buffer.
12. 0.5 M EGTA: 186.1 g of ethylene glycol-bis(β-aminoethyl ether)-*N*,*N*,*N*′,*N*′-tetraacetic acid (EGTA), and 800 mL of water. Stir with a magnetic spinner and adjust pH to 8.00 with NaOH (approximately 20 g NaOH pellets). Sterilize by filtering or autoclaving.
13. PerfeCTa SYBR Green FastMix ROX.
14. RNase-free water.

3 Methods

3.1 Design of sgRNA

1. For each *C. albicans* gene of interest (which is to be targeted by CRISPRi for genetic repression), obtain the gene sequence including ~1 kb upstream of the start codon using the Website *Candida* Genome Database (CGD; http://www.candidagenome.org) [32].
2. Take the 1 kb sequence and enter it into the sgRNA design tool Eukaryotic Pathogen CRISPR gRNA Design Tool (EuPaGDT, http://grna.ctegd.uga.edu) [33]. This tool will generate a number of sgRNA sequences located in the 1 kb region before the start codon of the gene of interest. sgRNAs are ranked based on their generated efficiency score and predicted specificity. As well, the distance from the start codon is taken into account. Previous research indicates that the optimal placement for the sgRNA in *C. albicans* is between ~55 bp and ~129 bp upstream from the ATG start codon [30] (*see* **Note 1**).
3. Based on the efficiency and specificity score, as well as the optimal targeting range, select sgRNAs for the gene of interest.
4. For each sgRNA, design two oligonucleotides: a forward and reverse complement version of the selected sgRNA, plus 3 bp overhangs for SapI cloning (shown underlined in the example below). A 'cga' motif should be added to the 5′ end of one oligonucleotide, and a 'aac' motif should be added to the reverse complement oligonucleotide.
   Example:

   sgRNA: TTTCCTAAGAAAAAGCAACG.

   Oligonucleotide 1: cgaTTTCCTAAGAAAAAGCAACG.
   Oligonucleotide 2: aacCGTTGCTTTTTCTTAGGAAA.
5. The forward and reverse complement oligonucleotide sequences of the sgRNA can be synthesized using Integrated DNA Technologies (https://www.idtdna.com). These sequences can be purchased already duplexed or in the single-stranded form which can be duplexed using the listed procedure below.

3.2 Duplexing of Forward and Reverse Oligos for the sgRNA

1. Reconstitute oligonucleotides (both the forward and the reverse sequence) using Nuclease-Free Duplexing buffer (stored at 4 °C) to a concentration of 100 μM.
2. Vortex the oligonucleotides well to ensure they have been distributed evenly in solution.
3. Heat the oligonucleotides at 94 °C for 1 min in the heating block. Suspend well by vortexing.
4. Add equal volumes (about 25–50 μL) of each complementary oligonucleotide into a microcentrifuge (*see* **Note 2**).

5. Heat the microcentrifuge tube with combined oligonucleotides to 94 °C for 2 min.
6. Place tubes at room temperature to slowly cool to room temperature to enable annealing of the two complementary oligonucleotides. Tubes can then be stored at −20 °C if not used immediately.

3.3 Plasmid Extraction from Bacterial Culture

1. Prepare streaked plates of the *C. albicans* CRISPRi backbone plasmid (aka pRS159; Fig. 2; *see* **Note 3**) from plasmid archives (plasmid can be purchased from Addgene, ref. No 122378, https://www.addgene.org/122378/). Using a wooden dowel, scrape a small portion of frozen culture and streak along an LB-AMP plate.
2. Incubate plates overnight at 37 °C.
3. Prepare overnight cultures from the streaked plates: In a culture tube, add 5 mL of LB broth and 5 μL of Ampicillin (100 mg/mL concentration).
4. Inoculate culture tubes with a loopful of culture from the streaked plates. Vortex tubes briefly to homogenize the mixture.
5. Grow up culture tubes at 37 °C in a shaking incubator at 250 rpm overnight (10–18h).
6. Centrifuge cultures at 1700 × *g* for 5 min.
7. Use a GeneJET Plasmid Miniprep Kit or equivalent, and follow manufacturer's protocol.
8. Measure the concentration of plasmid using a nanodrop spectrophotometer (*see* **Note 4**). Store purified plasmid at −20 °C.

3.4 Golden Gate Cloning of sgRNA into CRISPRi Plasmid Backbone

1. Use a type IIS restriction enzyme for this protocol (*see* **Note 5**).
2. For this protocol, prepare a master mix as follows for a 1× reaction: 10 μL of extracted plasmid DNA (previously done using GeneJET Plasmid Miniprep Kit above), 2 μL of Cutsmart buffer, 2 μL ATP, 1 μL of SapI (Type IIS Restriction enzyme), 1 μL of T4 DNA ligase, and 3 μL of Nuclease-free water (Fig. 3).
3. Do not vortex the mixture when enzymes have been added. Pipette up and down to mix. Prepare a master mix for the number of cloning reactions plus one time extra to account for pipetting errors. Transfer 19 μL of master mix to a PCR tube.
4. Add 1 μL of the duplexed sgRNA (prepared previously) to each PCR tube.
5. Run PCR tubes in a thermocycler under the following parameters: 37 °C for 2 min then 16 °C for 5 min: cycle 99 times;

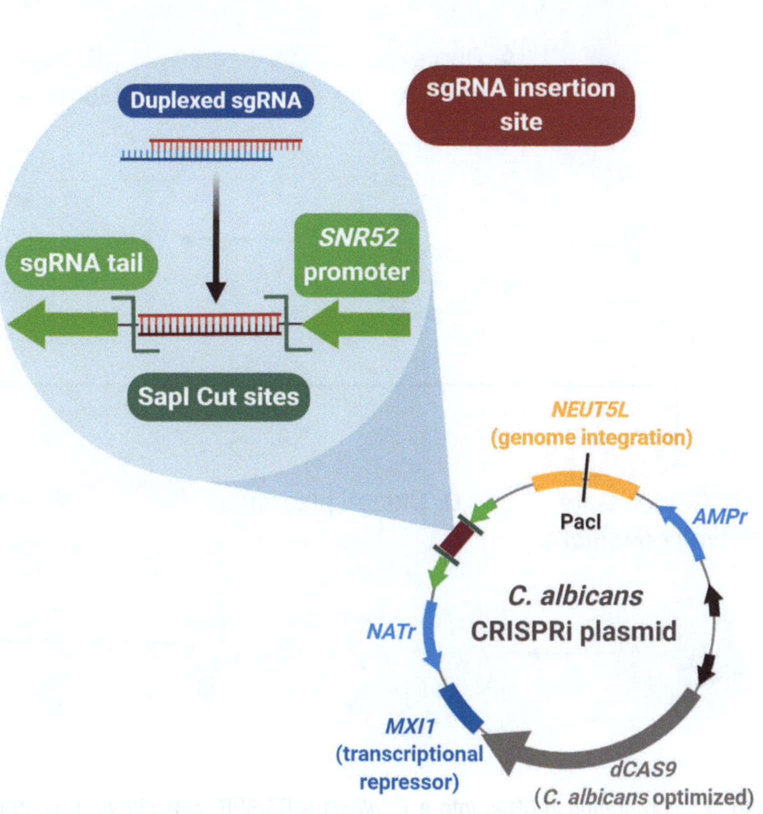

Fig. 2 CRISPRi plasmid for genetic repression in *C. albicans*. This CRISPRi plasmid (pRS159, Addgene No. 122378) has been optimized for use in *C. albicans*, as previously shown [30]. This plasmid contains all the required genetic material for targeted genetic repression in *C. albicans*, including *C. albicans* codon-optimized *dCAS9* fused to the transcriptional repressor *MXI1*, an *AMPr* resistance cassette for selection in bacterial transformations, an *NATr* resistance cassette for selection in *C. albicans* transformations, a *NEUT5L* site for *C. albicans* genomic integration with PacI restriction enzyme site for plasmid linearization, and the sgRNA cloning site which contains the *SNR52* RNA polymerase III promoter (guide RNA promoter) and a conserved sgRNA tail sequence, flanking SapI restriction enzyme sites for Golden Gate cloning of the sgRNA oligonucleotides

65 °C for 15 min and then 80 °C for 15 min. This will take approximately 13 h.

6. Once the cycling finishes, add an additional 1 μL of SapI to each PCR tube. Incubate in the thermocycler at 37 °C for 1 h. This ensures that all uncut plasmids are linearized and therefore will not be able to be transformed into the bacterial cells.
7. If the plasmids are not going to be used directly after this procedure, they can be stored at 4 °C for a short period of a few days.

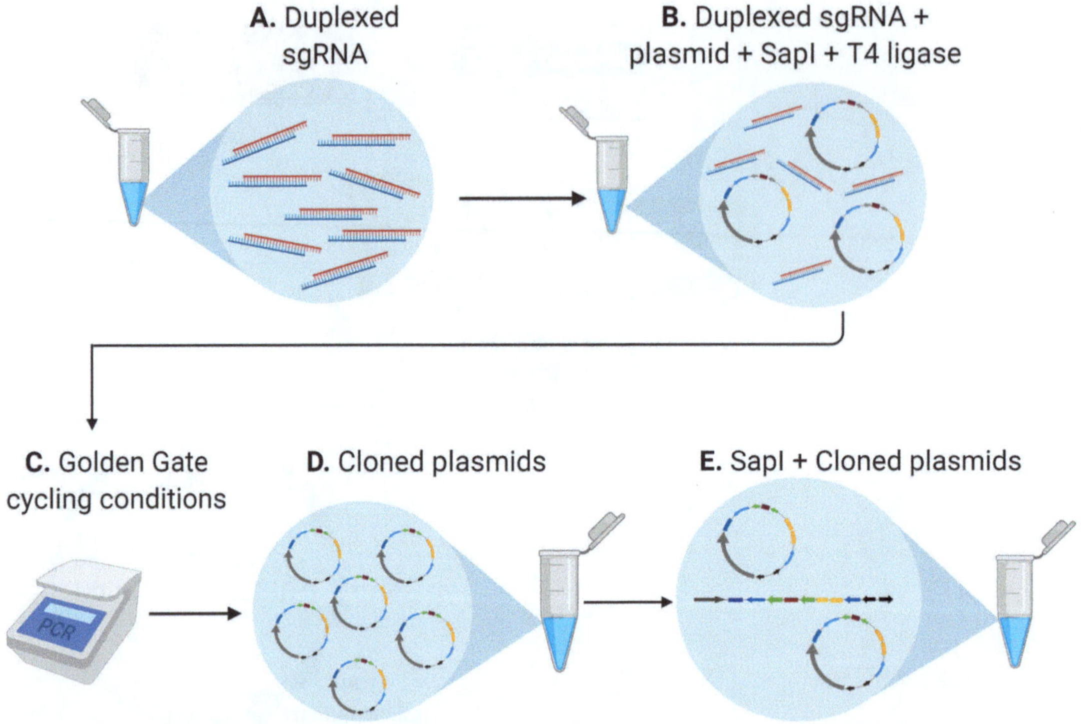

Fig. 3 Incorporating sgRNAs into a *C. albicans* CRISPRi plasmid via Golden Gate cloning. (**a**) A Golden Gate cloning strategy is used to clone the duplexed sgRNA oligonucleotides into the *C. albicans* CRISPRi plasmid. (**b**) This strategy exploits the use of a type IIS restriction enzyme, in this case SapI. By using this enzyme along with T4 ligase, one can cut and ligate the CRISPRi plasmid simultaneously. The duplexed sgRNA, SapI, and T4 ligase are incubated together, along with the CRISPRi plasmid backbone. (**c**) These components are then placed in a thermocycler under Golden Gate optimized cycling conditions for SapI cutting and T4 ligation. (**d**) After Golden Gate cycling is complete, circular plasmids will be formed that contain the sgRNA ligated into the CRISPRi plasmid backbone. Once the plasmid is cut and the duplexed sgRNA oligonucleotide is ligated into the plasmid, the circular plasmid cannot be cut with SapI again. (**e**) After a final SapI treatment, any unligated plasmid will be linearized, leaving intact (as a circular plasmid) only cloned plasmids containing the sgRNA for transformation into *E. coli*. Any CRISPRi plasmids that did not successfully incorporate the sgRNA will be linearized after this final SapI treatment

3.5 Transformation into Competent E. coli *Cells*

1. Remove chemically competent bacterial cell culture tubes from −80 °C freezer and place them on ice for 10 min.
2. Transfer the contents of the tube to a microcentrifuge tube. Add 5 μL of the cloned plasmid to the microcentrifuge tube. Allow tubes to sit on ice for 30 min.
3. Heat shock tubes in a water bath warmed to 42 °C for 30 s. Note, heat shocking for longer than 30 s can decrease the yield and efficiency of the transformation.
4. Place on ice and let stand for 5 min.

5. Add to tubes 950 μL of SOC media. Incubate on a shaker at 250 rpm for 1 h at 37 °C. During this incubation, place LB-AMP plates in a static incubator at 37 °C to warm up.
6. Pipette 100 μL of the transformation mixture to the LB-AMP plates along with 100 μL of sterile water. Using glass bead, spread the plates evenly. Place plates in a 37 °C static incubator overnight.
7. Remove from incubator when distinct colonies have formed.
8. Patch colonies on to a fresh LB-AMP plate to test for verification of cloned plasmids. Grow up patch plates overnight at 37 °C.
9. Remove from incubator and store at 4 °C until ready to perform PCR verification.

3.6 PCR Verification for Cloned Plasmids

1. For each cloning reaction and bacterial transformation, test a total of seven bacterial patches, plus one negative control. The negative control, in this case, is the uncloned CRISPRi plasmid pRS159.
2. In a PCR strip tube, pipette 100 μL of sterile water into each tube.
3. Using the tip of a sterile pipette tip, inoculate each tube with a sample from each of the patches (*see* **Note 6**).
4. In a microcentrifuge tube, prepare a PCR master mix reaction. A master mix will need to be prepared for each unique sgRNA used in the cloning process. Prepare a 10× reaction, seven for the patches one for the negative control and two extra to account for pipetting errors.
5. In a microcentrifuge tube, combine 100 μL of 2× Taq mix, 48 μL of nuclease-free water, 1 μL of 'Oligonucleotide 2' from the example in Subheading 3.1 (the reverse complement of the duplexed sgRNA: primer 1), and 1 μL of another primer that is complementary to the CRISPRi plasmid (primer 2). This second primer will be the same for testing all sgRNA integrations (*ACCCACTGAATTCTACATCGAAC*) and is 530 bp from the sgRNA (primer 2). Pipette up and down to mix.
6. Transfer 15 μL of PCR master mix into each PCR tube and then add 5 μL of the diluted colonies from **step 3**.
7. Place PCR tube strips in the thermocycler and use the following parameters: 94 °C for 3 min for 1 cycle, 94 °C for 30 s, Annealing temperature for the two primers (optimal temperature can be calculated using the NEB Website https://tmcalculator.neb.com/#!/main) (*see* **Note 7**) for 30 s, and 72 °C for 1 min cycled for 35 times, then cycled once at 72 °C for 10 min. The tubes can then be stored at 4 °C or −20 °C if not used immediately after.

8. Prepare a 1% agar gel for electrophoresis. In the first lane, add 8 μL of 1 kb ladder. Load the rest of the lanes with 8 μL of your PCR product.
9. Run the gel for 1 h at 95 V. Image gel using a gel imaging system. Observe for bands the size of the distance between your two primers 576 bp long. A band indicates that successful cloning occurred, and no band should be seen in the negative control PCR (empty CRISPRi backbone). Positive colonies can be archived for further use. Only one colony for each unique sgRNA needs to be archived.

3.7 Archiving Plasmids for Further Use

1. Start an overnight culture of colonies that PCR tested positive in Subheading 3.6: In a culture tube, add 5 mL of LB broth and 5 μL of Ampicillin (100 mg/mL concentration).
2. Inoculate culture tubes with a loopful of culture from the streaked plates. Vortex tubes briefly to homogenize the mixture. Grow overnight at 37 °C overnight.
3. In a cryotube labeled the assigned plasmid identification, add 900 μL of 40% glycerol and 900 μL of overnight culture. Invert the tube several times to mix well. Store prepared tubes at −80 °C.

3.8 Linearization of Cloned Plasmids

1. Remove archived plasmids from the −80 °C freezer. Perform plasmid extraction as described previously in Subheading 3.3.
2. Digest plasmids to obtain a linear plasmid form: pRS159 CRISPRi-based plasmids can be linearized by digestion with PacI (Fig. 2). In a microcentrifuge tube, add 88 μL of extracted plasmid, 2 μL of the PacI enzyme, and 10 μL of CutSmart buffer.
3. Incubate overnight at 37 °C.

3.9 Transformation of Cloned Plasmids into *C. albicans*

1. Prepare overnight cultures of the *C. albicans* strain you with to transform with the CRISPRi plasmid (Fig. 4). Prepare a culture tube with 5 mL of YPD broth. Inoculate tube with a loopful of *C. albicans* culture. Incubate overnight in a shaking incubator at 30 °C.
2. Measure the OD600 of the cultures using a spectrophotometer. In a 1-mm cuvette, add 900 μL of water and 100 μL of culture from overnight cultures (vortex cultures before adding culture into the cuvette). Pipet up and down to mix well. Measure the OD600 value.
3. Calculate the amount of culture needed for the transformation using the following formula. Volume of culture (in μL) = $\frac{10,000}{(\mathrm{OD600})(10)(1.5)}$.

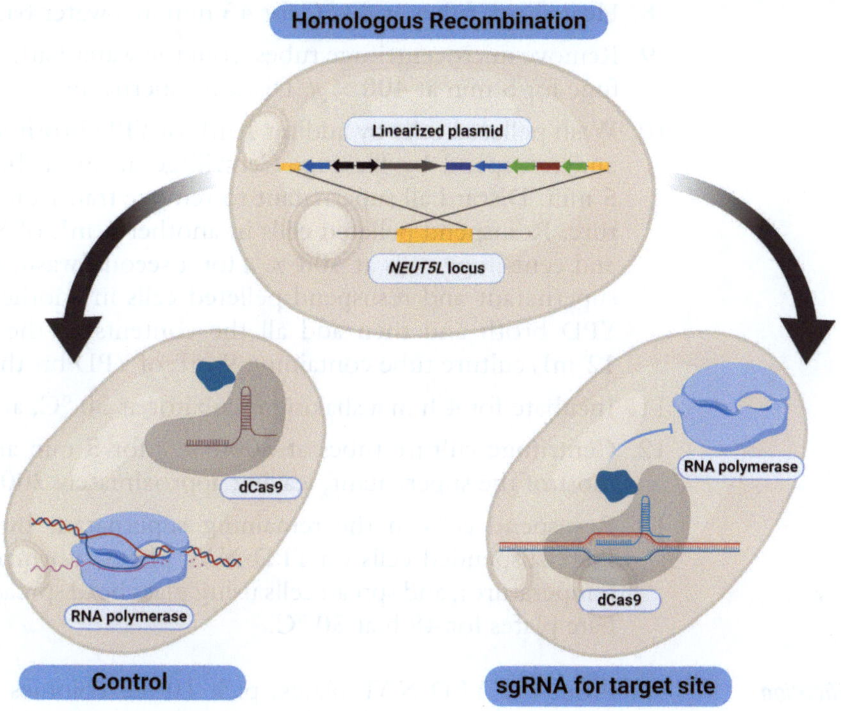

Fig. 4 Transformation and CRISPRi repression in *C. albicans*. The cloned CRISPRi plasmid containing the sgRNA is linearized and transformed into *C. albicans* for genomic integration at the *NEUT5L* locus. Two independent strains of *C. albicans* can be generated to monitor transcriptional repression from a gene of interest: a 'Control' strain, which is transformed with the empty CRISPRi plasmid backbone (containing no sgRNA), and the 'sgRNA for target site' strain, which contains the cloned CRISPRi plasmid containing a targeting sgRNA for the promoter of the gene of interest. The control strain contains the dCas9, but it will not be targeted to the gene of interest, and thus, this gene will remain expressed by RNA polymerase. The sgRNA targeted strain contains the dCas9 and the sgRNA to target it to the promoter of interest, and thus, transcription by RNA polymerase will be inhibited at this locus

4. Place calculated volume in a microcentrifuge tube and centrifuge at 400 × *g* for 5 min to form a pellet of cells. Remove most of the supernatant leaving approximately 50 μL of supernatant to resuspend cells in.
5. In a heating block heated to 100 °C, place a tube of salmon sperm DNA in the heating block and boil for 10 min.
6. Prepare transformation master mix for a 1× reaction by adding the following: 800 μL of 50% PEG 3350, 100 μL TE, 100 μL of 1 M LiAC, 40 μL of salmon sperm DNA, and 20 μL 1 M DTT. Prepare a reaction for each transformation plus an additional reaction to account for pipetting error.
7. To the microfuge tube containing the resuspended *C. albicans* cells from **step 4**, add 45 μL of the linearized plasmid from Subheading 3.8. Then, add 1060 μL of transformation master mix. Mix by inverting. Incubate for 1 h at 30 °C.

8. Heat shock tubes at 42 °C for 45 min in a water bath.
9. Remove microcentrifuge tubes from the water bath and centrifuge for 5 min at 400 × *g*. Discard supernatant.
10. Wash pelleted cells by adding 1 mL of YPD broth to the tube and resuspend by flicking. Centrifuge again at 400 × *g* for 5 min. Discard all supernatant to remove transformation mixture. Resuspend pelleted cells in another 1 mL of YPD broth and centrifuge cells at 400 × *g* for a second wash. Discard all supernatant and resuspend pelleted cells in another 1 mL of YPD broth and then add all the contents of the tube to a 12-mL culture tube containing 9 mL of YPD broth.
11. Incubate for 4 h in a shaking incubator at 30 °C, at 250 rpm.
12. Centrifuge culture tubes at 3000 × *g* for 5 min and remove most of the supernatant, leaving approximately 200 μL.
13. Resuspend cells in the remaining supernatant by pipetting. Plate suspended cells on YPD-NAT plates (warmed to room temperature), and spread cells using glass bead spreaders. Incubate plates for 48 h at 30 °C.

3.10 Verification of Integration into C. albicans Genome

1. From the YPD-NAT plates, pick 10–12 colonies and patch them onto another YPD-NAT plate. Allow plates to grow for 24 h at 30 °C.
2. Prepare the 2× Taq master mix as described in Subheading 3.6.
3. Use the following primers, which will validate integration of the CRISPRi plasmid into the *C. albicans* genome at the *NEUT5L* locus:

   Primer 1 (Forward): ACTATTAAAGAACGTGGACTC CAACGTCA.

   Primer 2 (Reverse, Neut5L): CAAGTTTG CACTCTTTTGTCTA
4. If plasmid is correctly integrated, PCR product should roughly 750 bp in length (*see* **Note 8**).
5. Positive PCR tests indicate that plasmid integration into *C. albicans* genome was successful.

3.11 qRT-PCR to Confirm Repression of Gene of Interest

1. (*See* **Notes 9** and **10**).
2. Prepare overnight cultures of archived fungal strains as stated in Subheading 3.9.
3. Using a spectrophotometer, take the OD600 value of the overnight cultures by preparing a 1/5 or 1/10 dilution with water. Blank spectrophotometer by preparing the same dilution concentration by combining YPD stock media with sterile water.

4. In a 250-mL flask, add 50 mL of YPD media. Prepare triplicate samples for each fungal strain (*see* **Note 11**).
5. From the OD600 value, calculate the volume of overnight culture that will need to be added to a 250-mL flask. Use the formula $(\text{OD600}_{\text{initial}})(V_1) = (\text{OD600}_{\text{final}})(V_2)$, where the final OD600 is 0.05 and the final volume (V_2) is 50,000 μL.
6. Add the calculated volume of overnight culture into the 50-mL conical tube.
7. Incubate for 30 °C in an incubating shaker at 250 rpm and incubate for 2 h to reach an OD600 value of 0.2 (*see* **Note 12**).
8. Transfer 50 mL culture to a 50-mL conical tube and centrifuge the 50-mL conical tubes for 5 min at 1700 × *g*. Decant the supernatant leaving approximately 300 μL of YPD to resuspend pelleted cells.
9. Transfer resuspended cells into a 1.5-mL microcentrifuge tube, and store at −80 °C overnight.
10. Use an RNA extraction kit for RNA extraction from fungal samples.
11. Prepare 200 U of zymolase solution (*see* **Note 13**) per RNA extraction reaction. Prepare an additional reaction to account for pipetting error.
12. Remove cells from the −80 °C freezer and thaw at room temperature for 10 min.
13. Centrifuge for 10 min at 2400 × *g* . Remove all supernatant.
14. Combine 590 μL of sorbitol buffer (from kit) from RNA extraction kit and 10 μL of 200 U zymolase solution and add to each tube. Resuspend cells in solution and incubate at 30 °C for 30 min.
15. Centrifuge cells for 10 min at 400 × *g*.
16. Discard the supernatant and resuspend pelleted cells in 300 μL of RB buffer (from kit) and 3 μL of β-mercaptoethanol (not from the kit) (combine in fume hood). Incubate at room temperature for 5 min.
17. Centrifuge at 17,000 × *g* for 2 min. Transfer supernatant to a fresh 1.5-mL microcentrifuge tube.
18. Add 500 μL of 70% ethanol (not in the kit) to mixture and pipet up and down to mix. Take 500 μL of the mixture and transfer to a column. Centrifuge at 17,000 × *g* for 1 min. Discard flow-through and then add the rest of the mixture and spin again. Discard flow-through.
19. Add 50 μL of DNase reaction mix (40 μL of DNase 1 and 360 μL of DNase 1 reaction buffer, not found in the kit). Incubate samples at room temperature for 15 min.

20. Perform an RNA wash by adding 400 μL of W1 buffer (found in the kit) to the sample columns. Centrifuge tubes at 17,000 × *g* for 1 min. Discard flow-through and add 600 μl of wash buffer and centrifuge at 17,000 × *g* for 1 min and discard flow-through. Repeat this wash step a second time. Discard flow-through 600 μl of wash buffer and centrifuge at 17,000 × *g* for 1 min and discard flow-through. Centrifuge again at 17,000 × *g* for 3 min to completely dry columns.
21. Elute RNA by placing the column in a fresh microcentrifuge tube. Add 50 μL of RNase-free water (found in the kit) to the center of the column matrix. Allow tubes to stand at room temperature for 3 min. Centrifuge tubes at 17,000 × *g* for 1 min.
22. Using the Nanodrop spectrophotometer, measure the RNA concentration. Transfer one 1 μg of each RNA sample to a new tube and treat with DNase to remove any residual DNA contamination. Do this by combining 1 μg pf RNA elution, 5 μL of DNase reaction buffer, and 0.5 μL of DNase 1 and then add nuclease-free water to reach a total volume of 50 μL.
23. Incubate tubes at 37 °C for 30 min.
24. To the tubes, add 1 μL of 20 mM EGTA (pH 8.0) to each sample tube to inhibit further DNase reaction. Incubate tubes in a water bath for 10 min at 65 °C.
25. Repeat RNA wash and elution steps until a 50 μL elution of DNase-treated RNA is obtained.
26. On a spectrophotometer check purity. An A_{260}/A_{280} ratio between 1.8 and 2.0 is considered acceptable.
27. Check for degraded RNA by preparing a 1% agarose and run to examine ribosomal RNA (rRNA) bands the upper rRNA band (28S for eukaryotic cells) should show twice the intensity of the lower rRNA band (18S in eukaryotic cells). Bands should be crisp and tight. If the intensity of the bands is equal, then some degradation has occurred. An mRNA band should run between the two rRNA bands. Some smearing may occur.
28. Generate cDNA from 800 ng RNA using High Capacity cDNA Reverse Transcription Kit (or equivalent kit). Follow the protocol steps provided by the kit.
29. For the primer design for qRT-PCR primers, use the Website *Candida* Genome Database (CGD; http://www.candidagenome.org). Look at the genomic DNA sequence for the gene of interest that is being repressed with CRISPRi. Locate the middle of the gene of interest and design a forward and reverse primers that are ~150 bp apart. The primers should be 20 bp in length and should have 40–60% GC content. As

well the housekeeping gene, *ACT1* will be used as a control to normalize against. Primers have been previously designed for this gene and are as follows:

Oligonucleotide 1 (Forward): GTTGGTGATGAAGCC CAATCC.

Oligonucleotide 2 (Reverse): CTGGATGTTCTTCTGGAG CAAC.

30. Prepare a 20 μL reaction by combining 10 μL of PerfeCTa SYBR Green FastMix ROX, 0.8 μL of forward and reverse primer mix at 5 μM (final concentration of primer at 200 nM), 4.2 μL of water, and 5 μL of diluted cDNA.
31. Using a real-time qPCR instrument, set the parameters for thermocycling to 95 °C for 30 s for 1 cycle followed by 40 times for the next 2 steps: 95 °C for 3 s and 60 °C for 30 s. Cycle once through 95 °C for 15 min, 60 °C for 1 min, and 95 °C for 15 min.
32. For data analysis, compare transcript expression levels of the gene of interest (being targeted by CRISPRi) relative to the *ACT1* housekeeping gene in CRISPRi repression *C. albicans* strains, compared with control strains that should have no CRISPRi repression (*see* **Note 14**).

4 Notes

1. Optimal range for placement of the sgRNA may not apply to every gene. Be sure to design several sgRNAs at varying distances from the transcription start site to determine which leads to the greatest repression.
2. For duplexing oligonucleotides, the volume used is dependent on the number of times the duplexed sgRNA will be used for cloning into a plasmid.
3. The *C. albicans* CRISPRi plasmid backbone (Fig. 2) contains all the required genetic information for transcriptional repression of targeted genes including *C. albicans*-optimized *dCAS9* fused to the transcriptional repressor *MXI1*, *AMPr* resistance cassette for bacterial selection, *NATr* resistance cassette for yeast selection, *NEUT5L* genomic integration site with PacI for plasmid linearization, and the sgRNA cloning site which contains the *SNR52* promoter and an sgRNA tail flanking two type IIS SapI sites for Golden Gate cloning of the duplexed sgRNA oligonucleotide.
4. Concentration of greater than 65 ng/μL with a 260/280 reading between 1.8 and 2 is acceptable.

5. For Golden Gate cloning to work, type IIS restriction enzymes are required for this protocol (here, we use SapI for Golden Gate cloning of the sgRNA at the sgRNA insertion site (Fig. 2)). By using this category of enzymes, one can cut and ligate simultaneously due to the enzyme ability to cut outside of the recognition sequence. Once the plasmid is cut and the duplexed oligonucleotide is ligated in, it cannot be cut out again. After a final SapI incubation, any unligated plasmid will be linearized, leaving intact only cloned plasmids containing the sgRNA, which can be used for transformation into *E. coli* (Fig. 3).
6. Using a sterile pipet tip allows you to inoculate the water and gives you the ability to mix thoroughly by pipetting up and down.
7. Annealing temperatures range between 45 °C and 65 °C.
8. Positive PCR tests indicate that plasmid integration into *C. albicans* genome was successful. Archive one colony for each plasmid integration using the procedure listed previously in Subheading 3.7.
9. Using qRT-PCR is a quick and accurate way for us to test the repression capacity of the CRISPRi mechanism with the target sgRNA.
10. This is a 3-day experiment.
11. The fungal strains that should be used here include the one you are looking to test, and and a control fungal strain containing the "sgRNA empty CRISPRi plasmid" that has been previously transformed (Fig. 4).
12. This indicates that the cells are still in the logarithmic growth phase.
13. Zymolase is not included in the kit and can be bought from vendor.
14. Repression of the gene of interest can be seen through the reduction of transcripts produced for the gene as compared to the housekeeping gene.

Acknowledgments

We would like to thank members of the Shapiro laboratory for helpful discussions. This research is supported by a Natural Sciences and Engineering Research Council (NSERC) Discovery Grant, and NSERC Discovery Accelerator Supplement. Figures were created with BioRender (https://biorender.com/).

References

1. Bongomin F, Gago S, Oladele R, Denning D (2017) Global and multi-national prevalence of fungal diseases—estimate precision. J Fungi 3:57
2. Geddes-McAlister J, Shapiro RS (2018) New pathogens, new tricks: emerging, drug-resistant fungal pathogens and future prospects for antifungal therapeutics. Ann N Y Acad Sci 1435(1):57–78. https://doi.org/10.1111/nyas.13739
3. Webb BJ, Ferraro JP, Rea S et al (2018) Epidemiology and clinical features of invasive fungal infection in a US health care network. Open Forum Infect Dis 5:ofy187
4. Pfaller MA, Diekema DJ (2007) Epidemiology of invasive candidiasis: a persistent public health problem. Clin Microbiol Rev 20:133–163
5. Bassetti M, Righi E, Montravers P, Cornely OA (2018) What has changed in the treatment of invasive candidiasis? A look at the past 10 years and ahead. J Antimicrob Chemother 73: i14–i25
6. Perfect JR (2017) The antifungal pipeline: a reality check. Nat Rev Drug Discov 16:603–616
7. Enloe B, Diamond A, Mitchell AP (2000) A single-transformation gene function test in diploid *Candida albicans*. J Bacteriol 182:5730–5736
8. Nobile CJ, Mitchell AP (2009) Large-scale gene disruption using the UAU1 cassette. Methods Mol Biol 499:175–194
9. Segal ES, Gritsenko V, Levitan A et al (2018) Gene essentiality analyzed by in vivo transposon mutagenesis and machine learning in a stable haploid isolate of *Candida albicans*. MBio:9. https://doi.org/10.1128/mBio.02048-18
10. Roemer T, Jiang B, Davison J et al (2003) Large-scale essential gene identification in *Candida albicans* and applications to antifungal drug discovery. Mol Microbiol 50:167–181
11. Shapiro RS, Chavez A, Collins JJ (2018) CRISPR-based genomic tools for the manipulation of genetically intractable microorganisms. Nat Rev Microbiol 16:333–339
12. Ng H, Dean N (2017) Dramatic improvement of CRISPR/Cas9 editing in *Candida albicans* by increased single guide RNA expression. mSphere 2(2):e00385–e00316. https://doi.org/10.1128/mSphere.00385-16
13. Vyas VK, Guy Bushkin G, Bernstein DA et al (2018) New CRISPR mutagenesis strategies reveal variation in repair mechanisms among fungi. mSphere 3:e00154–e00118
14. Vyas VK, Barrasa MI, Fink GR (2015) A *Candida albicans* CRISPR system permits genetic engineering of essential genes and gene families. Sci Adv 1:e1500248
15. Min K, Ichikawa Y, Woolford CA, Mitchell AP (2016) *Candida albicans* gene deletion with a transient CRISPR-Cas9 system. mSphere 1: e00130–e00116
16. Nguyen N, Quail MMF, Hernday AD (2017) An efficient, rapid, and recyclable system for CRISPR-mediated genome editing in *Candida albicans*. mSphere 2:e00149–e00117. https://doi.org/10.1128/mSphereDirect.00149-17
17. Shapiro RS, Chavez A, Porter CBM et al (2018) A CRISPR-Cas9-based gene drive platform for genetic interaction analysis in *Candida albicans*. Nat Microbiol 3:73–82
18. Halder V, Porter CBM, Chavez A, Shapiro RS (2019) Design, execution, and analysis of CRISPR-Cas9-based deletions and genetic interaction networks in the fungal pathogen *Candida albicans*. Nat Protoc 14:955–975
19. Román E, Prieto D, Alonso-Monge R, Pla J (2019) New insights of CRISPR technology in human pathogenic fungi. Future Microbiol 14:1243–1255
20. Qi LS, Larson MH, Gilbert LA et al (2013) Repurposing CRISPR as an RNA-guided platform for sequence-specific control of gene expression. Cell 152:1173–1183
21. Dominguez AA, Lim WA, Qi LS (2015) Beyond editing: repurposing CRISPR–Cas9 for precision genome regulation and interrogation. Nat Rev Mol Cell Biol 17:5–15
22. La Russa MF, Qi LS (2015) The new state of the art: Cas9 for gene activation and repression. Mol Cell Biol 35:3800–3809
23. Sato'o Y, Hisatsune J, Yu L et al (2018) Tailor-made gene silencing of *Staphylococcus aureus* clinical isolates by CRISPR interference. PLoS One 13:e0185987
24. Peters JM, Colavin A, Shi H et al (2016) A comprehensive, CRISPR-based functional analysis of essential genes in bacteria. Cell 165:1493–1506
25. Peters JM, Koo B-M, Patino R et al (2019) Enabling genetic analysis of diverse bacteria with Mobile-CRISPRi. Nat Microbiol 4:244–250. https://doi.org/10.1038/s41564-018-0327-z
26. Liu X, Gallay C, Kjos M et al (2017) High-throughput CRISPRi phenotyping identifies

new essential genes in *Streptococcus pneumoniae*. Mol Syst Biol 13:931

27. Tan SZ, Reisch CR, Prather KLJ (2018) A robust CRISPR interference gene repression system in pseudomonas. J Bacteriol 200. https://doi.org/10.1128/JB.00575-17
28. Rock JM, Hopkins FF, Chavez A et al (2017) Programmable transcriptional repression in mycobacteria using an orthogonal CRISPR interference platform. Nat Microbiol 2:16274
29. Choudhary E, Thakur P, Pareek M, Agarwal N (2015) Gene silencing by CRISPR interference in mycobacteria. Nat Commun 6:6267
30. Wensing L, Sharma J, Uthayakumar D et al (2019) A CRISPR interference platform for efficient genetic repression in *Candida albicans*. mSphere 4. https://doi.org/10.1128/mSphere.00002-19
31. Román E, Coman I, Prieto D et al (2019) Implementation of a CRISPR-based system for gene regulation in *Candida albicans*. mSphere 4. https://doi.org/10.1128/mSphere.00001-19
32. Skrzypek MS, Binkley J, Binkley G et al (2017) The *Candida* genome database (CGD): incorporation of assembly 22, systematic identifiers and visualization of high throughput sequencing data. Nucleic Acids Res 45:D592–D596
33. Peng D, Tarleton R (2015) EuPaGDT: a web tool tailored to design CRISPR guide RNAs for eukaryotic pathogens. Microb Genom 1: e000033

Chapter 5

Identification of Essential Genes Using Sequential CRISPR and siRNA Screens

Luke DeHart, Oliver P. Yockey, and Jesse Bakke

Abstract

Genome-wide CRISPR and siRNA screening methodologies are powerful tools that are aptly suited to the discovery of essential genes. In this chapter, we outline our methods to conduct sequential CRISPR and siRNA screens to quickly and efficiently identify essential genes within a collection of cell lines. The utilization of both screening methodologies provides a pipeline that minimizes costs and time while enabling the robust detection of candidate genes.

Key words CRISPR, siRNA, Essential genes

1 Introduction

The discovery of essential genes is important to deepen our basic understanding of biology as well as to uncover novel therapeutics for diseases such as cancer. There are a collection of essential genes that are required by most cells in the human body, and there are tissue- and disease-specific essential genes. Genome-wide CRISPR screens revealed that there are over 1500 core essential genes that are likely required in most cell types [1]. In an effort to understand tissue and disease essential genes, many groups have utilized an array of screening methodologies, such as CRISPR, siRNA, and shRNA [2–7].

Genome-wide CRISPR screening is a powerful pool based screening methodology that has proven to be robust, reproducible, and efficient [8, 9]. This screening tool has been used to screen for an array of genes required for oxidative phosphorylation [10], cancer metastasis [11], immune regulatory networks [12], cell invasion [13], and many others.

Whole-genome siRNA screening has been historically used to identify genes contributing to a specific phenotype. These screens are single-well based and traditionally utilize small-form factor

Ren Zhang (ed.), *Essential Genes and Genomes: Methods and Protocols*, Methods in Molecular Biology, vol. 2377,
https://doi.org/10.1007/978-1-0716-1720-5_5,

plates, such as 384-well plates. They require robotic decks to function to capacity due to the small volumes and the large array of plates. As one would expect, these screens are typically expensive and time-consuming. However, small siRNA validation screens are quick due to the availability of commercially available siRNA and the speed of transfection.

In this method, we paired a CRISPR screen with siRNA screens to identify and validate essential genes for a specific disease or tissue. We outline the process from library preparation of the CRISPR screen to deconvolution screens by siRNA. Through this process, elimination of broadly essential genes and elimination of tissue-specific essential genes can be elucidated from disease-specific essential genes.

2 Materials

2.1 Genome-Wide Crispr Screen

There are many CRISPR screening libraries available, and they are continually being improved. We currently use the "Brunello" library with a two plasmid system utilizing the lentiguide-Puro backbone [9, 14]. This library consists of 76,441 gRNAs targeting 19,114 genes and includes 1000 control gRNAs.

1. Brunello Library (Addgene).
2. Lentiguide-Puro Plasmid (Addgene).

2.2 Library Preparation

1. 1600 ng Brunello Library plasmid DNA (Addgene).
2. 400 μL electrocompetent cells. Any high efficiency ($\geq 10^9$ cfu/μg) cells may be used.
3. 40–50 mL of SOC outgrowth medium.
4. 12 untreated 245-mm-Square BioAssay dishes.
5. Plasmid maxi kit. 10 columns are required.
6. LB Agar powder.
7. Ampicillin.
8. Disposable cell spreader.
9. 12 electroporation cuvettes, 0.1 cm gap or equivalent.
10. Electroporator or equivalent.
11. 14-mL round-bottom tubes.

2.3 Lentivirus Production

1. Low-generation 293 T cells (ATCC).
2. T-175 Tissue Culture Flasks. Total of 16 for transfection.
3. Cell maintenance media: Dulbecco's Modification of Eagle's Medium (DMEM), 10% Fetal Bovine Serum (FBS), 2 mM L-Glutamine, Penicillin–Streptomycin (Optional).

4. Opti-MEM reduced serum media.
5. Lipofectamine 3000 transfection reagent (or equivalent).
6. DNA Plasmids: Plasmid DNA, VSV-G plasmid (Addgene), and psPAX2 plasmid (Addgene).
7. Takara Lenti-X GoStix (Takara Bio).
8. Microcentrifuge tubes.
9. 50-mL conical tubes
10. 1-L bottles.
11. 0.45-μm PVDF filters.

2.4 Determination of Infection Efficiency

1. Cells that are going to be screened. Cas9 positive cells if you are utilizing the two-vector system. We strongly recommend using the actual cell line in which the screening will be conducted due to the fact that transduction efficiency may vary among cell lines.
2. Complete growth media for the cells.
3. 9–10 mL of library virus (Ideally high titer, as indicated by Lenti-X GoStix)
4. 10 mg/mL Polybrene in water.
5. 10 mg/mL Puromycin in water.
6. Cell counter or hemocytometer.
7. 6-well plates.
8. Pipettes.
9. 50-mL conical tubes.
10. PBS.
11. Trypsin.

2.5 Library Transduction

1. Cells that are going to be screened. Cas9 positive cells if you are utilizing the two-vector system.
2. Complete growth media for the cells.
3. library virus (amount calculated below).
4. Polybrene 10 mg/mL in water.
5. Puromycin 10 mg/mL in water.
6. Cell counter or hemocytometer.
7. T-225 flasks.
8. Pipettes.
9. 50-mL conical tubes.
10. PBS.
11. Trypsin.

2.6 Genomic DNA Extraction

Many kits provide efficient genomic DNA extraction. We aim for at least 100 μg of genomic DNA.

1. Genomic DNA extraction kit.
2. Vortex.
3. Centrifuge.
4. 96–100% Ethanol.
5. Optional, but recommended: RNAase, dissolved in water 20 mg/mL.

2.7 PCR and Purification for Sequencing

1. >100 μg genomic DNA per sample.
2. Premix ex*taq*HS (Takara Bio) or equivalent.
3. Ultrapure Water (molecular biology grade).
4. 10 μM Equal molar mix of P5 primers for each sample barcode (Fig. 1). Do not mix barcode primers with other barcode primers, just those with the same sample barcode.
5. 10 μM P7 primer.
6. Thermocycler.
7. AMPure XP (Beckman Coulter) or equivalent.
8. 96-well Magnetic Plate.
9. Microcentrifuge tubes.
10. PCR plates.
11. 96-well round-bottom plates.
12. 70% ethanol (molecular biology grade).
13. TE buffer.

2.8 Pooled siRNA Validation Screen

1. Cells of interest. Cells should not have CAS9 or other genetic alterations.
2. "Normal/control" cells.
3. CellTiter-Glo 2.0 Cell Viability Assay.
4. Pools of four individual siRNA in a 96-well plate diluted to 2 μM with siRNA resuspension buffer. Include nontargeting controls. We prefer multiple individual controls rather than pooled siRNA controls. We also like to include a known essential gene from the literature as a positive control.
5. Lipofectamine RNAiMax Transfection Reagent.
6. Tissue culture 96-well solid white microplates.
7. Complete cell growth media.
8. Opti-MEM reduced serum media.
9. Multichannel pipette(s) capable of pipetting 2–100 μL.
10. Plate reader capable of reading luminescence.

Primers

F1	**AATGGACTATCATATGCTTACCGTAACTTGAAAGTATTTCG**
R1	CTTTAGTTTGTATGTCTGTTGCTATTATGTCTACTATTCTTTCC

Vector priming site

P5 Forward Primer sets (barcoded)

Barcode #1:

Barcode1_F2_1	AATGATACGGCGACCACCGAGATCTACACTCTTTCCCTACACGACGCTCTTCCGATCTNAACGTGATTCTTGTGGAAAGGACGAAACACCG
Barcode1_F2_3	AATGATACGGCGACCACCGAGATCTACACTCTTTCCCTACACGACGCTCTTCCGATCTNNNAACGTGATTCTTGTGGAAAGGACGAAACACCG
Barcode1_F2_6	AATGATACGGCGACCACCGAGATCTACACTCTTTCCCTACACGACGCTCTTCCGATCTNNNNNNAACGTGATTCTTGTGGAAAGGACGAAACACCG
Barcode1_F2_9	AATGATACGGCGACCACCGAGATCTACACTCTTTCCCTACACGACGCTCTTCCGATCTNNNNNNNNNAACGTGATTCTTGTGGAAAGGACGAAACACCG

P5 flow-cell sequence, Illumina sequencing primer, Stagger, Barcode, Vector priming site

Barcode #2-6

Replace Barcode with the following 8 nt sequences. The rest of the sequence remains the same

Example: Sample 1 (Pre-selection event/control) = barcode #1, Sample 2 (Post-selection event/variable)

Barcode #2 AACGCATT; Barcode #3 ACTGTATC; Barcode #4 CATGATCG; Barcode #5 CGTTACCA; Barcode #6 CGCAGATT

For example- Barcode #2

Barcode1_F2_1	AATGATACGGCGACCACCGAGATCTACACTCTTTCCCTACACGACGCTCTTCCGATCTNAACGCATTTCTTGTGGAAAGGACGAAACACCG
Barcode1_F2_3	AATGATACGGCGACCACCGAGATCTACACTCTTTCCCTACACGACGCTCTTCCGATCTNNNAACGCATTTCTTGTGGAAAGGACGAAACACCG
Barcode1_F2_6	AATGATACGGCGACCACCGAGATCTACACTCTTTCCCTACACGACGCTCTTCCGATCTNNNNNNAACGCATTTCTTGTGGAAAGGACGAAACACCG
Barcode1_F2_9	AATGATACGGCGACCACCGAGATCTACACTCTTTCCCTACACGACGCTCTTCCGATCTNNNNNNNNNAACGCATTTCTTGTGGAAAGGACGAAACACCG

P5 flow-cell sequence, Illumina sequencing primer, Stagger, Barcode, Vector priming site

P7 Reverse Primer

P7_R2_1	CAAGCAGAAGACGGCATACGAGATGTGACTGGAGTTCAGACGTGTGCTCTTCCGATCTTCTACTATTCTTTCCCCTGCACTGT

P7 flow-cell sequence, Illumina sequencing primer, Vector priming site

Must order F1, R1, P5 forward primer sets (1 set/barcode/sample), and Reverse P7

Fig. 1 Primer set for nested PCR of integrated gRNAs utilizing sample-specific barcodes. These primers are sufficient for Illumina Sequencing

2.9 Deconvolution Screen

1. Multiple Cell lines of interest. One should be the original cell line and 3–4 additional cell lines.
2. 10–20 siRNA targets each with four individual siRNA sequences in a 96-well plate format (2 μM). The four individual siRNA sequences are the same four from the pooled siRNA screen.

3 Methods

This next section details the protocols required to conduct a genome-wide CRISPR dropout screen and siRNA screen to identify and validate essential genes (Fig. 2).

3.1 Library Preparation

This protocol has been adapted from the broad institute GPP Web Portal (https://portals.broadinstitute.org/gpp/public/resources/protocols). It is important to remember that maintaining library diversity is of the utmost importance and is the reason this protocol requires large amounts of DNA and bacteria.

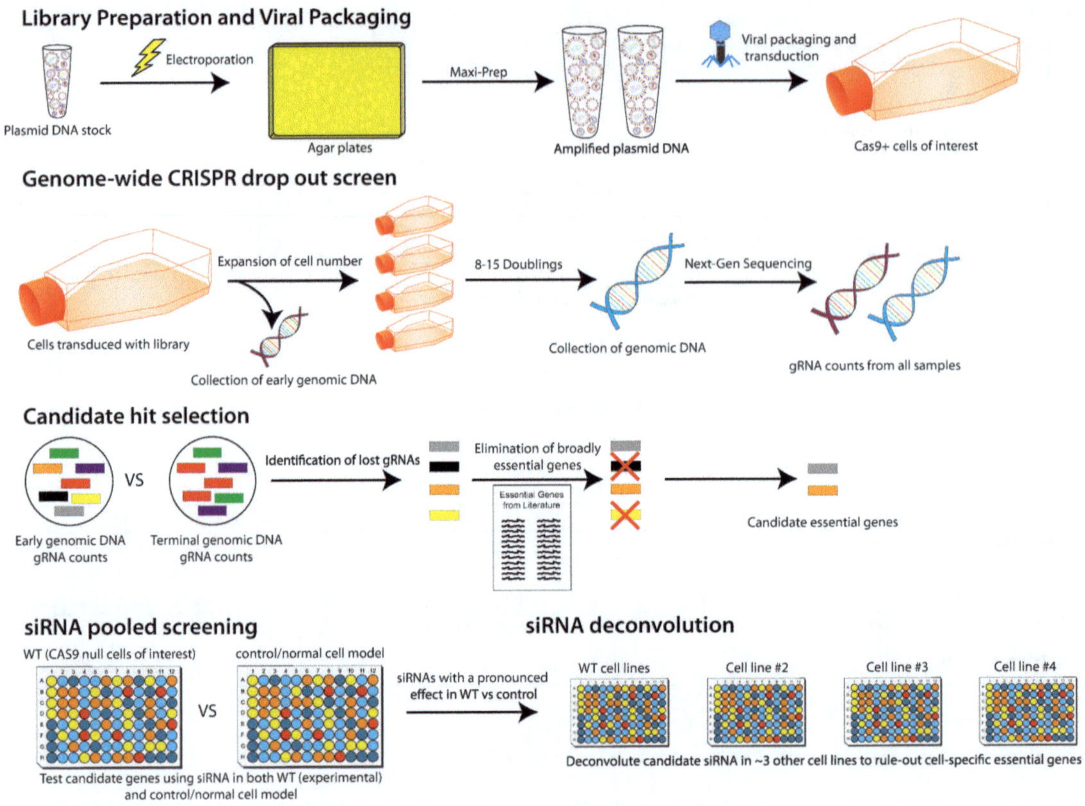

Fig. 2 Overview schematic of the entire workflow, including library preparation and viral packaging, genome-wide CRISPR dropout screening, candidate hit selection, and siRNA screening

3.1.1 Preparation of LB Agar Plates

1. Add 70 grams LB Agar, Lennox (follow manufacturer's guidelines) to two liters of distilled water.
2. Adjust pH to 7.2 and autoclave.
3. Once the LB Agar solution is cooled to ~50 °C, add 2 mL ampicillin (100 mg/mL) for a final concentration of 100 μg/mL ampicillin.
4. Pour mixture into twelve 245-mm plates about 13 mm deep. Allow to completely cool and solidify (3–6 h).

3.1.2 Electroporation and Bacterial Growth

1. Thaw 400 μL competent bacteria on wet ice for 10–20 min.
2. Mix 1600 ng Brunello Library plasmid DNA with 400 μL competent bacteria. Keep on wet ice for 10 min.
3. Add 30 μL of this bacteria/plasmid DNA mixture to a chilled cuvette and electroporate at 1.8 kV. Add 450 μL of cold SOC outgrowth medium, and transfer to 14-mL round-bottom tube containing 2.5 mL cold SOC outgrowth medium. Keep on wet ice.

4. Repeat **step 3** for a total of 12 cuvettes. At this point, you should have a total of 12 round-bottom tubes containing ~3 mL of SOC outgrowth medium on wet ice.
5. Shake tubes at 30–33 °C for 30 min to 1 h.
6. Using disposable cell spreaders, spread 3 mL from each round-bottom tube on a prewarmed 245 mm LB agar with ampicillin plate for a total of 12 plates.
7. Allow the medium to be fully absorbed into the agar (~1 h), flip upside down, and place into an incubator at 30–33 °C for 16 h.

3.1.3 Plasmid DNA Extraction

1. Remove plates from incubator and allow the plates to cool to room temperature (1 h).
2. Using a cell spreader, scrape the plates using 25 mL of cold LB media (25 mL/plate). Collect the LB media into a sterile 500-mL beaker. There should be a total of more than 300 mL of LB media.
3. Divide the total by 10 and aliquot into individual 50-mL conical tubes.
4. Spin down the tubes at 1500 × *g* for 20 min at 4 °C. Decant supernatant.
5. Purify the plasmid DNA according to the plasmid maxi kit instructions.
6. Using the purified plasmid DNA, PCR amplifies amplicons to submit to illumina sequencing. See PCR amplification protocol in Subheading 2.1, **item 6**. This step is necessary to make sure library representation has been maintained.

3.2 Lentivirus Production

Large-scale viral production is required to maintain library representation and to generate high-titer virus capable of infecting up to 2.5×10^8 cells. We recommend against the use of antibiotics in all cell culture steps as they may mask bacterial contamination. Additionally, follow all BSL-2 guidelines at your institution.

3.2.1 Day 1: Cell Seeding

1. Seed 17.5×10^6 293T cells into 16 T-175 flasks with a total media volume of 30 mL. Make sure cells are healthy and have never been above 80% confluence during growth. Health of the cells is vital to efficient transfection of plasmid DNA.

3.2.2 Day 2: 293T Cell Preparation for Transfection

1. Warm Opti-MEM in a 37 °C water bath.
2. Gently remove complete maintenance media and replace with 25 mL of Opti-MEM 1 h before adding transfection mixture (next steps). Cells should be 60–70% confluent.

3.2.3 Day 2: Transfection Master Mix Formulation

1. Thaw plasmid DNA, VSV-G, and psPAX2 at room temperature.
2. In a 50-mL conical tube, generate a plasmid master mix by adding the DNA and bringing up the total volume to 51 mL with Opti-Mem.
   (a) Plasmid DNA 714 μg (42 μg/flask).
   (b) psPAX2 DNA 714 μg (42 μg/flask).
   (c) VSV-G DNA 115 μg (6.75 μg/flask).
   (d) 3060 μL P3000 reagent (180 μL/flask).
3. In a second 50-mL conical tube, add 1.7 mL Lipofectamine 3000 to 48 mL Opti-MEM.
4. Using 16 separate 15-mL conical tubes, mix 3 mL of the DNA mixture with 3 mL of the Lipofectamine 3000 mixture and incubate at room temperature for 25 min.

3.2.4 Day 2: 293T Plasmid DNA Transfection

1. Gently remove the 16 T-175 flasks from the incubator that have had 25 mL of Opti-MEM for 30 min to 1 h. It is best to work with four at a time. Make sure not to tilt or cause the media to be agitated.
2. With the flasks stacked and flat in your biological safety hood, gently pipette 6 mL from each 15-mL flask on to one T-175 flask. Add this mixture drop wise on to as much of the surface area as possible. Repeat for all 16 flasks.
3. 6–8 h posttransfection, gently remove the flasks and change the media to 50 mL complete maintenance media (optional penicillin–streptomycin).

3.2.5 Day 4: Virus Media Collection and Purification

See **Note 1**.

1. These next steps should be done 36–40 h after the addition of complete maintenance media on day 2.
2. Gently remove the 16 flasks from the incubator and pipette the media into 50-mL conical tubes.
3. Centrifuge the tubes (with protections from aerosols) at $1500 \times g$ for 5 min.
4. Filter 50-mL tubes using 0.45-μm PVDF filter. Centrifugation and filtering eliminate 293T cell contamination with minimal losses to lentiviral particle number.
5. In an effort to make a homogenous population, decant each 50-mL tube into a 1-L sterile bottle.
6. Optional: You may use Takara Lenti-X GoStix o quickly check for the presence of lentiviral particles and to test the titer. This is a 10-min readout that we highly recommend before moving on.

7. Aliquot 10 mL from the homogenous solution into 15-mL tubes. Freeze at −80 °C. Concentrate the virus if required.

3.3 Determination of Infection Efficiency

A critical aspect of genome-wide CRISPR screening is the assumption that there is 1 gRNA per cell. To combat the odds of having multiple genes knocked-out, we need to limit the efficiency of infection to less than 0.5, ideally the lower the better. However, this must be weighed against the magnitude of cell numbers needed to maintain coverage of the gRNA library. To balance these requirements, an infection efficiency of 0.3–0.5 is recommended. We personally aim for an MOI of 0.3–0.4. For nonadherent cells, a spin-fection (virus transduction aided by centrifugation) is preferred; however, this protocol details transduction for adherent cells (no centrifugation required). Since we originally designed this protocol, a more detailed and robust protocol has been developed and is available at the Broad Institute's protocol collection (GPP Web Portal).

3.3.1 Before Starting Protocol

1. Determine cell seeding density that yields 90% cell confluence in *6 days*.
2. Conduct a puromycin killing curve to determine the optimal puromycin concentration to achieve 100% cell death at day 3.

3.3.2 Day 1—Cell Seeding

1. Seed the predetermined optimal number of cells (from above) in seven 6-well plates with 2 mL of complete growth media per well. These cells should be in logarithmic growth. To minimize well variability, it is best to intermittently mix the cells prior to seeding them in the six-well plates.

3.3.3 Day 2—Viral Transduction

1. 24 h after cell seeding, thaw 9–10 mL of library lentivirus at room temperature. Do not thaw at 37 °C.
2. While the lentivirus is thawing, take one of the 6-well plates and wash it with PBS, trypsinize, and count the cells in each well. These cell counts will be averaged to determine the number of cells being infected. There should be six 6-well plates remaining.
3. Using Table 1, mix complete growth media (no penicillin-streptomycin or other antibiotics) with polybrene (stock concentration 10 mg/mL) and the library lentivirus. Polybrene final concentration may be determined for your cell type; however, here we are using 6 μg/mL. The media should be similar temperature to the library lentivirus; do not warm to 37 °C.
4. Label the remaining six of the 6-well plates as shown in Fig. 3.
5. Each of the 6-well labels corresponds to the amount of library lentivirus is in each well and to the 15-mL tubes generated in **step 2**.

Table 1
Amounts of growth media, polybrene, and the library lentivirus

Tube Label (15 mL conical)	Growth Media (mL)	Library Lentivirus (μL)	Polybrene (μL)
0	14	0	8.4
40	13.7	280	8.4
80	13.4	560	8.4
160	12.9	1120	8.4
320	11.8	2240	8.4
640	9.5	4480	8.4

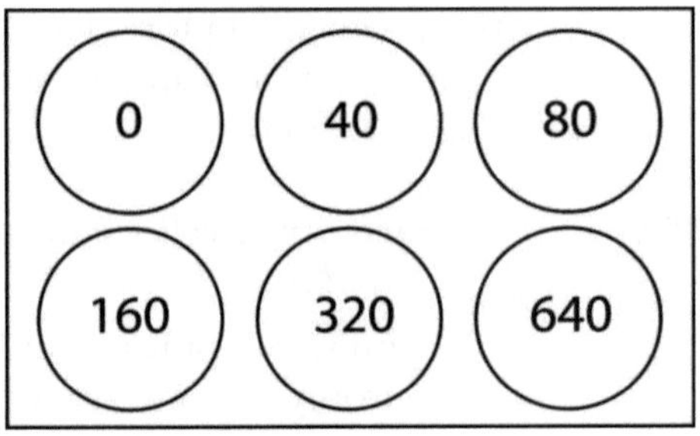

Fig. 3 6-well plate layout for Subheading 3.3

6. Add 2 mL from tube 0 to all six wells labeled 0; repeat for each label until all six 6-well plates have media.
7. Gently, put the plates back in the incubator.

3.3.4 Day 3—Drug Selection

1. 24 h after viral transduction, gently aspirate viral media from three 6-well plates and replace with complete growth media *with* the predetermined puromycin concentration to achieve 100% cell death.
2. Gently aspirate the media from the remaining three 6-well plates and replace with 2 mL complete growth media *without* puromycin. At this time, three 6-well plates should have media *with puromycin* and three 6-well plates *without puromycin.*

3.3.5 Day 6—Calculation of Infection Efficiency

1. Working with one plate, wash the wells with PBS and trypsinize the cells. Count the cells and record the amount per well. Repeat with all six plates. Making sure to keep note of the plates with and without puromycin.
2. Calculate your puromycin killing efficiency. Average of three 0 wells *with* puromycin divided by the average of the three 0 wells *without* puromycin. Subtract it from 100 to get puromycin killing efficiency of nontransduced cells.

3. Calculate the infection efficiency for each viral load. Average of three 40-μL wells with puromycin divided by the average of the three 40-μL wells without puromycin. Multiply this number by 100 to get a percent. Repeat for 80 μL, 160 μL, 320 μL, and 640 μL virus wells. Example: 650,000/1,000,000 = 0.65, $0.65 \times 100 = 65\%$ efficiency of infection.
4. Plot the calculated infection efficiencies, and using a linear regression line, determine the amount of virus needed to achieve 35% infection efficiency. *Do not extrapolate data, if 35% efficiency falls outside the range of the tested virus amounts; redo the experiment with a new viral range.*
5. Calculate the amount of virus required to infect 1×10^6 cells. Use the average values from the 6-well plate found in Subheading 3.3.3 Day 2 **step 2**. Example: 120 μL of virus/ $6 \times 10^5 = X$ μL of virus/1×10^6.

3.4 Library Transduction

Correct mathematical calculations are required to ensure adequate coverage of your gRNA library. The starting cell numbers are relatively large due to the number of the gRNAs, the limited infection efficiency, and an ideal coverage of more than 300 cells/ gRNA.

3.4.1 Required Calculations

1. Calculate number of required cells for appropriate gRNA coverage

   76,441 (# of gRNAs in library) × 500 (minimum # of cells/ gRNA)/0.35 (infection efficiency) = $\mathit{1.09 \times 10^8}$.
2. Calculate the number of T-225 flasks required to seed the amount of cells calculated in **step 1** above. Keep in mind that this number ($\mathit{1.09 \times 10^8}$) is the amount of cells required at the time of infection, not at seeding. Average cell count from Day 2 (Subheading 3.3.3) divided by Day 1 (Subheading 3.3.2) seeding cell density is the growth rate. 1.09×10^8 divided by the growth rate is the total number of cells to seed. Day 1 seeding cell density in a 6-well plate multiplied by the scaling factor (23.44) is the seeding density for a T-225 flask. The total number of cells to seed (calculated above) divided by the seeding density for a T-225 flask is the number of T-225 flasks required for optimal coverage. Round up the number of flasks.
3. Calculate the amount of library lentivirus to infect 1.09×10^8 cells. Use the value (μL of virus/1×10^6 cells) calculated in Day 6 **step 5** of Subheading 3.3.5. X μL of virus × 109 = total μL of library lentivirus required to infect 1.09×10^8 cells.

3.4.2 Day 1—Seeding T-225 Flasks

1. Warm complete growth medium to 37 °C.
2. Seed the calculated number of cells (**step 1**, Subheading 3.4.1) into the calculated number T-225 flasks (**step 2**, Subheading 3.4.1) with 50 mL of growth medium/flask.

3.4.3 Day 2—Library Lentivirus Transduction

1. Warm complete growth medium (without penicillin-streptomycin or other antibiotics) at room temperature.
2. Thaw library lentivirus at room temperature. The amount was calculated in Subheading 3.4.1, **step 3**.
3. Thaw polybrene (10 mg/mL) at room temperature.
4. Prepare master mix of media, virus, and polybrene.
   Total complete growth medium = # of T-225 flasks × 50 mL.
   Calculate required volume of polybrene ($C_1V_1 = C_2V_2$) (*see* **Note 2**).
   In a large sterile bottle, mix the media, virus, and polybrene to create a master mix and to ensure equal distribution of the library lentivirus.
5. Gently aspirate media from T-225 flasks, working in manageable batches.
6. Add 50 mL from the master mix (growth media, lentivirus, and polybrene) to each T-225 flask of cells.
7. While minimizing the agitation of the media, put the flasks back in the incubator.

3.4.4 Day 3—Puromycin Selection

1. Warm complete growth medium (penicillin-streptomycin is optional) at 37 °C. Add puromycin to the media at your predetermined dosage that yields 100% cell death in nontransduced cells in 3 days.
2. Working in batches, aspirate media and replace with 50 mL of growth media + puromycin.

3.4.5 Day 6–11—Cell Maintenance

1. At this point, nontransduced cells should be dead. Perform media changes until cells are 80% confluent. We suggest maintaining the puromycin concentration throughout cell passages.
2. Maintain at least 1000x gRNA coverage. With a library of 76,441 gRNAs this equates to 7.6×10^7 cells. Cells may be banked in batches for future experiments without the need to redo transduction. It is important to note that each subsequent experiment must start with a minimum of 7.6×10^7 to maintain coverage.
3. During these initial passages, it is important to pellet at least 7.6×10^7 cells and store at −80 °C. Genomic DNA will be extracted and will be used to determine the starting

representation of the library of gRNAs prior to a selection event (*see* **Note 3**).

3.5 Dropout Screen

A genome-wide CRISPR dropout screen is also known as a negative selection screen. Dropout screens are used to identify gRNAs that are "missing" after a selection event. The overall goal of this chapter is to lay out the necessary steps to identify essential genes. In addition, a dropout screen is appropriate to achieve this goal because we would expect cells with a gRNA targeting an essential gene to be removed from our genomic DNA samples due to cellular death.

We and others [1, 7, 15, 16] opted to allow the cells to grow for 8–18 doublings before collecting genomic DNA to determine which gRNAs are lost. This is because over time, some gRNAs will outcompete other gRNAs due to growth properties and not due to targeting essential genes.

3.5.1 Day 0—Maintain gRNA Library Diversity for 8–18 Doublings

1. Thaw or maintain cells from Subheading 2.1, **item 4** with at least 7.6×10^7 total cell number or 1000× gRNA coverage. Dropout screens require more sequencing read counts, and as such, it is vital to maintain representation of the gRNA library during cell passages.
2. Recommended: Every three-five doubling store 7.6×10^7 cells that may be used to determine a time-lapse loss of specific gRNAs.
3. We typically recommend a terminal endpoint of 12–15 doublings. At this time, pellet and store a minimum of 7.6×10^7 at −80 °C.

3.6 Genomic DNA Extraction

We follow the protocol according to the manufacturer. However, researchers at the Broad Institute have published improvements to the protocol (GPP Web Portal). These improvements include:

1. Extend the incubation of the cell pellet, PBS, Proteinase K, and lysis buffer from 10 to 15 min at 70 °C to *overnight* at 70 °C.
2. Following this overnight incubation at 70 °C, they add 20 μL RNAase (stock 20 mg/mL) and incubate for 5 min at room temperature.

3.7 PCR and Purification for Sequencing

Utilizing the genomic DNA, PCR is used to amplify the integrated gRNA sequences. These sequences are used identify the gRNAs and to determine the number of reads of each gRNA. We use PCR to add sample-specific barcodes (to allow sample pooling) and to add Illumina adapter sequences (P5 and P7). It is imperative to use multiple primers (with stagger) to maintain diversity across the flow-cell.

PCR bias is a problem and may arise from preferential amplification of products during early cycles. In an effort to minimize these effects, we aimed to keep PCR cycles as low as possible. The following protocol was largely adapted from Autry et al. 2020 [17] and uses a nested PCR design. Like many other protocols, other researchers have streamlined the process. The Broad Institute has a collection of protocols, which has a single step PCR protocol (GPP Web Portal), that includes modifications to PCR design, *Taq* DNA polymerase, and primers.

3.7.1 PCR Round 1 (Nested PCR)

1. Make sure to working on a clean surface or hood reserved for PCR work to avoid contamination.
2. Make a master mix of the following components with one genomic DNA sample at a time.
   (a) 1 mL Premix eq*Taq*HS.
   (b) 120 μL of 10 μM of F1 primer.
   (c) 120 μL of 10 μM of R1 primer.
   (d) 80–200 μg genomic DNA from one sample.
   (e) Add ultrapure water up to 2 mL.
3. Repeat **step 2** for each sample of genomic DNA. Make sure to clean the area between sample preparations.
4. Pipette 100 μL/well across 20 wells of a 96-well PCR plate for each genomic DNA sample. For two samples, 40 wells are required.
5. PCR program:
   (a) 2 min at 95 °C.
   (b) 30 s at 95 °C.
   (c) 30 s at 54 °C.
   (d) 30 s at 72 °C.
   (e) Go to **step 2** 14×.
   (f) Hold at 4 °C.

3.7.2 PCR Round 2

1. Working in a clean area, mix the PCR products from the 20 reactions/sample into a single tube.
2. Make a master mix of the following components:
   (a) 1 mL Premix eq*Taq*HS.
   (b) 120 μL of an equal molar mix (10 μM) of P5 Forward Primer set (1 barcode set/genomic DNA sample).
   (c) 120 μL of 10 μM of P7 Reverse Primer.
   (d) 100 μL from the aggregated PCR products from PCR round 1.
   (e) Add ultrapure water up to 2 mL.

3. Repeat **step 2** for each sample of genomic DNA. Make sure to clean the area between sample preparations.
4. Pipette 100 μL/well across 20 wells of a 96-well PCR plate for each genomic DNA sample. For two samples, 40 wells are required.
5. PCR program:
   (a) 2 min at 95 °C.
   (b) 30 s at 95 °C.
   (c) 30 s at 62.3 °C.
   (d) 30 s at 72 °C.
   (e) Go to **step 2** 18×.
   (f) Hold at 4 °C.

   See **Note 4**.

3.7.3 Purification of PCR Products

1. Mix the PCR products from round 2 of PCR, making sure to keep the samples from different genomic DNA separate.
2. Follow the manufactures protocol, AMPure XP (Beckman Coulter), using 100 μL/well in a 96-well round-bottom plates (Costar).
3. Optional: Run the purified product on a gel to confirm size and quantity. Or measure amount and quantity using a spectrophotometer.

3.8 Illumina Sequencing and Data Analysis

We highly recommend working with a sequencing core facility to determine the best options for sequencing. This discussion should include read counts, data deconvolution, and analysis of data quality. For dropout screens, it is imperative to have adequate read depth; we generally aim for upwards of 4×10^8 reads. For sequencing of the plasmid library, $\sim 8 \times 10^7$ is adequate.

Additionally, there are multiple options for data analysis. We [7, 13] opted to use the highly cited Mageck-Vispr software [18, 19]. Complete instructions are available at https://bitbucket.org/liulab/mageck-vispr/src/master.

3.9 Validation Using siRNA Screening

It is important to note that genome-wide CRISPR screens have been more robust and reliable screening tools compared to siRNA/shRNA screening [1, 20]. Genome-wide screening with siRNA is expensive and tedious compared to CRISPR pooled screens. However, *targeted* siRNA screening does offer some benefits. First, siRNA knockdown is fast, easy, and relatively homogenous among the cell population. Secondly, siRNA knockdown tests the robustness of the phenotype. Diverging phenotypes (siRNA vs. CRISPR) are not unexpected. A gene knocked-down ~90% of its endogenous level may not confer lethality, whereas the same gene knocked-out it may be lethal. A siRNA screen will confirm hits that upon knockdown share the same phenotype as the CRISPR screen. Thus, a small siRNA screen of the top ~100 candidate genes from a genome-wide CRISPR screen is a cost efficient, robust, and fast method of validating potential hits.

3.9.1 Pooled siRNA Validation Screen

A pooled siRNA screen utilizes four individual siRNAs targeting the same gene. This method offers the most robust response without the need to validate on-target knockdown efficiency. However, off-target phenotypes are common and a subsequent deconvolution screen is required.

The main goal of this siRNA validation step is twofold: to validate the phenotype from the CRISPR screen and to test the hits against another "normal/control" cell lines. Identification of disease-state specific essential genes requires the elimination of broadly cell essential genes, which can be done by comparison with other publications, and the elimination of tissue-specific essential genes. For example, say we are interested in testing ovarian cancer cells for essential genes. To do so, we must compare the resulting genes with normal healthy ovarian cells (or the best model available) to eliminate genes that are essential for both disease and nondisease state cells.

To assess gene essentiality, many different methods are used. We use CellTiter-Glo from Promega to quickly assess ATP levels using a spectrophotometer, with the assumption that dead cells have low ATP. Another method we have successfully used is live imaging to track cellular confluence.

1. Dilute the siRNA 96-well plate to 2 μM with siRNA resuspension buffer. Make sure to work under sterile conditions and avoid splashing between wells.
2. Using an 8-channel pipette transfer 8 μL of Opti-MEM to four tissue culture white 96-well plates.
3. Using an 8-channel pipette, transfer 2 μL of the stock siRNA (2 μM) solutions to the four 96-well plates (refer to Fig. 4). Column 1 and 12 are controls. Plate in quadruplicate among four experimental plates.

   Reverse Transfection

   Reverse Transfection is the addition of the lipid/RNA prior to the addition of cells.
4. Create a master mix of RNAiMax and opti-MEM. Mix 0.3 μL RNAiMax and 10 μL Opti-MEM per well. For a total for four plates, 120 μL RNAiMax and 4 mL of Opti-MEM are required.
5. Using a multichannel pipette, add 10 μL of the RNAiMax/Opti-MEM mastermix to each well of the four experimental plates.
6. Add 80 μL of complete growth media with 3000–10,000 cells to each well of the four experimental plates. *Growth kinetics of*

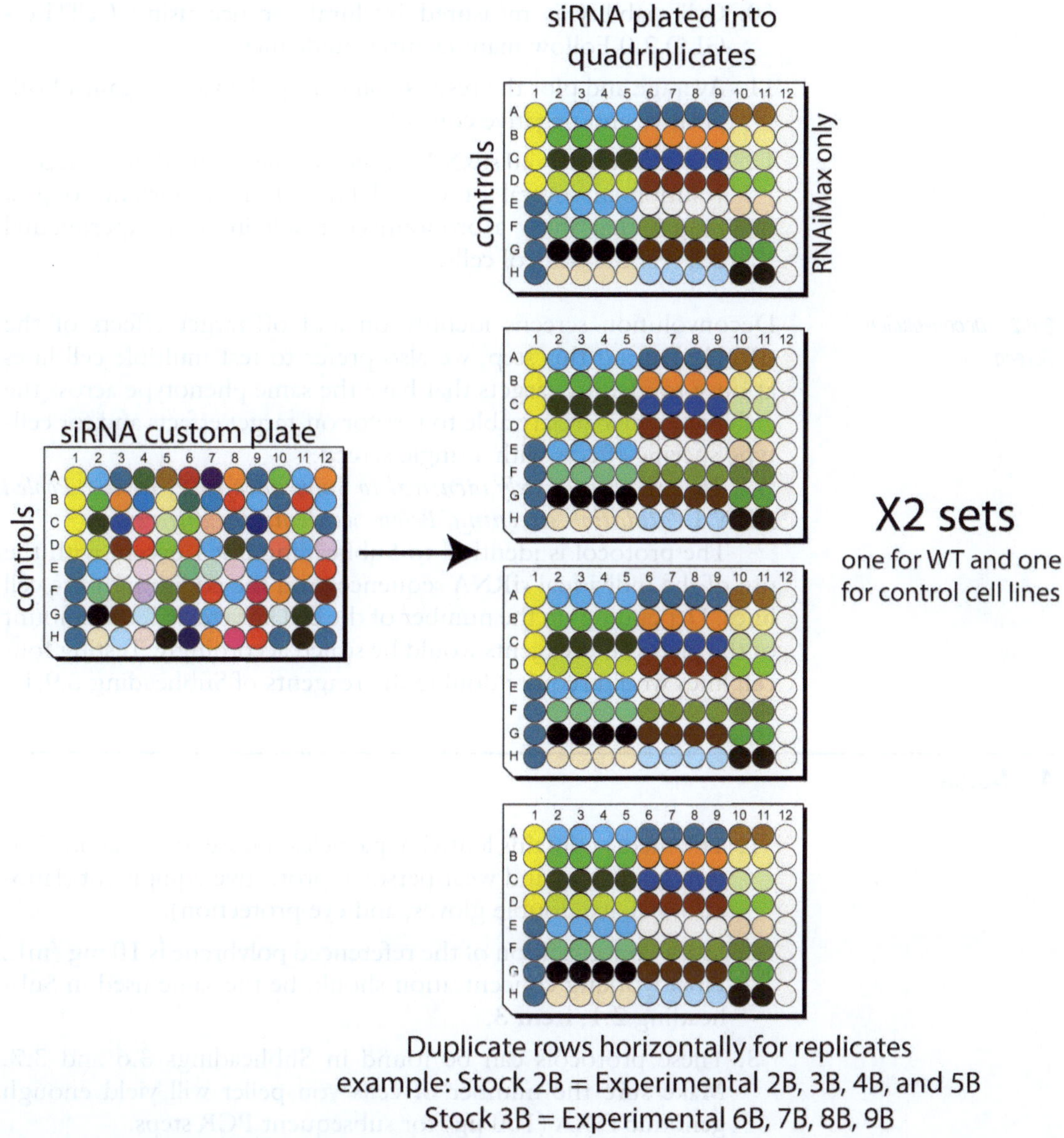

Fig. 4 96-well plate setup and design for Subheading 3.9.1

the cell lines should be used to determine optimal seeding density. Nontargeting and RNAiMax cells should be 95% confluent at day 4–5.

7. Repeat **steps 2–6** for the control cell line. At this stage, you should have at least eight 96-well plates in the incubator.
8. Change media ~48 h after transfection to complete growth media.
9. Monitor cells for 4–5 days.

Viability Assay

10. Cell viability is measured by luminescence using CellTiter-GLO 2.0.Follow manufactures guidelines.
11. Average and plot the results comparing the values against both positive and negative controls.
12. Pick the top 10–20 siRNA targets to validate further by deconvolution in several other cell lines. It is important to pick targets that show a pronounced result in your experimental cells vs. the control cells.

3.9.2 Deconvolution Screen

Deconvolution screens identify on and off-target effects of the siRNA. During this step, we also prefer to test multiple cell lines at once to identify targets that have the same phenotype across the cell lines. Thus, we are able to test for on-target effects and for cell-type specific effects with a single screen.

The process is largely identical to Subheading 2.2, ***item 1*** *Pooled siRNA Validation screening. Below, we detail the differences.*

The protocol is identical to Subheading 3.9.1 except with the use of the individual siRNA sequences and the addition of the cell lines. Depending on the number of the cell lines tested, the amount of the plates and reagents would be scaled accordingly. Testing four cell lines would require double the reagents of Subheading 3.9.1.

4 Notes

1. The media contains lentiviral particles. Please follow all institution guidelines and wear personal protective equipment (laboratory coats, double gloves, and eye protection).
2. Stock concentration of the referenced polybrene is 10 mg/mL. Final working concentration should be the same used in Subheading 2.1, **item 3**.
3. These protocols can be found in Subheadings 3.6 and 3.7. Make sure the number of cells you pellet will yield enough genomic DNA (100 μg) for subsequent PCR steps.
4. For sequencing of the plasmid DNA, the above steps are identical except the amount of plasmid DNA is 100 pg/reaction for a total of <2 μg among all ~ 20 reactions. (10 reactions may be adequate.)

Acknowledgements

We would like to thank the many scientists at St. Jude Children's Research Hospital, specifically Drs. Taosheng Chen, William E. Evans, and RJ Autry.

References

1. Hart T, Chandrashekhar M, Aregger M et al (2015) High-resolution CRISPR screens reveal fitness genes and genotype-specific cancer liabilities. Cell 163:1515–1526. https://doi.org/10.1016/j.cell.2015.11.015
2. Marcotte R, Brown KR, Suarez F et al (2012) Essential gene profiles in breast, pancreatic, and ovarian cancer cells. Cancer Discov 2:172–189. https://doi.org/10.1158/2159-8290.CD-11-0224
3. McCleland ML, Adler AS, Shang Y et al (2012) An integrated genomic screen identifies LDHB as an essential gene for triple-negative breast cancer. Cancer Res 72:5812–5823. https://doi.org/10.1158/0008-5472.CAN-12-1098
4. Sun S, Cheng S, Zhu Y et al (2016) Identification of PRKDC (protein kinase, DNA-activated, catalytic polypeptide) as an essential gene for colorectal cancer (CRCs) cells. Gene 584:90–96. https://doi.org/10.1016/j.gene.2016.03.020
5. Fei T, Chen Y, Xiao T et al (2017) Genome-wide CRISPR screen identifies HNRNPL as a prostate cancer dependency regulating RNA splicing. PNAS 114:E5207–E5215. https://doi.org/10.1073/pnas.1617467114
6. Bakke J, Wright WC, Zamora AE et al (2017) Transcription factor ZNF148 is a negative regulator of human muscle differentiation. Sci Rep 7:1–10. https://doi.org/10.1038/s41598-017-08267-5
7. Bakke J, Wright WC, Zamora AE et al (2019) Genome-wide CRISPR screen reveals PSMA6 to be an essential gene in pancreatic cancer cells. BMC Cancer 19:23. https://doi.org/10.1186/s12885-019-5455-1
8. Shalem O, Sanjana NE, Hartenian E et al (2014) Genome-scale CRISPR-Cas9 knockout screening in human cells. Science 343:84–87. https://doi.org/10.1126/science.1247005
9. Sanjana NE, Shalem O, Zhang F (2014) Improved vectors and genome-wide libraries for CRISPR screening. Nat Methods 11:783–784. https://doi.org/10.1038/nmeth.3047
10. Arroyo JD, Jourdain AA, Calvo SE et al (2016) A genome-wide CRISPR death screen identifies genes essential for oxidative phosphorylation. Cell Metab 24:875–885. https://doi.org/10.1016/j.cmet.2016.08.017
11. Chen S, Sanjana NE, Zheng K et al (2015) Genome-wide CRISPR screen in a mouse model of tumor growth and metastasis. Cell 160:1246–1260. https://doi.org/10.1016/j.cell.2015.02.038
12. Parnas O, Jovanovic M, Eisenhaure TM et al (2015) A genome-wide CRISPR screen in primary immune cells to dissect regulatory networks. Cell 162:675–686. https://doi.org/10.1016/j.cell.2015.06.059
13. Oladimeji PO, Bakke J, Wright WC, Chen T (2020) KANSL2 and MBNL3 are regulators of pancreatic ductal adenocarcinoma invasion. Sci Rep 10:1–8. https://doi.org/10.1038/s41598-020-58448-y
14. Doench JG, Fusi N, Sullender M et al (2016) Optimized sgRNA design to maximize activity and minimize off-target effects of CRISPR-Cas9. Nat Biotechnol 34:184–191. https://doi.org/10.1038/nbt.3437
15. Cross BCS, Lawo S, Archer CR et al (2016) Increasing the performance of pooled CRISPR–Cas9 drop-out screening. Sci Rep 6:1–8. https://doi.org/10.1038/srep31782
16. Lenoir WF, Lim TL, Hart T (2018) PICKLES: the database of pooled in-vitro CRISPR knockout library essentiality screens. Nucleic Acids Res 46:D776–D780. https://doi.org/10.1093/nar/gkx993
17. Autry RJ, Paugh SW, Carter R et al (2020) Integrative genomic analyses reveal mechanisms of glucocorticoid resistance in acute lymphoblastic leukemia. Nat Cancer 1:329–344. https://doi.org/10.1038/s43018-020-0037-3
18. Li W, Xu H, Xiao T et al (2014) MAGeCK enables robust identification of essential genes from genome-scale CRISPR/Cas9 knockout screens. Genome Biol 15:554. https://doi.org/10.1186/s13059-014-0554-4
19. Li W, Köster J, Xu H et al (2015) Quality control, modeling, and visualization of CRISPR screens with MAGeCK-VISPR. Genome Biol 16:281. https://doi.org/10.1186/s13059-015-0843-6
20. Evers B, Jastrzebski K, Heijmans JPM et al (2016) CRISPR knockout screening outperforms shRNA and CRISPRi in identifying essential genes. Nat Biotechnol 34:631–633. https://doi.org/10.1038/nbt.3536

Chapter 6

A Simple Method that Combines CRISPR and AID to Quickly Generate Conditional Knockouts for Essential Genes in Various Vertebrate Cell Lines

Kohei Nishimura and Tatsuo Fukagawa

Abstract

Cells with a loss-of-function mutation in a gene (knockout cells) are powerful tools for characterizing the function of such gene product. However, for essential genes, conditional knockout cell lines must be generated. The auxin-inducible degron (AID) technique enables us to conditionally and rapidly deplete a target protein from various eukaryotic cell lines. A combination of CRISPR-/Cas9-based gene editing and AID technique allows us to generate AID-based conditional knockout cell lines. Using these two techniques, we recently proposed a simple and quick way to generate conditional knockout cells for essential genes. In this chapter, we introduce the reader to the experimental procedures to generate these AID-based conditional knockout cell lines.

Key words Auxin-Inducible Degron (AID), CRISPR/Cas9, Essential genes, Conditional knockout

1 Introduction

The disruption of a target gene (gene knockout) is the most reliable method to understand its function. The CRISPR/Cas9-based gene-editing technology allows us to easily generate knockout cells for a gene of interest in various cell lines [1–4]. However, some genes are essential for cell viability; therefore, conditional gene knockout cell lines are useful for characterizing the functions of these essential gene products. There are various ways to generate conditional gene knockout cell lines, including Cre-loxP-mediated gene targeting [5, 6] or tetracycline-dependent transcriptional silencing [7–9]. As these methods inactivate the gene itself or its mRNA, it takes 1 or 2 days to deplete the target gene product in cells when either one of these methods has been used. In contrast, auxin-inducible degron (AID) technology [10] enables a direct degradation of the target protein. It utilizes the ubiquitin–proteasome system, which is well conserved in eukaryotes, and a

Ren Zhang (ed.), *Essential Genes and Genomes: Methods and Protocols*, Methods in Molecular Biology, vol. 2377, https://doi.org/10.1007/978-1-0716-1720-5_6,

plant-specific component of the E3 ubiquitin ligase that responds to the plant hormone auxin. In cells expressing the auxin perceptive F-box protein TIR1, any target protein fused to an AID-tag is degraded upon the addition of auxin. Using AID technology, conditional knockout cell lines have been generated in various vertebrate cells [10, 11]. Natsume et al. [11] generated cells expressing the rice TIR1 (OsTIR1) and inserted AID-tag sequences into endogenous genes using the CRISPR/Cas9 technology. In these cells, the AID-fused target product was expressed under the control of the endogenous promoter and conditionally degraded after the addition of auxin.

Combining AID technology with CRISPR/Cas9 gene editing is quite a powerful tool; however, these experiments sometimes need complicated procedures to generate final conditional knockout cell lines. To overcome this complication, we have developed a simple and quick method to generate AID-based conditional knockout chicken DT40 cell lines [12] in which a donor vector for homologous recombination is not needed. In this chapter, we describe those experimental procedures needed to obtain an AID-based conditional knockout cell line using our simple method and demonstrate the possibilities of applicating this procedure to other mammalian cell lines.

2 Materials

2.1 Plasmids

For the generation of an AID-based conditional cell line by our method [12], a pAID plasmid containing a clone of the target gene must be integrated into the genome (Fig. 1). The pAID plasmid encodes the rice F-box protein TIR1 (OsTIR1) and an AID-tagged protein (*see* **Note 1**). Any cDNA from a target protein can be cloned into an empty pAID plasmid. Both for gene knockout or linearization of the pAID plasmid, we use the pX330 plasmid, which contains a Cas9 nuclease gene and the U6 promoter. The sgRNA sequence for each target gene can be cloned into the pX330 plasmid. These plasmids can be obtained from Addgene.

The full list of plasmids and the materials needed for their manipulation are as follows:

1. pX330 plasmid (Addgene: #42230) for directing target gene knockout.
2. pX330 plasmids for linearization of pAID plasmids (Addgene: #140609, #140610).
3. pAID plasmids (Addgene: #105985; #105986, #140601, #140602, #140603, #140604, #140605, #140606, #140607, #140608).

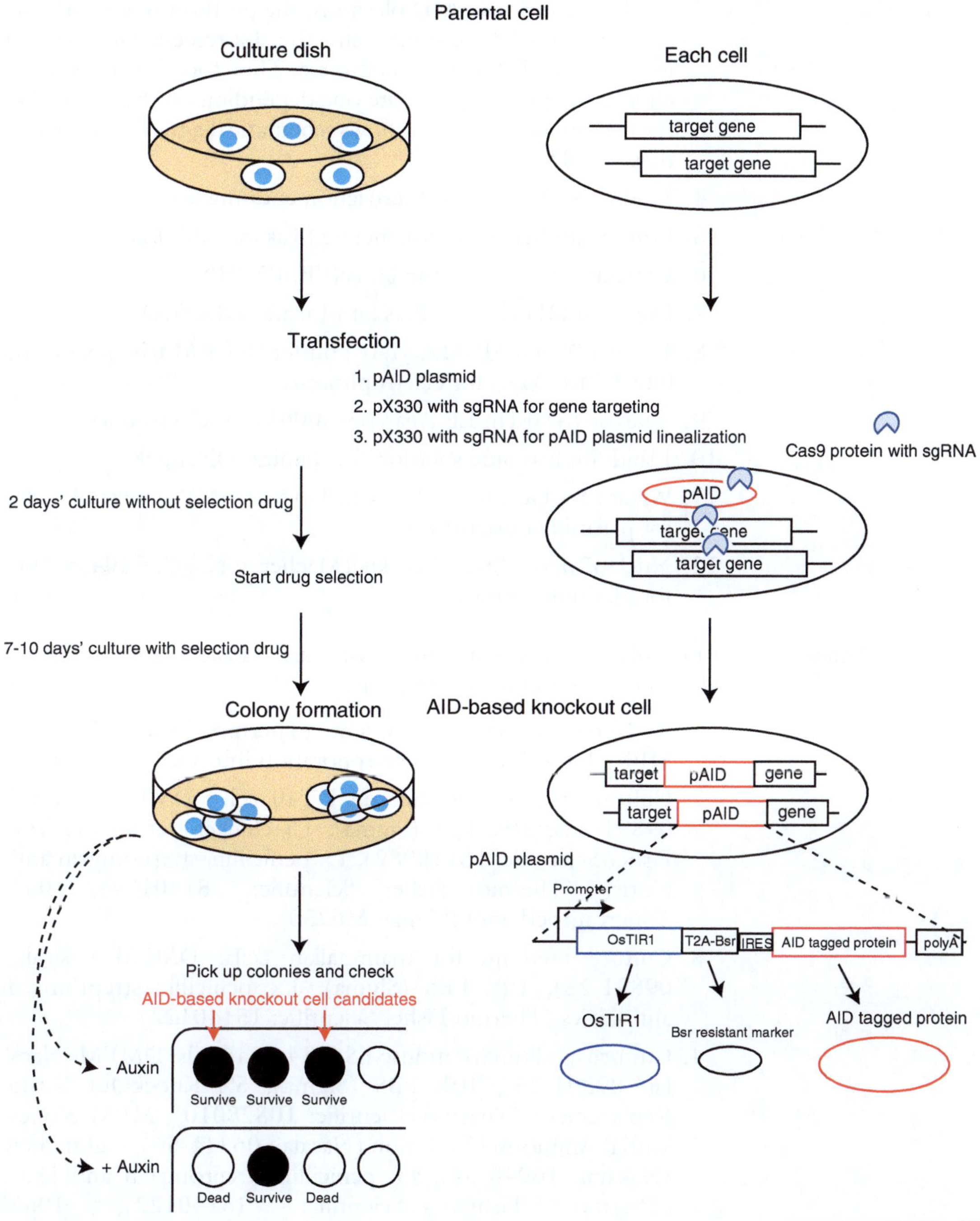

Fig. 1 Experimental scheme for generating AID-based conditional knockout cell lines. Three kinds of plasmids, pAID, pX330 with cloned sgRNA for gene targeting, and pX330 with cloned sgRNA for the linearization of the pAID plasmid are transfected to cultured cells. After drug selection, colonies are picked and cultured both in the presence and absence of auxins. Those colonies grown in the absence of auxin but not in its presence are

In the various pAID plasmids, the position of the AID tag (N-terminal or C-terminal) and the fluorescent tag (GFP or mCherry) included are also diverse. Therefore, it is possible to choose the most appropriate one depending on the purpose of the experiment. Figure 2 shows schematic diagrams for some of these pAID plasmids.

4. *Eco*RV (NEB: R0195S) restriction endonuclease.
5. PrimeStar GXL PCR polymerase (Takara: R050A).
6. Gibson assembly cloning kit (NEB: E5510S).
7. InFusion HD cloning Plus kit (Takara: Z8909N).
8. 1× Tris-Acetate-EDTA (TAE) buffer (0.04 M tris-acetate and 0.001 M EDTA) for electrophoresis.
9. Agarose (SeaKem LE Agarose: 50001) for electrophoresis.
10. Ethidium bromide solution for staining DNA gels.
11. Wizard SV Gel and PCR Clean-Up System (Promega: A9281) for plasmid construction.
12. NucleoBond® Xtra Maxi kit (Macherey-Nagel: 740414.100) for plasmid isolation.

2.2 Transfection

Once plasmids are constructed, they are transfected into cultured cells using the following materials:

1. A Neon transfection system (Thermo Fisher Scientific: MPK5000S) for DNA electroporation into various cell lines.
2. Culture medium for chicken DT40 cells: DMEM (Nacalai: 09891-25), 10% FBS (Sigma), 1% chicken serum (Thermo Fisher Scientific: 16110082), 1× penicillin–streptomycin antibiotics (Thermo Fisher Scientific: 15140122), 10μM 2-mercaptoethanol (Sigma: M6250).
3. Culture medium for mammalian cells: DMEM (Nacalai: 09891-25), 10% FBS (Sigma), 1× penicillin–streptomycin antibiotics (Thermo Fisher Scientific: 15140122).
4. Culture medium for mouse ES (E14tg2a) cells: DMEM (Nacalai: 09891-25), 10% FBS (Sigma), 5% KnockOut Serum Replacement (Thermo Scientific: 10828010), MEM Nonessential Amino Acid solution (Nacalai: 06344-56), L-glutamate (Nakarai: 16948-04), 1× penicillin–streptomycin antibiotics (Thermo Fisher Scientific: 15140122), 10μM

Fig. 1 (continued) potential candidates for conditional knockout lines. In successful conditional knockout cells, the target gene is disrupted and both OsTIR1 and the AID-tagged target protein are expressed. However, the AID-tagged target protein is degraded in the presence of auxin via the OsTIR1-dependent ubiquitin–proteasome pathway

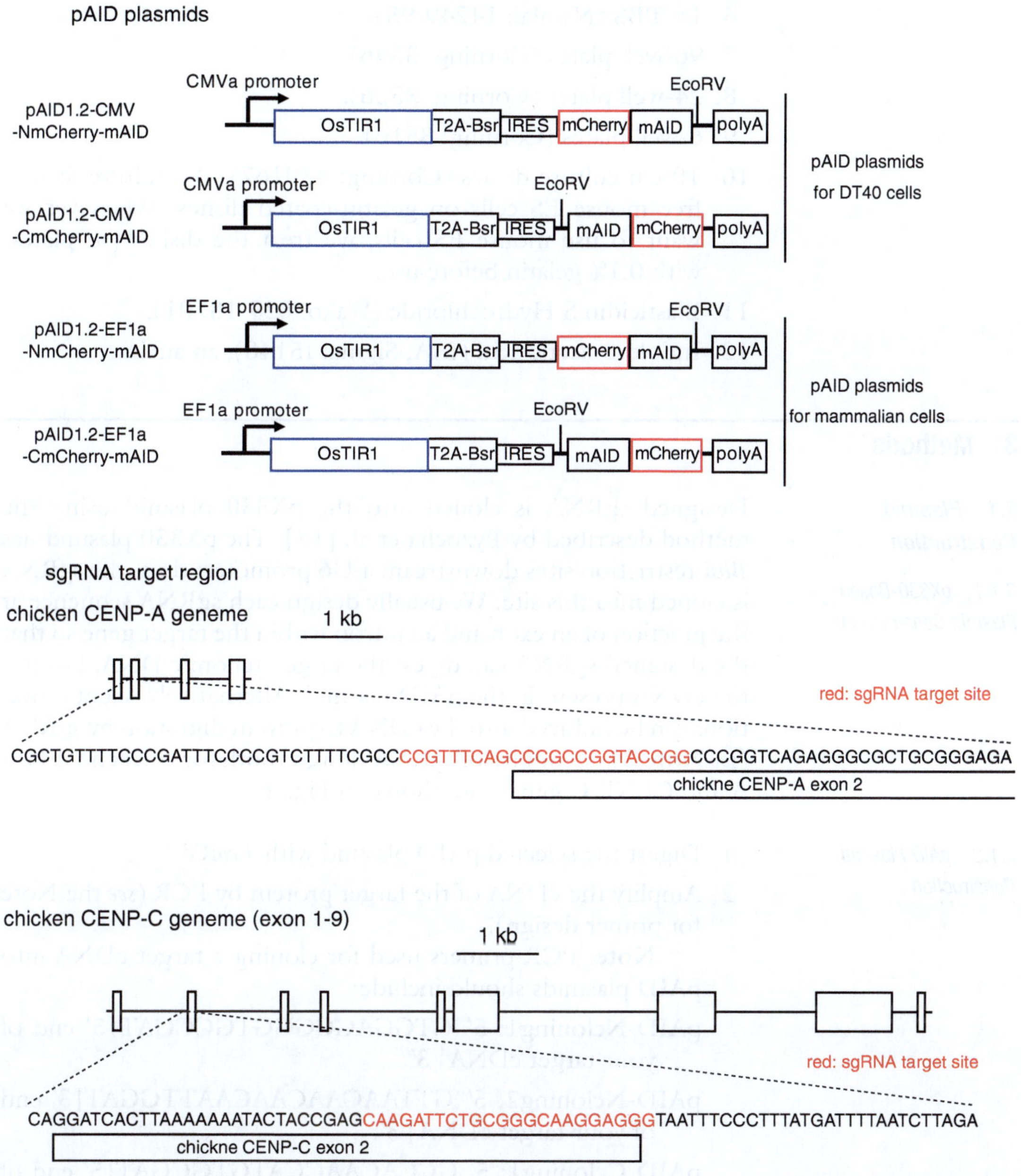

Fig. 2 Schematic diagrams of pAID plasmids and sgRNA sequence information of target chicken CENP-A and CENP-C genes. The pAID plasmid encodes OsTIR1 and an mCherry (or GFP)-mAID protein. Every plasmid has an *Eco*RV site useful for cloning any target cDNA. The position of the AID-tag (either N-terminal or C-terminal) is sometimes critical for the function of the target protein. The domain organization of target proteins should be considered for designing optimal constructs

2-mercaptoethanol (Sigma: M6250), 0.1% culture medium containing LIF overexpressed in 293T cells.

5. Polyethylenimine (Polysciences: 23966-1). We use 1 mg/mL polyethylenimine solution for plasmid transfection.

6. 1× PBS (Nacalai: 14249-95).
7. 96-well plates (Corning: 3596).
8. 24-well plates (Corning: 3526).
9. 6-well plates (Corning: 3516).
10. 10-cm culture dishes (Corning: 430167). We culture feeder-free mouse ES cells on gelatin-coated dishes. Whenever we want to use mouse ES cells, we treat the dishes (or plates) with 0.1% gelatin before use.
11. Blasticidin S Hydrochloride (Wako: 029-18701).
12. Indole-3-acetic acid (IAA, Sigma: I5148), an auxin.

3 Methods

3.1 Plasmid Construction

3.1.1 pX330-Based Plasmid Construction

Designed sgRNA is cloned into the pX330 plasmid using the method described by Pyzocha et al. [13]. The pX330 plasmid has *Bbs*I restriction sites downstream a U6 promoter; thus, the sgRNA is cloned into this site. We usually design each sgRNA sequence at the junction of an exon and an intron within the target gene so that the designed sgRNA can digest the target genomic DNA, but not the cDNA present in the pAID plasmid. Alternatively, silent mutations can be induced into the cDNA to prevent digestion by sgRNA (*see* **Note 2**). Examples of sgRNA design (both for chicken CENP-A and CENP-C genes) are shown in Fig. 2.

3.1.2 pAID Plasmid Construction

1. Digest the selected pAID plasmid with *Eco*RV.
2. Amplify the cDNA of the target protein by PCR (*see* the Note for primer design).

   Note: PCR primers used for cloning a target cDNA into pAID plasmids should include:

   pAID-Ncloning1: 5′ CTGGAGCGGGTGCCGAT[5′ end of your target cDNA] 3′.

   pAID-Ncloning2: 5′ GTTAACAACAACAATTGGAT[3′ end of your target cDNA] 3′.

   pAID-Ccloning1: 5′ GCCACAACCATGTGCGAT[5′ end of your target cDNA] 3′.

   pAID-Ccloning2: 5′ GCCTGCACCAGCTCCGAT[3′ end of your target cDNA] 3′.
3. Purify both the linearized pAID plasmid and the PCR fragment using the Wizard SV Gel and PCR Clean-Up System.

   Note: We always load DNA fragments into agarose gel wells for electrophoresis and perform gel purification of the desired bands in order to remove nonspecific PCR amplification products.

4. Assemble the PCR fragment and linearized pAID plasmid using Gibson assembly mix (NEB) or InFusion cloning kit (Takara).
5. Transform assembled DNA into *E. coli* cells.
6. Pick up colonies and isolate the corresponding cloned plasmids.
7. Confirm the sequence of cloned plasmids.

3.2 Transfection of Plasmids

For transfection, pure plasmid DNAs must be prepared. We usually use the NucleoBond® Xtra Maxi kit for pure plasmid isolation. The whole experimental scheme is shown in Fig. 1.

3.2.1 Transfection into a Chicken DT40 Cell Line

1. Prepare 5μg of the pAID plasmid, 5μg of the pX330 plasmid with cloned sgRNA for the linearization of the pAID plasmid, and 10μg of pX330 with cloned sgRNA for gene targeting. Precipitate all plasmids using 100% ethanol, wash them with 70% ethanol, and resuspend them in 130μL (total volume) of R buffer for Neon transfection (R buffer is included in the Neon transfection system kit).
2. Prepare 2×10^5 DT40 cells for each transfection.
3. Harvest and wash cells with 1x PBS once.
4. Resuspend cells in the 130μL mix containing R buffer and plasmid DNAs (described in **step 1**).
5. Electroporate cells using the Neon transfection system (1400 V, 5 ms, 6 pulses).
6. Suspend cells in 10 mL of DT40 medium immediately after Neon transfection.
7. Culture cells for 48 h, without drug selection, at 38.5 °C.
8. Take 3 mL, 0.6 mL, and 120μL from each DT40 cell culture and separately dilute them using 40 mL of fresh DT40 medium containing 25μg/mL of Blasticidin S (*see* **Note 4**). Split cells into two 96-well plates (200μL per well) and culture them under selective conditions for 1–2 weeks at 38.5 °C.
9. Pick up isolated colonies into 24-well plates both in the presence and absence of IAA (500μM) (*see* **Note 5**).
10. Colonies that grow in the absence of IAA but die in its presence are conditional knockout candidates of the target essential gene.
11. Analyze candidate colonies by PCR. In addition, degradation of the target protein must be confirmed by immunoblot analysis.

3.2.2 Transfection into Human Cell Lines

1. Prepare 5μg of the pAID plasmid, 5μg of the pX330 plasmid with cloned sgRNA for the linearization of the pAID plasmid, and 10μg of the pX330 plasmid with cloned sgRNA for gene targeting. Precipitate all plasmids with 100% ethanol, wash them with 70% ethanol, and resuspend them in 130μL of R buffer for Neon transfection (R buffer is included in the Neon transfection system kit) (*see* **Note 3**).
2. Prepare 2 × 10^5 human cells (HeLa-S3, HT1080, or RPE1 cells) for each transfection.
3. Harvest and wash cells with 1× PBS once.
4. Resuspend cells in the 130μL mix containing R buffer and plasmid DNAs (described in **step 1**).
5. Electroporate using the Neon transfection system (1400 V, 20 ms, 1 pulse).
6. Suspend cells into 10 mL of mammalian culture medium and plate 3 mL, 0.6 mL, and 120μL, separately, in 10-cm dishes containing 10 mL of mammalian culture medium.
7. Culture cells for 48 h without any drug selection at 37 °C.
8. Aspirate medium off, add 10 mL of new culture medium containing 4μg/mL (for HeLa-S3 cells) or 10μg/mL (for HT1080 or RPE1 cells) of Blasticidin S, and select cells for 1–2 weeks at 37 °C (*see* **Note 4**).

   Note: This method is applicable to other mammalian cell lines. However, the effective concentration of Blasticidin S on each cell line may vary; therefore, checking for the optimal concentration of the antibiotic for each cell line is needed to ensure a successful selection.
9. Transfer isolated colonies into 24-well plates both in the presence and absence of IAA (500μM) (*see* **Note 5**).
10. Colonies that grow in absence of IAA but die in its presence are conditional knockout candidates for the target essential gene.
11. Analyze candidate colonies by PCR and immunoblot analysis.

3.2.3 Transfection into Mouse ES Cell Line (E14tg2a)

1. Prepare 1μg of the pAID plasmid, 1μg of the pX330 plasmid with cloned sgRNA for the linearization of the pAID plasmid, and 2μg of the pX330 with cloned sgRNA for gene targeting (*see* **Note 3**).
2. Prepare two tubes containing 50μL of serum-free DMEM. Add the DNA to one of the tubes and 18μL of 1 mg/mL PEI solution to the other.
3. Mix both the DNA solution and the PEI solution by vortexing.
4. Prepare 1 × 10^6 ES cells for each transfection.
5. Harvest cells by trypsinization and wash them with 1× PBS.

6. Suspend cells in 300μL of serum-free DMEM.
7. Add the DNA/PEI solution to the cell suspension and mix by vortexing.
8. Incubate at room temperature (20–25 °C) for 30 min.
9. Plate cells on 6-well plates using different cell dilutions (1/5, 1/25, 1/125, 1/625, or even more diluted).
10. Culture cells in the absence of any selection drug for 24 h at 37 °C.
11. Add Blasticidin S to a final concentration of 10μg/mL to each well (*see* **Note 4**).
12. Select for 5–7 days.
13. Pick colonies into 24-well plates both in the presence and absence of IAA (500μM) (*see* **Note 5**).
14. Colonies that grow in absence of IAA but die in its presence are conditional knockout candidates for the target essential gene.
15. Analyze candidate colonies by PCR and immunoblot analysis.

Because we are focusing on the generation of conditional knockout cell lines for essential genes, cells should die after the addition of auxins. Therefore, we first screen for colonies that do not grow in the presence of IAA but grow in its absence. Disruption of the target genomic region must be characterized by PCR and DNA sequencing for candidate knockout colonies. Once gene disruption is confirmed, the expression and degradation of AID-tagged proteins after auxin addition should be analyzed by immunoblot analysis. In our system, the expression levels of AID-tagged proteins may be variable. Therefore, several clones should be characterized in parallel.

In Fig. 3, we show two examples of conditional knockout cell lines (CENP-A and CENP-C knockout cells) that were done using DT40 cells. Both CENP-A and CRNP-C are critical proteins for kinetochore assembly and are required for mitotic progression [14]. AID-fused proteins were not detected by immunoblot analysis within 1 h after addition of auxins to the medium (Fig. 3). The growth curves of these cells demonstrated that both lines die after the addition of the auxin (Fig. 3). Furthermore, we performed FACS analysis to know cell cycle distribution of cells. As both CENP-A and CENP-C are required for mitotic progression, disruption of these proteins causes accumulation of G2/M cells in the population (Fig. 3).

Our method is applicable to other cell lines. In this chapter, we introduce the specifications for its application in HeLa-S3, HT1080, and RPE1 human cell lines, and in mouse ES cells. We show data for CENP-H (a kinetochore protein) conditional knockout lines using all these cell lines (Fig. 4). Based on immunoblot

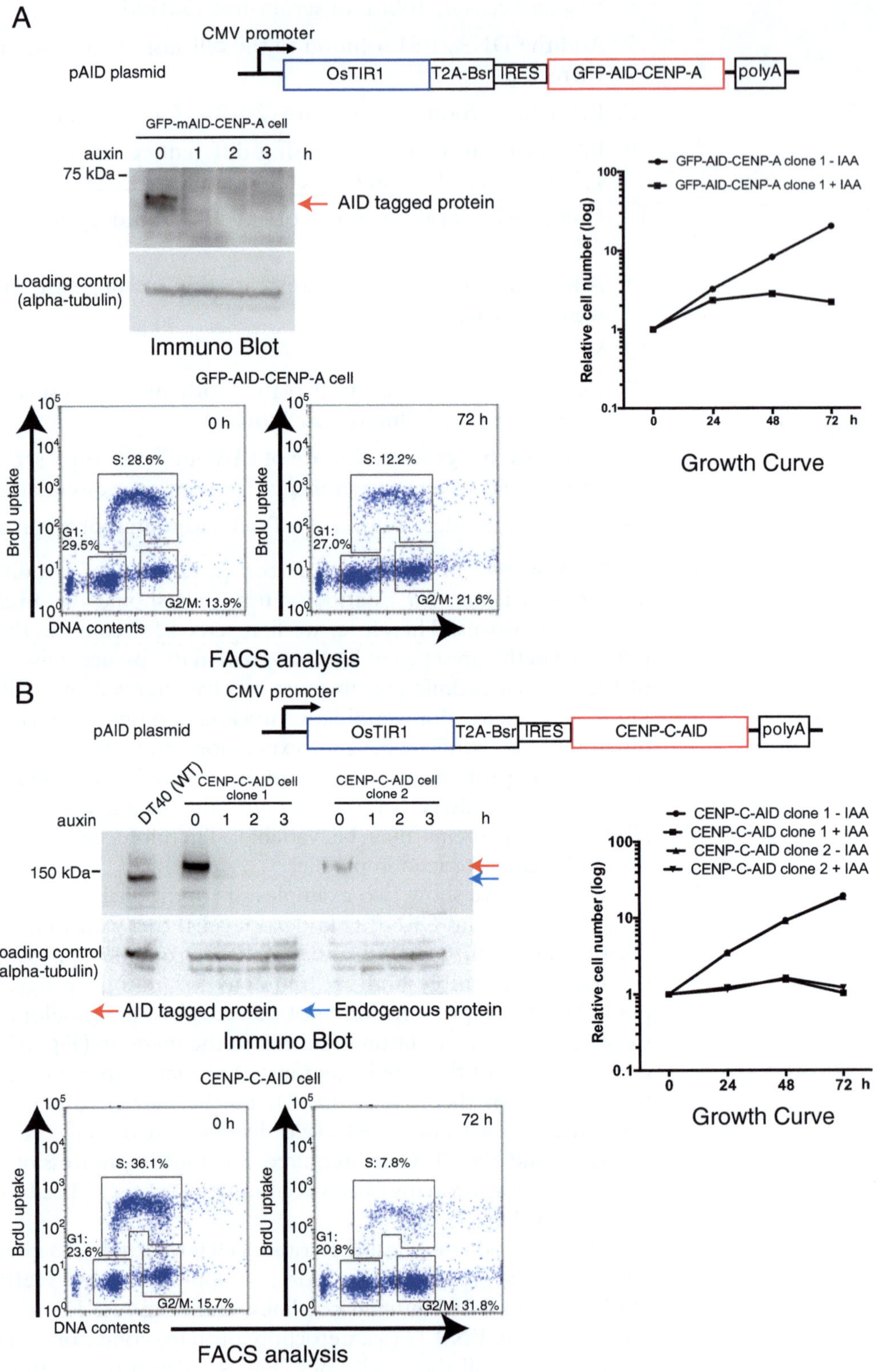

Fig. 3 Generation of AID-based conditional knockout DT40 cell lines. (**A**) CENP-A conditional knockout DT40 cell line. Immunoblot analysis indicated that GFP-mAID-CENP-A could not be detected within 1 h after auxin

analysis and microscopy observations, we confirmed the degradation of the AID-fused CENP-H protein after auxin addition in all tested cell lines. We also examined cell viability of these CENP-H conditional knockout lines based on colony formation assays. When same numbers of cells were plated in the absence or presence of auxin, colonies were clearly detected in the absence of the auxin, but not in its presence (Fig. 4). All these data indicate that AID-based CENP-H conditional knockout lines were successfully generated using tested cell lines confirming that this method is applicable to any cell line and is a powerful tool to characterize the function of essential gene products.

4 Notes

1. We designed two types of pAID plasmids containing either CMV promoter or EF1α promoter. We usually use the pAID plasmid containing CMV promoter for chicken DT40 cells and one containing EF1α promoter for other mammalian cell lines.
2. We usually design a sgRNA sequence at the intron–exon boundary of a target gene to avoid digestion of cDNA on the pAID plasmid. It is preferable to check the DNA sequence of the target site in your cell lines. The target sequence is sometimes changed due to point mutations or SNPs in some cell lines.
3. Although we have used transfection with Polyethylenimine or the Neon transfection system, other methods (such as a lipofection method) would be applicable to the transfection for mammalian cell lines. It is important to use transfection methods with low toxicity, which will not inhibit cell growth after the transfection.
4. The effective concentration of Blasticidin S depends on cell types. It is preferable to check the effective concentration of Blasticidin S on your cell lines, before the drag selection.
5. We have picked up 12 to 24 colonies for each transfection. These clones are cultured in growth media with DMSO (control) or 500μM IAA. It is preferable to use parental cells for

Fig. 3 (continued) (IAA) addition. Growth analysis demonstrated that this line did not grow in the presence of auxin. Furthermore, FACS analysis showed an accumulation of G2/M cells within the population after addition of auxin to the medium. (**B**) CENP-C conditional knockout DT40 cell lines. In this case, analyses of two independent clones are shown. Immunoblot analysis indicated that CENP-C-AID could not be detected within 1 h after auxin (IAA) addition in any of the clones. Growth analysis also demonstrated that these lines did not grow in the presence of auxin. Furthermore, FACS analysis showed an accumulation of G2/M cells within the population after adding auxin to the medium

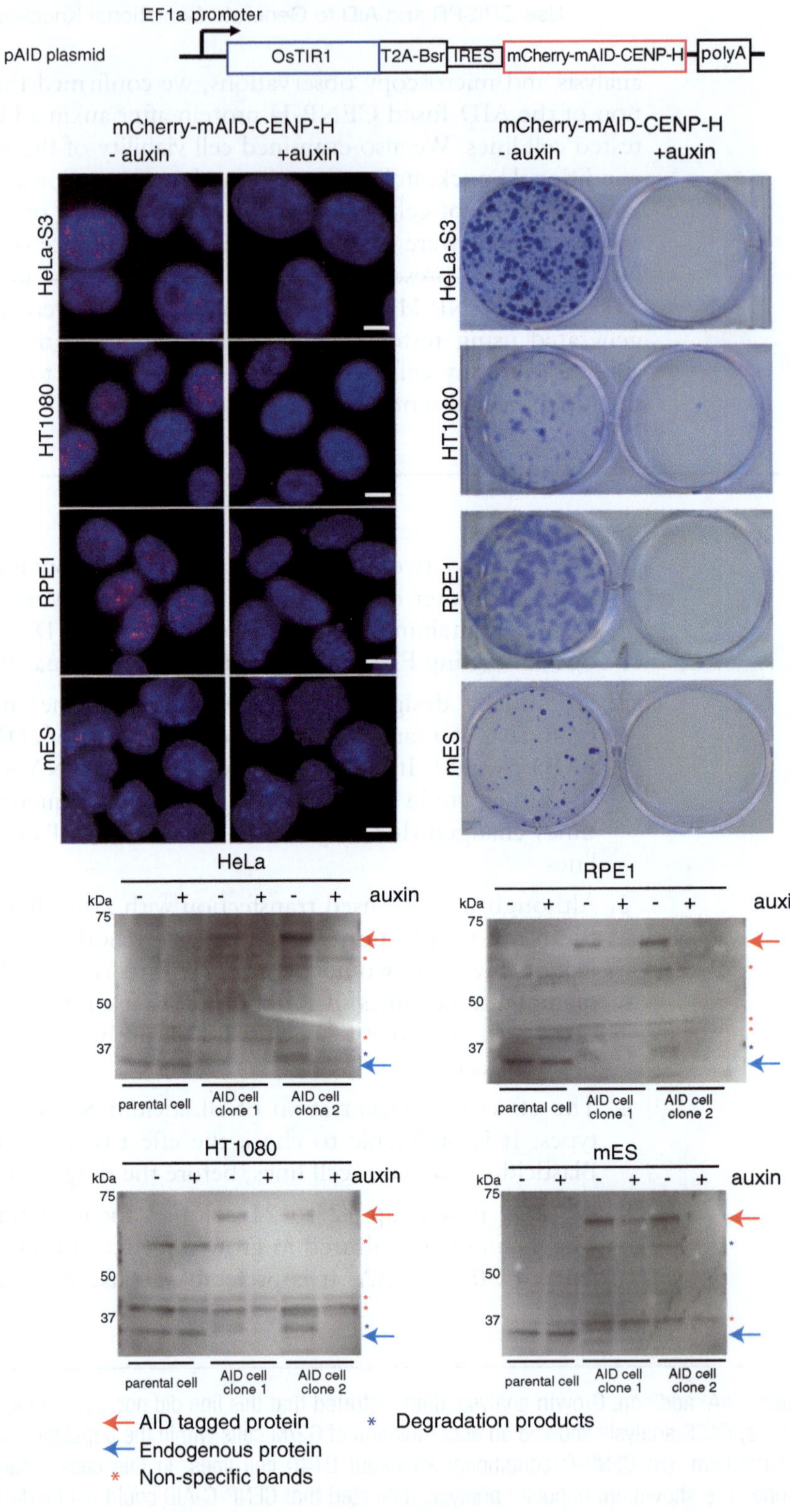

Fig. 4 Generation of AID-based conditional knockout mammalian cell lines. Using human HeLa-S3, human HT1080, human RPE1, and mouse ES cell lines, AID-based CENP-H conditional knockout lines were generated. Immunoblot analyses and microscopy observations showed the degradation of mCherry-mAID-CENP-H proteins 3 h after auxin addition in all tested cell lines. Scale bar: 10μm; red arrowhead: mCherry-mAID-CENP-H; blue arrowhead: endogenous CENP-H; red asterisk: nonspecific bands; blue asterisk: degradation products. Colony formation assays demonstrated that introduced mutations were lethal in all these cells in the presence of auxin

growth control. AID-based conditional knockout cells for essential gene showed cell death or slow growth in the presence of 500μM of IAA. After checking IAA sensitivity, target gene disruption must be checked by DNA sequencing and immunoblot analyses.

Acknowledgments

We are very grateful to R. Fukuoka and Y. Fukagawa for technical assistance. OsTIR1 cDNA was given from Prof. Masato Kanemaki (National Institute of Genetics). This work was supported by JSPS KAKENHI Grant Numbers 25221106, 15H05972, and 17H06167 to TF and JSPS KAKENHI Grant Number 17K15041 and 19K06611 to KN.

References

1. Cong L, Ran FA, Cox D, Lin S, Barretto R, Habib N, Hsu PD, Wu X, Jiang W, Marraffini LA, Zhang F (2013) Multiplex genome engineering using CRISPR/Cas systems. Science 339(6121):819–823. https://doi.org/10.1126/science.1231143
2. Mali P, Yang L, Esvelt KM, Aach J, Guell M, DiCarlo JE, Norville JE, Church GM (2013) RNA-guided human genome engineering via Cas9. Science 339(6121):823–826. https://doi.org/10.1126/science.1232033
3. Shalem O, Sanjana NE, Hartenian E, Shi X, Scott DA, Mikkelsen TS, Heckl D, Ebert BL, Root DE, Doench JG, Zhang F (2014) Genome-scale CRISPR-Cas9 knockout screening in human cells. Science 343(6166):84–87. https://doi.org/10.1126/science.1247005
4. Wang T, Wei JJ, Sabatini DM, Lander ES (2014) Genetic screens in human cells using the CRISPR-Cas9 system. Science 343 (6166):80–84. https://doi.org/10.1126/science.1246981
5. Gu H, Marth JD, Orban PC, Mossmann H, Rajewsky K (1994) Deletion of a DNA polymerase beta gene segment in T cells using cell type-specific gene targeting. Science 265 (5168):103–106. https://doi.org/10.1126/science.8016642
6. Lakso M, Pichel JG, Gorman JR, Sauer B, Okamoto Y, Lee E, Alt FW, Westphal H (1996) Efficient in vivo manipulation of mouse genomic sequences at the zygote stage. Proc Natl Acad Sci U S A 93(12):5860–5865. https://doi.org/10.1073/pnas.93.12.5860
7. Gossen M, Bonin AL, Freundlieb S, Bujard H (1994) Inducible gene expression systems for higher eukaryotic cells. Curr Opin Biotechnol 5(5):516–520. https://doi.org/10.1016/0958-1669(94)90067-1
8. Sonoda E, Sasaki MS, Buerstedde JM, Bezzubova O, Shinohara A, Ogawa H, Takata M, Yamaguchi-Iwai Y, Takeda S (1998) Rad51-deficient vertebrate cells accumulate chromosomal breaks prior to cell death. EMBO J 17(2):598–608. https://doi.org/10.1093/emboj/17.2.598
9. Fukagawa T, Mikami Y, Nishihashi A, Regnier V, Haraguchi T, Hiraoka Y, Sugata N, Todokoro K, Brown W, Ikemura T (2001) CENP-H, a constitutive centromere component, is required for centromere targeting of CENP-C in vertebrate cells. EMBO J 20 (16):4603–4617. https://doi.org/10.1093/emboj/20.16.4603
10. Nishimura K, Fukagawa T, Takisawa H, Kakimoto T, Kanemaki M (2009) An auxin-based degron system for the rapid depletion of proteins in nonplant cells. Nat Methods 6 (12):917–922. https://doi.org/10.1038/nmeth.1401
11. Natsume T, Kiyomitsu T, Saga Y, Kanemaki MT (2016) Rapid protein depletion in human cells by auxin-inducible Degron tagging with short homology donors. Cell Rep 15 (1):210–218. https://doi.org/10.1016/j.celrep.2016.03.001
12. Nishimura K, Fukagawa T (2017) An efficient method to generate conditional knockout cell

lines for essential genes by combination of auxin-inducible degron tag and CRISPR/Cas9. Chromosome Res 25(3-4):253–260. https://doi.org/10.1007/s10577-017-9559-7

13. Pyzocha NK, Ran FA, Hsu PD, Zhang F (2014) RNA-guided genome editing of mammalian cells. Methods Mol Biol 1114:269–277. https://doi.org/10.1007/978-1-62703-761-7_17

14. Fukagawa T, Earnshaw WC (2014) The centromere: chromatin foundation for the kinetochore machinery. Dev Cell 30(5):496–508. https://doi.org/10.1016/j.devcel.2014.08.016

Chapter 7

New Method for Genome-Scale Functional Genomic Study in Bacteria with Superior Performance: CRISPR Interference Screen

Xihao Liao, Xin-Hui Xing, and Chong Zhang

Abstract

High-throughput genetic screens based on CRISPR/Cas9 technology are powerful tools to genome-wide identify gene function and genotype–phenotype association. Here, we describe a detailed protocol for conducting and evaluating pooled CRISPR screens interfering with gene expression in *Escherichia coli*. We provide step-by-step instructions for guide RNA library design and construction, genome-scale screening and next-generation sequencing data processing. This tool outperforms transposon sequencing (Tn-seq) with similar library sizes and short gene length. The workflow can be used in follow-up studies implemented in other bacteria systems.

Key words CRISPRi, High-throughput screen, Next-generation sequencing, Computational biology, Microbial functional genomics

1 Introduction

Due to the numerous nodes and complex interactions of the system genetic network in bacteria, it is unattainable to accumulate a large number of genotype–phenotype quantitative data without high-throughput perturbation methods. High-throughput functional genomics uses highly parallelized experiments to perturb the basic units in the genome and detect the corresponding changes in phenotype, thereby correlating the genotype with the phenotype, which would lay a solid foundation for the description of microbial life phenomena at system level [1]. With that, the numbers of high-quality gene functional data can be greatly expanded by utilizing computer technology for functional annotation of genetic elements through genome-wide genotype–phenotype associations, which could provide high-quality and functionally diverse genetic elements for microbial study. On the other hand, deep understanding of genotype-phenotype relationships would also provide theoretical

Ren Zhang (ed.), *Essential Genes and Genomes: Methods and Protocols*, Methods in Molecular Biology, vol. 2377,
https://doi.org/10.1007/978-1-0716-1720-5_7,

guidance or direct gene targets for engineering complex phenotypes.

Genome-wide arrayed collections of single-gene knockout mutants have been constructed in several important model microorganisms, such as *Escherichia coli* [2] and *Bacillus subtilis* [3], by using the homologous recombination-based methods. These collections provide a gold standard for studying the functional genomics of these microorganisms and have produced many important results and datasets [4]. However, the manipulation of such a library containing thousands of different mutants, which requires expensive automated equipment, is time-consuming and labor-intensive. Moreover, this approach is difficult to apply to a wide range of species because of the specificity of homologous recombination among different organisms. Another most widely used method named as transposon sequencing (Tn-seq) depends on random transposon insertion-derived knockout libraries, which enables quantitative gene-phenotype mapping by mixing a large number of transposon mutants and using next-generation sequencing (NGS) to monitor abundance in competitive growth cultures in a pooled manner [5]. Although there are many successful examples [6–8], Tn-seq suffers from the random insertion of transposons into genomes, leading to a bias with long coding regions. Thus, the statistical robustness is poor while assigning functions to short genes, for example, small but important ncRNAs [9, 10]. Besides, Tn-seq is only applicable to the whole genome rather than specific regions in the genome, which will increase labor and cost when only focusing on a subset of genes. An alternative approach depends on homologous recombination and subsequent pooled screening, such as trackable multiplex recombineering (TRMR) [11, 12] or CRISPR-enabled trackable genome engineering (CREATE) [13]. These methods have been demonstrated in a variety of genome-scale engineering applications in *E. coli* with the ability to link the effects of mutations in promoter elements or coding regions in highly parallel manner. However, similar problem is faced when applying these strategies to a wider range of microorganisms, as the efficient recombination tools have been established in only a limited number of strains.

In order to solve the shortcomings of these methods, we proposed to combine the recently developed CRISPR genome interference technology (CRISPRi) [14, 15] [16] with pooled screening and next-generation sequencing (NGS) to develop a novel approach of high-throughput functional genomics [17]. As shown in Fig. 1, the CRISPRi system consists of two basic molecular components: the nuclease-activity-free Cas9 mutant protein (dCas9) acting as a roadblock and binds the targeted gene to inhibit gene expression; and the single guide RNA (sgRNA) providing targeting specificity to the target gene with complementary pairing of 20 base variable sequences. This highly programmable feature

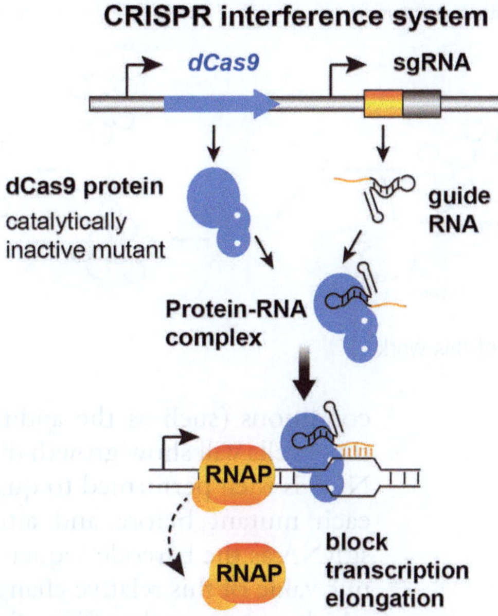

Fig. 1 Diagram of CRISPRi working mechanism

allows us to easily design sgRNA libraries to target hundreds or thousands of genes in the genome. CRISPR genome editing technology has good host adaptability and has been proven reliable in many hosts from bacteria to human cells [15, 18–21]. Therefore, it can theoretically be applied to any microorganism with basic genetic manipulation tools. In addition, the sgRNA encodes a target-specific portion of only 20 bases, which is well compatible with current DNA synthesis and sequencing (NGS) technology, greatly facilitating the synthesis of sgRNA libraries and the subsequent preparation of NGS libraries. Moreover, compared with randomly inserted transposon mutation libraries, sgRNA libraries can be customized for specific gene subsets, thereby reducing the cost and operational difficulty of small-scale studies with preliminary assumptions.

We have established a universal, rapid and high-throughput functional genomic platform in *E. coli* based on CRISPRi and NGS technology [17]. An overview of this platform is shown in Fig. 2. The sgRNA library was designed by using known genomic sequences according to the target genes. With the help of DNA chip synthesis [22] and DNA in vitro assembly technology [23, 24], sgRNA libraries in plasmid form can be rapidly prepared. Then the library is transformed into cells expressing the dCas9 protein to obtain a mutant library in cell form. Each of these cells containing different sgRNA represents a specific genotype. According to the phenotype of interest, such as the tolerance of microorganisms, the resulting library is cultured with the corresponding

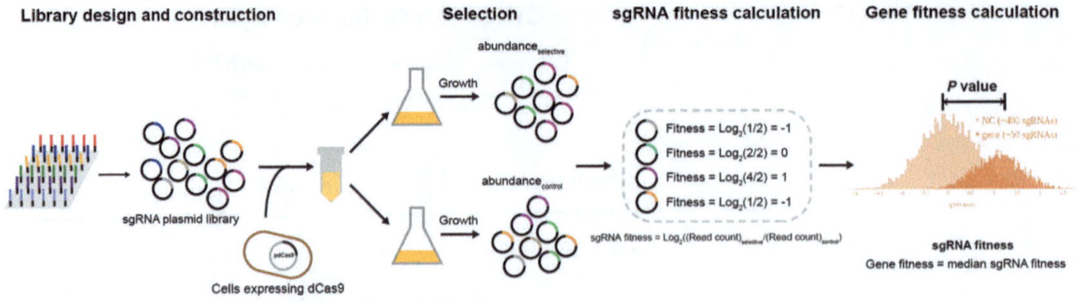

Fig. 2 Overview of this work

conditions (such as the addition of toxic chemical), and each of these cells will show growth differences due to different tolerances. NGS is then performed to quantitatively analyze the abundance of each mutant before and after screening by using the 20-base sgRNA as the barcode sequence of each mutant cell. The logarithmic value of this relative change is defined as the fitness of sgRNA, which represents the effect of sgRNA+dCas9 on the phenotype of a gene after its expression is repressed. In order to further improve the reliability of the method and reduce the interference of experimental noise, multiple sgRNAs are designed for each gene, and the median is taken as the fitness of the gene to characterize the effect of the gene on the target phenotype after being suppressed. At the same time, control sgRNAs that do not target any sites of the genome can provide statistically significant information.

2 Materials

1. Strains. *E. coli* K-12 MG1655 from ATCC (700926) was used as wild type. *E. coli* s17–1 sfGFP (super fold GFP) was a gift from Guoqiang Chen's laboratory at Tsinghua University [25]. We amplified the chloramphenicol expression cassette from pKM154 plasmid [26] and then cloned it into the *smf* locus of *E. coli* K-12 MG1655 by λ/RED recombineering [27] to produce the MCm strain which was used in the screening experiments. We inserted a *sacB* expression cassette from pKM154 [26] into the *smf* locus of *E. coli* K-12 MG1655 by CRISPR/Cas9 recombineering [28] to construct the *E. coli* strain Msac. *E. coli* lyc001 containing *crtEIB* gene cluster for overproducing lycopene [29] was stored in our laboratory. *E. coli* DH10B (BioMed, China) was used as the cloning host for plasmid construction.
2. Plasmids. Plasmid pdCas9-cons expressing dCas9 was generated by replacing the inducible promoter and resistance marker of Addgene plasmid #44249 [16] with a wild-type promoter

for Cas9 expression from *Streptococcus pyogenes* and the kanamycin marker from pCas (Addgene # 62225) [28], respectively. Then, a set of dCas9 plasmids pdCas9-J23109, 111, 112, 113, and 116 with different expression degree were constructed by using a series of iGEM Anderson promoter. To construct the plasmid pKana_NC which is used as the negative control in essential gene identification experiments, we amplified the pdCas9-J23111 to remove the dCas9 cassette. By replacing the spectinomycin marker with an ampicillin cassette (pTrc99a [30]) lacking the *Bsa*I restriction site, the vector expressing sgRNA was derived from the pTargetF plasmid (Addgene #62226) [28]. Moreover, the promoter region was substituted with an IPTG-induced promoter $P_{LlacO\text{-}1}$ [31] and *LacI* repressor cassette (pTrc99a [30]), resulting in pTargetF_lac. To promote sgRNA library insertion, pTargetF_lac_preLib was constructed through introducing two *Bsa*I sites in opposite direction between the promoter and the dCas9 binding region of pTargetF_lac.

3. Media: Luria-Bertani (LB), MOPS (Sigma-Aldrich, USA) [32].
4. Antibiotics: 50 mg/L kanamycin, 100 mg/L ampicillin, and 7 mg/L chloramphenicol.
5. Chemicals: glucose, sucrose, glycerol, lycopene, casamino acid, furfural, isobutanol, calcium chloride, and magnesium chloride.
6. Reagents for molecular biology: agarose for gel electrophoreses, KAPA HiFi HotStart polymerase (KAPA Biosystems, USA), deoxyribonucleotide triphosphates (dNTPs, 2.5 mM each nucleotide), Mag-Bind Plasmid Maxi Kits (Omega Bio-Tek, USA), Gel Extraction Kit (Omega Bio-Tek, USA), *Bsa*I restriction enzyme (New England Biolabs, USA), T4 DNA ligase (New England Biolabs, USA), and Gibson Assembly master mix (New England Biolabs, USA).
7. 0.2-cm electroporation cuvette for High-Throughput ECM™ 630 system (Avantor, USA).

3 Methods

The outline of our method is as follows: (1) construction and characterization of CRISPRi system in *E. coli*, (2) design and preparation of sgRNA library, (3) pooled screening experiments, (4) NGS library preparation and sequencing, (5) NGS data processing, (6) performance comparation with existing methods, (7) software tool.

3.1 Construction and Characterization of CRISPRi System in E. coli

In order to reduce the noise introduced during screening, we constructed a series of constitutive dCas9 expression plasmids and an inducible sgRNA expression plasmid for investigating the optimal CRISPRi system. However, due to expression leakage, we were failed to observe any inducible profile in a test inhibiting *sfGFP* expression in *E. coli*, instead found that the repression activity generally depended on the strength of dCas9 expression [17]. According to these findings, we used the pdCas9-J23111 driven by the strongest promoter among the five constructs and pTargetF_lac plasmids for the following work. The plasmid maps of them are deposited at following link:

https://benchling.com/s/seq-UtUqPXinL0XxdjVCAPHG

https://benchling.com/s/seq-ht6aO2giauphixQSGbP8

For characterization of CRIPSRi system, we firstly used the *sfGFP* as the reporter in *E. coli* s17–1. We designed multiple sgRNAs that target different sites of the *sfGFP* as well as control sgRNAs that do not target any site of the genome. The *E. coli* s17-1 strain carrying dCas9 and the corresponding sgRNA was inoculated into 5 mL of LB medium with 50 mg/L kanamycin and 100 mg/L ampicillin in a 15-mL conical tube and cultured overnight. The seed solution was transferred to 10 mL of the fresh LB medium with 50 mg/L kanamycin and 100 mg/L ampicillin in a 50-mL shake flask at an initial $OD_{600} = 0.02$ and cultured for 12 h. Fluorescence was measured with a Hitachi Fluorescence Reader. The excitation and emission wavelengths for *sfGFP* were 488 nm and 510 nm, respectively. At the same time, the Amersham Bioscience Spectrophotometer was used to determine the OD_{600} of the cells to normalize the measured fluorescence values. The activity of the CRISPRi system was calculated by using Eq. (1).

$$\text{Repression Ratio} = 1 - \frac{\text{Fluorescence(Target)}/\text{OD}_{600}\text{(Target)}}{\text{Fluorescence(Control)}/\text{OD}_{600}\text{(Control)}} \quad (1)$$

Second, CRIPSRi system was used to repress the expression of *crtE* by characterizing lycopene accumulation. Overnight, LB cultures (with ampicillin and kanamycin) of *E. coli* lyc001 carrying dCas9 and the corresponding sgRNA were individually 10 mL of the fresh LB medium with 50 mg/L kanamycin and 100 mg/L ampicillin in a 50-mL shake flask at an initial $OD_{600} = 0.02$ and cultured for 24 h. Subsequently, lycopene was measured according to the previous method [29], and the titer was normalized to the culture OD_{600} value. Third, CRIPSRi system was performed for repressing the expression of *sacB* gene by the growth testing of *E. coli* Msac. 1.25 × sodium chloride-free LB medium was autoclaved and then was mixed with 500 g/L filter-sterilized sucrose stock solution at 1:4 (v:v) to prepare a screening plate by adding

50 mg/L kanamycin and 100 mg/L ampicillin. A single colony of *E. coli* Msac carrying the dCas9 protein and the corresponding sgRNA was streaked and cultivated at 30 °C for 24 h, and the growth phenotype was measured.

3.2 Design and Preparation of Genome-Wide sgRNA Library

3.2.1 Design Process

Designing sgRNA library was based on the genome file (NC_000913.3) of *E. coli* K-12 MG1655, including annotation files of protein-coding gene (NC_000913.ptt and NC_000913.fna) and RNA-coding gene (NC_000913.rna and NC_000913.frn). In order to maximize the activity of the selected sgRNA for each target gene, possible sgRNA candidates (20-mer) were searched within the first 5% of the coding region from the 5′ end of open reading frame (ORF) on the nontemplate strand [17]. The SeqMap package [33] was used to check potential off-target sites of the designed sgRNAs by searching for all potential N20NG(A)G 23-mers with a tolerance setting of five mismatches, and customized scoring method was shown in Fig.3 to calculate the score of off-target sites.

In short, N20-mer is divided into three parts in line with the distance to the PAM site, which include 7, 5, and 8 nt from the 3′ end to the 5′ end as Region I, II and III, respectively. According to the literature, the seed region closed to PAM (Region I, Fig. 3) has the greatest effect on sgRNA activity. Therefore, an off-target site scoring system was designed. If the PAM site of the off-target 23-mer is "NGG," the mismatch penalty in the three regions was set as 8, 4.5 and 2.5, respectively, and the penalty set for "NAG" was 10, 7, and 3, respectively. Otherwise, the penalty was automatically set to be 100. For the genome-wide sgRNA library, when the total score is less than the threshold 21, the off-target site is considered to be significant, and corresponding sgRNA candidates were excluded. Moreover, the computer program further checked the GC content of the N20-mer to preclude sgRNA candidates with extreme GC content (<25% or >75%).

After the above quality control, once 15 sgRNAs were designed within the first 5% of the coding region, the program stopped and moved to the next gene. If the designed sgRNAs in the particular region were insufficient to 15, sgRNA design continued to uniformly select the additional sgRNAs in the remaining part of the coding region, until 15 sgRNAs were designed for target gene. Meanwhile, if sufficient sgRNAs were still not found by searching to the stop codon, the number of sgRNAs currently designed is standard (Fig. 4). In addition, 400 negative control sgRNAs were designed by searching random 23-mers with proper GC content, which were less than 5 mismatches with any N20NGG encoded on the *E. coli* genome (*see* **Note 1**).

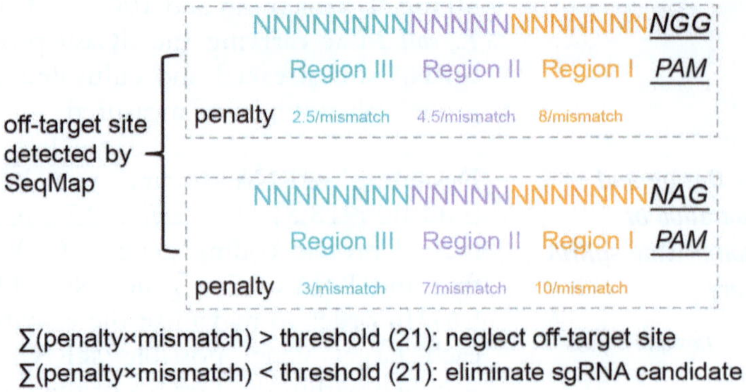

Fig. 3 Scoring matrix performed to measure the significance of sgRNA off-target hits

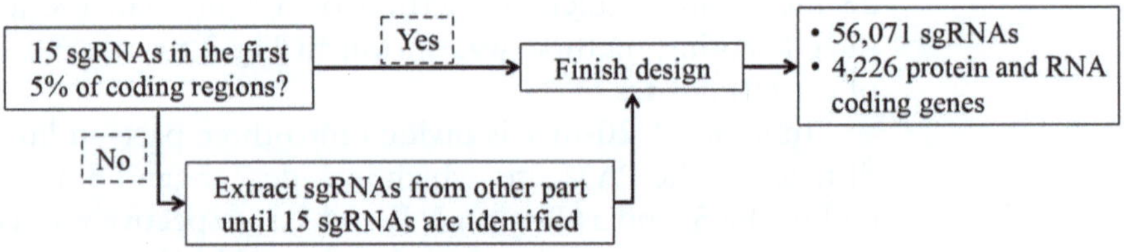

Fig. 4 Design algorithm of genome-wide sgRNA library

3.2.2 Preparation of sgRNA Library

The designed sgRNA library in N20 format was converted into oligomers to be synthesized on a microarray (N82, flanking region + *Bas*I (GGTCTCA) + promoter overhang (GCAC) + N20 + Cas9 binding region overhang (GTTT) + *Bas*I (GGAGACC) + flanking region). The obtained oligonucleotide library was amplified by PCR with primers on the flanking region, ligated into the pTarget-F_lac_pre plasmid using the Golden Gate Assembly, transformed into *E. coli* DH10B, and coated on an LB plate 100 mg/L ampicillin for overnight culture. The specific process is shown in Fig. 5. All sgRNA libraries are available from Addgene (https://www.addgene.org/browse/article/28196099/) (*see* **Note 2**).

3.3 Pooled Screening Experiments

3.3.1 Preparation of E. coli Electroporation-Competent Cells

The preparation of *E. coli* MCm/pdCas9-J23111 or pKanaNC (control) electroporation-competent cells was performed according to the previous methods [2] with proper modification as below.

1. Transfer *E. coli* MCm/pdCas9-J23111 or pKanaNC cultures into 100 mL LB medium with 50 mg/L kanamycin and 7 mg/L chloramphenicol in a 500-mL shake flask and incubate the flask at 37 °C with agitation.
2. Measure the OD_{600} of growing culture. When it reaches 0.6, rapidly transfer the culture to two 50-mL conical tubes and incubate in an ice-water bath for 15 min.

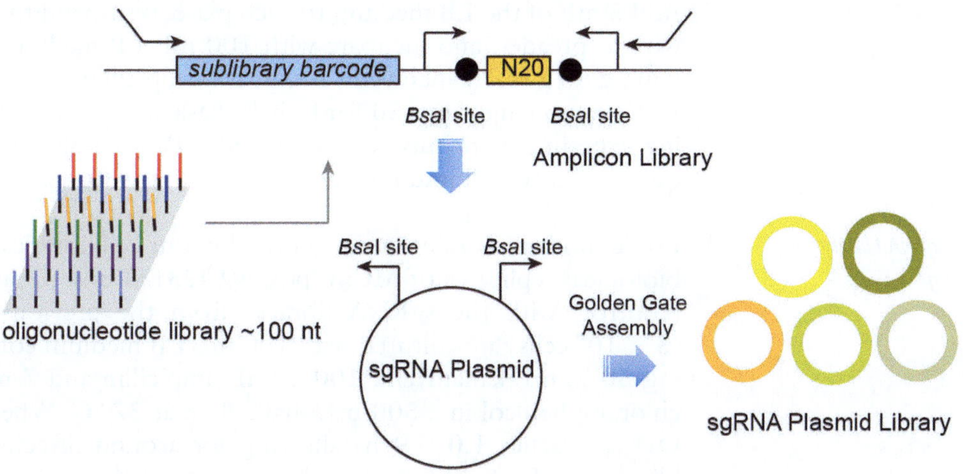

Fig. 5 This work developed a method for preparing plasmid-form sgRNA library. After the variable region of sgRNA was synthesized by the chip, it was amplified by PCR and assembled to form the sgRNA plasmid library using Golden Gate Assembly method

3. Harvest the cells by centrifugation at 8000 × *g* for 5 min at 4 °C. Remove the supernatant and resuspend the cell pellet in 5 mL of prechilled, sterilized H_2O.
4. Transfer the suspension to a 1.5-mL centrifuge tube with 1-mL per tube. Harvest the cells by centrifugation at 8000 × *g* for 2 min at 4 °C. Resuspend the cells in 1 mL prechilled, sterilized H_2O. Repeat this step for five times.
5. Harvest the cells by centrifugation at 8000 × *g* for 2 min at 4 °C. Resuspend all cells in 6 mL of ice-cold 15% (v:v) glycerol. Store at −80 °C (*see* **Note 3**).

3.3.2 Plasmids Transformation

The library plasmids were independently transformed twice into either *E. coli* MCm/pdCas9-J23111 or MCm/pKanaNC (negative control), providing two biological replicates for each.

1. Add 500 ng sgRNA library plasmid for per mL of competent cells, mix gently, and transfer to the 5 × 5 0.2-cm electroporation cuvette (*see* **Note 4**).
2. Electroporation is performed by using High-Throughput ECM™ 630 system at 2.1 kV with 25 μF and 1 kΩ (*see* **Note 5**).
3. Immediately followed by addition of 1:4 (v:v) volume of LB medium and incubation for 1 h at 37 °C.
4. The recovered cells are spread uniformly on the LB plate with 50 g/mL kanamycin, 100 g/mL ampicillin, and 7 mg/L chloramphenicol. Culture overnight and calculate electrical transformation efficiency and library coverage (*see* **Note 6**).

5. Add 3 mL of the LB medium to each plate, resuspend the cells with a spreader, and incubate with 100 mL LB medium containing 50 g/mL kanamycin, 100 g/mL ampicillin, and 7 mg/L chloramphenicol in a 500-mL shake flask at 37 °C for screening experiments or mix with sterilized 50% (v:v) glycerol and quickly freeze in liquid nitrogen and store at −80 °C.

3.3.3 sgRNA Library Screening

1. For identification of essential genes, the four cell libraries (two biological replicates of MCm/pdCas9-J23111 and pKanaNC (control) with the sgRNA library) from the previous step (3×10^6 cells) are cultured with 100 mL LB medium containing 50 g/mL kanamycin, 100 g/mL ampicillin and 7 mg/L chloramphenicol in a 500-mL shake flask at 37 °C. When the OD_{600} reaches 1.0 (~9 h) allowing for around fifteen doublings, 5 mL of cultures are taken to extract the plasmids for NGS library construction (*see* **Note 7**).
2. For identification of auxotrophic genes (or tryptophan biosynthesis genes), the cultures from the previous step representing two biological replicates of MCm/pdCas9-J23111 with sgRNA library are washed with MOPS medium (1:1, v/v). Transfer the cell libraries into 100 mL LB, MOPS medium (or MOPS + 0.5 g/L casamino acid for identification of L-Trp biosynthesis genes) containing 50 g/mL kanamycin, 100 g/mL ampicillin, and 7 mg/L chloramphenicol with initiate $OD_{600} = 0.03$ in a 500-mL shake flask at 37 °C, respectively. When the OD_{600} reaches 1.0, 5 mL of cultures are taken to extract the plasmids for NGS library construction (*see* **Note 8**).

For identification of genes carrying toxic chemical tolerance, the cultures from the previous step representing two biological replicates of MCm/pdCas9-J23111 with sgRNA library are washed with MOPS medium (1:1, v/v). The remaining cultures are mixed together, serving as the initial library for the following phenotypes to be tested (Table 1). Transfer the cell libraries into 100 mL MOPS + 0.4 g/L furfural, MOPS + 4 g/L isobutanol medium containing 50 g/mL kanamycin, 100 g/mL ampicillin, 7 mg/L chloramphenicol with initiate $OD_{600} = 0.03$ in a 500-mL shake flask at 37 °C, separately. When the OD_{600} reaches 1.0, 5 mL of cultures are taken to extract the plasmids for NGS library construction (*see* **Note 8**).

3.4 NGS Library Preparation and Sequencing

To construct the NGS library for amplicon sequencing, the robust maintenance of both dCas9 and sgRNA expression plasmids were verified by gel electrophoresis after plasmid extraction.

3.4.1 PCR Amplification

The N20 regions of plasmid library were amplified using the extracted plasmids as a template with the primer pair of

Table 1
The phenotypes studied in this work

Phenotype	Selective condition	Control condition
Essentiality	dCas9, LB	Empty plasmid, LB
Auxotrophy	MOPS	LB
L-Trp biosynthesis	0.5 g/L casamino acid, MOPS	LB
Furfural tolerance	0.4 g/L furfural, MOPS	Initial
Isobutanol tolerance	4 g/L isobutanol, MOPS	Initial

PF_pTargetLacNGS_PE150 (TGCGCCGACATCATAACGGTTCTG) and PR_pTargetLacNGS_PE150 (CGACTCGGTGCCACTTTTTCAAGTTG).

1. 50 ng template.
2. 1.5 μM of each primer.
3. 2 mM dNTPs.
4. 1 U KAPA HiFi HotStart polymerase.
5. Fill up with H_2O to 50 μL.
6. Reactions were "hot started" at 95 °C for 3 min.
7. 98 °C for 20 s, 67.5 °C for 15 s, 72 °C for 30 s for 20 cycles.
8. Additional 1 min at 72 °C.

3.4.2 Library Construction and Sequencing

The sequencing library of the sgRNA library was prepared following the manufacturer's protocol (TruSeq DNA Nano Library Prep Kit for Illumina).

1. Treat the amplified product with End Prep Mix for end repair according the manufacturer's instructions.
2. Purify the amplified product using Sample Purification Beads (SPBs) for 5′ end phosphorylation and purification according the manufacturer's instructions.
3. The purified fragments are treated with A-tailing Mix for adenylated 3′ end, and followed by ligation to the adapters indexed with "T" overhang.
4. Purify the amplified product using SPBs again according the manufacturer's instructions.
5. PCR amplify the DNA library for eight cycles using the primer pair of P5 and P7.
6. Clean up the amplified product using SPBs again.
7. Validate the purified DNA library using Agilent 2100 Bioanalyzer (Agilent Technologies, Palo Alto, USA) and quantify with

a Qubit 2.0 Fluorometer (Invitrogen, Carlsbad, USA) according the manufacturer's instructions.

8. Load the DNA library with index on an Illumina HiSeq instrument according to the manufacturer's instructions (Illumina, San Diego, USA).
9. Sequencing is carried out using a 2 × 150 paired-end configuration, and image analysis and base calling were conducted with the HiSeq Control Software (HCS) + OLB + GAPipeline-1.6 (Illumina) on the HiSeq instrument. Yield approximately 10^6 reads per library.

3.5 NGS Data Processing

Generally, the data analysis consists of three steps: read count mapping, sgRNA fitness calculation, and gene fitness calculation.

3.5.1 Read Mapping

The raw data were demultiplexed, and the adapter was removed from the sequencing reads to collect clean data for each library. Each of the two pairs was merged by FLASH script [34] for pair-end data, and those reads without detected pairs were removed. Subsequently, we removed the read longer than 194 nucleotides and from each end of read trimmed those nucleotides with the base quality below 25. After that, we performed a filter step based on a customized cutoff (Q10 < 1, Q20% > 85%, and Q30% > 60%) to enrich the high-quality reads. Customized Python script was carried out to search each read for "GCACN20GTTT" 28_mer in the sequencing reads, where the "GCAC" is the last four bases of the promoter region of the sgRNA expression plasmid (pTargetF_lac) immediately adjacent to the 20-mer variable region, and the "GTTT" is the four starting bases of sgRNA scaffold for the dCas9 binding. In this way, each sequencing read was clipped to N20 and matched with the designed library N20 sequence to get the read count of each mutant (sgRNA) in a specific sequencing library. Since the depth of each sequencing library was not the same, the normalization factor (n sequencing libraries) for each sequencing library was calculated using Eq. (2). The sgRNAs with less than 20 read counts in the initiate library were excluded for increasing statistical robustness. The normalized read count data of two biological replicates were integrated using geometric averaging.

$$\text{Normalication factor}_i = \text{Read count}_i / \sum\nolimits_{i=1}^{n} \text{Read count}_i / n \quad (2)$$

3.5.2 Calculation of sgRNA Fitness

For each phenotype to be studied, we defined a selective condition and a control condition for it (Table 1). For example, for the phenotype of essentiality, the dCas9-expressing condition was regarded as the selective condition (MCm/pdCas9-J23111), and the dCas9-free condition was treated as the control condition

(MCm/pkana_NC). The fitness of each sgRNA for a phenotype was calculated by dividing the number of reads for this sgRNA in the corresponding selective condition by the number of reads in the relevant control condition and taking the $\log_2$ value (Eq. 3). Then, we figured the median of fitness for all negative control sgRNAs and used it to normalize the sgRNA fitness calculated in Eq. (3) to generate the final sgRNA fitness shown in Eq. (4).

$$\text{Fitness}'_{\text{sgRNA}} = \log_2\left(\frac{(\text{Read count})_{\text{selective}}}{(\text{Read count})_{\text{control}}}\right) \tag{3}$$

$$\text{Fitness}_{\text{sgRNA}} = \text{Fitness}'_{\text{sgRNA}} - \text{median}\left(\text{Fitness}'_{\text{NC sgRNA}}\right) \tag{4}$$

To calculate the Z score of each sgRNA for assessment of its activity, we fit the fitness for all negative control sgRNAs to a normal distribution producing the standard deviation (σ). Then, the Z score for each sgRNA is determined by dividing the sgRNA fitness by the σ value of all negative control sgRNAs (Eq. 5).

$$Z_{\text{sgRNA}} = \text{Fitness}_{\text{sgRNA}}/\sigma_{\text{normalized distribution of NC sgRNA}} \tag{5}$$

3.5.3 Calculation of Gene Fitness

To determine the fitness for each gene and calculate the statistical significance based on the fitness of all sgRNAs belonging to this gene, we applied a framework similar to that described recently [35] with an improved version of the algorithm via a sgRNA sampling approach.

First, a strategy of constructing quasi-gene was used to calculate the false-positive rate (FPR) associated a single gene with a particular phenotype (such as essentiality). sgRNA subsets were used to calculate related parameters of genotype–phenotype association, because different genes contain diverse numbers of sgRNA. We explored 400 negative control sgRNAs to construct 15 quasi-gene sets by random sampling, each consisting of 10,000 members with 1–15 sgRNAs per member, which were referred to as QGi (i = 1–15) (QG1 has only 400 members). For each quasi-gene in QGi, the P value was calculated using a two-tailed Mann–Whitney U (MWU) test with reference to the distribution of fitness of all negative control sgRNAs. Then, a score was defined for each quasi-gene, which is the product of the absolute value of the fitness of the gene and $-\text{Log}_{10}P$, incorporating both the effect size and the P values as Eq. (6). The quasi-gene in each QGi set was sorted according to the score, and the FPR of each quasi-gene was calculated using Eq. (7). The hypothesis here was that when a quasi-gene derived from control sgRNAs was misidentified as a positive gene, it naturally represented a false positive. As a result, the proportion of false positives under different score threshold conditions, that is FPR, could be calculated through a large-scale (10,000) quasi-gene dataset. Since the score–FPR relationship obtained by Eq. (7) was discrete, this relationship was converted into a continuous function of FPR that varied with scores using a simple linear interpolation method (Eq. 8). In this way, a total of 15 equations describing the quantitative change of FPR with scores were obtained, each describing the relationship between the potential FPR and score when calculating the fitness of a gene under the corresponding phenotype using i sgRNAs (*see* **Note 9**).

$$\text{Score} = \left|\text{median}\left(\text{Fitness}_{\text{sgRNA}}\right)\right| \times \left(-\log_{10}P\right) \tag{6}$$

$$FPR(\lambda)_i(i = 1, 2, \ldots, 15) = \frac{\{score_j > \lambda, j = 1, 2, \ldots, N_i\}}{N_i(N_1 = 400\ and\ N_{2\tilde{1}5} = 10,000)} \tag{7}$$

$$FPR(score)_i(i = 1, 2, \ldots, 15) = \text{linear interpolation (data by equation 7)} \tag{8}$$

By correlating FPR with the gene score of genes with different numbers of sgRNAs, we calculated the fitness of each gene based on the sgRNA sampling method, because we found that sgRNAs located within the 5′ region of the sequence showed better activities [17]. Suppose that N sgRNAs belong to a specific gene. We sorted these sgRNAs based on their relative positions within the coding region (from the one closest to the 5′ end to that most proximal to the 3′ end). Then, we extracted the first M sgRNAs and calculated the fitness based on the median sgRNA fitness and *P* values using two-tailed MWU test against all negative control sgRNAs, followed by using Eqs. (6) and (8) to obtain score and the FPR of this gene based on the current sgRNA subset with *M* members ($M = 1, 2, \ldots, N$). The sgRNA subset with the smallest FPR value was selected as the sgRNA set to determine all indicators of the gene. The fitness of the gene to the studied phenotype was then the median fitness of all sgRNAs belonging to the selected subset. We used the classic Storey–Tibshirani method [36] to convert the FPR value (a particular type of *P* value) to a *Q* value, which was used as the threshold for hit-gene calling (FDR < 0.05), such as predicting *E. coli* essential genes. Essential genes of *E. coli* identified by CRISPRi were excluded (fitness ≤ -4, FDR ≤ 0.01) for other screening experiments of genome-wide sgRNA libraries except for gene essentiality (*see* **Note 10**).

3.6 ROC-AUC to Compare Performance of Different Methods for Identification of Essential Genes

Compared to the classical genetic methods for functional studies of a single-gene, high-throughput method inevitably has a false-positive rate. To compare the performance of different methods to identify the essential genes of *E. coli*, we trained these methods as binary classifiers as described below.

For the CRISPRi dataset in this work, according to the fitness and statistical significance (FPR) of each knockdown gene under eutrophic conditions, we converted the screening results into prediction scores for the essentiality of each gene (Eq. 9).

$$\text{prediction}_{\text{gene}} = (1 - \text{FRP}_{\text{gene}}, \text{if fitness}_{\text{gene}} < 0) \text{ or } (0, \text{if fitness}_{\text{gene}} \geq 0) \quad (9)$$

For the dataset of transposon-based genetic footprinting method [37] (https://www.genome.wisc.edu/Gerdes2003/table_s1.txt), the authors have divided the genes into "essential" or "nonessential" and assigned a *P* value to each gene according to the footprinting profile. Therefore, we simply use Eq. (10) to convert it into a binary classifier of gene essentiality to produce the prediction score.

$$\text{prediction}_{\text{gene}} = (1 - P_{\text{gene}}, \text{if essential}_{\text{gene}} = \text{yes}) \text{ or } (0, \text{if essential}_{\text{gene}} = \text{no}) \quad (10)$$

For the Tn-seq dataset [38] (http://genomics.lbl.gov/supplemental/rbarseq/html/Keio/all.poolcount), an optimized strategy [6] was applied to generate the prediction score for each gene to reveal its essentiality since the original paper did not divide the essential genes. The transposon insertion index was calculated for each gene annotated in the genome *E. coli* NC000913.2, which was the unique insertion number in the coding region of this gene divided by gene length (Eq. 11). Barquist et al. proposed the distribution of the insertion index is bimodal [6], corresponding to the required model (mode 0) and the nonrequired model [17]. Thus, we called a gene as required using a cutoff of the insertion index <0.00125. Then, we used the Eq. (12) to assign a prediction score for each required gene to quantify its essentiality (*see* **Note 11**).

$$\text{insertion index}_{\text{gene}} = \text{number of unique transposon insertions/gene length} \quad (11)$$

$$\left(1 - e^{-(\text{local insertion index}_{\text{gene}} \times \text{gene length})}, \text{if insertion index}_{\text{gene}} < 0.00125\right)$$
$$\text{or } (0, \text{if insertion index}_{\text{gene}} \geq 0.00125) \quad (12)$$

Therefore, the three binary classifiers were trained, and the ROC-AUC analysis was performed based on the scikit-learn (0.19.0) Python package with the Keio collection essential genes as the gold standard.

3.7 Software Tools

The CRISPRi functional genomic method requires a lot of computer processing, including sgRNA library design and NGS data analysis. These requirements are not friendly for the researchers who have traditionally engaged in wet laboratory. In order to solve this problem, the Python script written in this work was integrated and packaged. Users only need to provide the most basic genomic files or original sequencing files, complete the concise configuration file, and run the entire data processing process in one click. The integrated software toolkit including detailed instructions for users has been loaded on GitHub (https://github.com/zhangchonglab/CRISPRi-functional-genomics-in-prokaryotes).

4 Notes

1. For the size of the sgRNA library, it depends on the purpose of the experiment and the cost of oligonucleotide synthesis. For example, before the design of genome-wide sgRNA library, we designed a tiling sgRNA library composed of 2281 members targeting 44 genes with known phenotypes to evaluate the activities of these sgRNAs. We used a threshold of 11 for the tiling library design to maximize the number of sgRNAs designed for each gene (about 50 sgRNAs for each gene).
2. We confirmed the high quality of the library by NGS performed as described above to profile the genome-scale library, which was consistent with the reported error rate of the current massively parallel DNA oligomer synthesis technology [12] (Table 2).
3. For freshly prepared competent cells, the transformation efficiency decreases with longer storage and can be stored for up to 3 months at −80 °C. We recommend the competent cells were prepared just before use or stored in 15% glycerol stock not exceeding 1 week.
4. For the plasmid-form library used to transform into the particular host for phenotype study, fresh plasmid extraction should be used or maintained at 4 °C within 1 week.
5. Transformation parameters should be optimized: plasmid amount and applied voltage. The optimal parameters depend on host cell and plasmid (generally ~50 ng plasmid seems OK for *E. coli*). In addition, we typically obtained around 10^5 colonies per well. To achieve a proper coverage for the library (~60,000 members), we electroporated 32 wells of cells for each biological replicate.
6. We took 50 μL of the resulting culture to test the real transformation efficiency by dilution, streaking the diluted culture onto LB agar plates with kanamycin and ampicillin and counting the colonies after overnight incubation at 37 °C. The results confirmed that the coverage for each replicate was at least 20-fold.
7. 10^5 colonies per well × 30 wells = 3 × 10^6 cells of initial incubation. Cultivation to OD_{600} ~ 1.0 results in around 10^9 cells/mL × 100 mL = 10^{11} cells. $\log_2(10^{11}/(3 \times 10^6))$ = ~15. Theoretically, doubling of 15 generations

Table 2
Summary of genome-wide sgRNA library metrics

4140 protein-coding genes	In silico	Plasmid	NC1[a]	NC2[a]
Genes with at least 1 sgRNA ≥20 reads (ratio vs. all genes %)	4084 (98.6)	4079 (98.5)	4080 (98.6)	4081 (98.6)
Genes with at least 3 sgRNA ≥20 reads (ratio vs. all genes %)	4006 (96.8)	3999 (96.6)	3994 (96.5)	3996 (96.5)
178 RNA-coding genes	In silico	*Plasmid*	*NC1*	*NC2*
Genes with at least 1 sgRNA ≥20 reads (ratio vs. all genes %)	142 (79.8)	142 (79.8)	139 (78.1)	139 (78.1)
Genes with at least 3 sgRNA ≥20 reads (ratio vs. all genes %)	87 (48.9)	86 (48.3)	86 (48.3)	85 (47.8)

[a]Transformation with 20-fold coverage in two biological replicates (NC1 and NC2)

can expand the growth difference of 10% per generation to about 400%.

8. Similar to the identification experiments of essential gene, cultivation to an OD_{600} of about 1.0 allowed the cells in each experiment to multiply approximately five times from 5 mL of each culture ($Log_2(1/0.03) = \sim 5$). Five times during the screening experiment was designed to reduce the risk of undetected de novo mutations in the bacterial genome, which could affect growth and cause noise in the final data. At the same time, a reliable fivefold doubling is reliable for hits of auxotrophic genes. In principle, five times can double the 15% fitness advantage (1.15^5–2), which is enough for NGS to detect.
9. Our results showed that the FPR function obtained by taking 5 or 10 sgRNAs for each quasi-gene was little different from that when taking more sgRNAs, indicating that 5 to 10 sgRNAs could reliably detect the corresponding genotype-phenotype association [17].
10. We have also tried alternative methods to calculate FPR values and compare them with the current method. For example, we tested the performance by using a simple two-tailed Student's *t*-test between sgRNA fitness belonging to the extracted subset and the fitness of the negative control sgRNA to determine the FPR value. In terms of identification of essential gene, current algorithm was slightly better than *t*-test method.
11. In Eq. (11), considering the bias of transposon insertion, the global average insertion index [6] was replaced with the local average insertion index [37] in the transposon-based genetic footprinting method. The specific calculation method is the insertion index in the range of 10 kb region around the gene, instead of the previous insertion index in the whole genome.

Compared with the original processing method, the new processing method proposed here performed better (ROC-AUC: 0.878 vs. 0.876 all genes; 0.773 vs. 0.766 is less than 400 bp genes).

Acknowledgments

The authors thank Tianmin Wang and Jiahui Guo for their helpful suggestions. The research is supported by the National Key Research and Development Program (Grant No. 2018YFA0901500) and National Natural Science Foundation of China (Grant No. 21938004).

References

1. Shalem O, Sanjana NE, Zhang F (2015) High-throughput functional genomics using CRISPR–Cas9. Nat Rev Genet 16(5):299–311
2. Baba T, Ara T, Hasegawa M et al (2006) Construction of *Escherichia coli* K-12 in-frame, single-gene knockout mutants: the Keio collection. Mol Syst Biol 2(1)
3. Koo B-M, Kritikos G, Farelli JD et al (2017) Construction and analysis of two genome-scale deletion libraries for Bacillus subtilis. Cell Systems 4(3):291–305. e297
4. Nichols RJ, Sen S, Choo YJ et al (2011) Phenotypic landscape of a bacterial cell. Cell 144 (1):143–156
5. Van Opijnen T, Bodi KL, Camilli A (2009) Tn-seq: high-throughput parallel sequencing for fitness and genetic interaction studies in microorganisms. Nat Methods 6(10):767
6. Barquist L, Langridge GC, Turner DJ et al (2013) A comparison of dense transposon insertion libraries in the salmonella serovars Typhi and typhimurium. Nucleic Acids Res 41(8):4549–4564
7. Lluch-Senar M, Delgado J, Chen WH et al (2015) Defining a minimal cell: essentiality of small ORFs and ncRNAs in a genome-reduced bacterium. Mol Syst Biol 11(1):780
8. Ibberson CB, Stacy A, Fleming D et al (2017) Co-infecting microorganisms dramatically alter pathogen gene essentiality during polymicrobial infection. Nat Microbiol 2(8):17079
9. Wagner EGH, Romby P (2015) Small RNAs in bacteria and archaea: who they are, what they do, and how they do it. In: Advances in genetics. Elsevier, Amsterdam, pp 133–208
10. Fröhlich KS, Vogel J (2009) Activation of gene expression by small RNA. Curr Opin Microbiol 12(6):674–682
11. Warner JR, Reeder PJ, Karimpour-Fard A et al (2010) Rapid profiling of a microbial genome using mixtures of barcoded oligonucleotides. Nat Biotechnol 28(8):856
12. Freed EF, Winkler JD, Weiss SJ et al (2015) Genome-wide tuning of protein expression levels to rapidly engineer microbial traits. ACS Synth Biol 4(11):1244–1253
13. Garst AD, Bassalo MC, Pines G et al (2017) Genome-wide mapping of mutations at single-nucleotide resolution for protein, metabolic and genome engineering. Nat Biotechnol 35 (1):48
14. Jinek M, Chylinski K, Fonfara I et al (2012) A programmable dual-RNA–guided DNA endonuclease in adaptive bacterial immunity. Science 337(6096):816–821
15. Cong L, Ran FA, Cox D et al (2013) Multiplex genome engineering using CRISPR/Cas systems. Science 339(6121):819–823
16. Qi LS, Larson MH, Gilbert LA et al (2013) Repurposing CRISPR as an RNA-guided platform for sequence-specific control of gene expression. Cell 152(5):1173–1183
17. Wang T, Guan C, Guo J et al (2018) Pooled CRISPR interference screening enables genome-scale functional genomics study in bacteria with superior performance. Nat Commun 9(1):1–15
18. DiCarlo JE, Norville JE, Mali P et al (2013) Genome engineering in Saccharomyces cerevisiae using CRISPR-Cas systems. Nucleic Acids Res 41(7):4336–4343

19. Jiang W, Bikard D, Cox D et al (2013) RNA-guided editing of bacterial genomes using CRISPR-Cas systems. Nat Biotechnol 31(3):233
20. Hwang WY, Fu Y, Reyon D et al (2013) Efficient genome editing in zebrafish using a CRISPR-Cas system. Nat Biotechnol 31 (3):227–229
21. Wang H, Yang H, Shivalila CS et al (2013) One-step generation of mice carrying mutations in multiple genes by CRISPR/Cas-mediated genome engineering. Cell 153 (4):910–918
22. Kosuri S, Church GM (2014) Large-scale de novo DNA synthesis: technologies and applications. Nat Methods 11(5):499
23. Gibson DG, Young L, Chuang R-Y et al (2009) Enzymatic assembly of DNA molecules up to several hundred kilobases. Nat Methods 6 (5):343–345
24. Engler C, Gruetzner R, Kandzia R et al (2009) Golden gate shuffling: a one-pot DNA shuffling method based on type IIs restriction enzymes. PLoS One 4(5):e5553
25. Lv L, Ren Y-L, Chen J-C et al (2015) Application of CRISPRi for prokaryotic metabolic engineering involving multiple genes, a case study: controllable P (3HB-co-4HB) biosynthesis. Metabolic Eng 29:160–168
26. Murphy KC, Campellone KG, Poteete AR (2000) PCR-mediated gene replacement in Escherichia coli. Gene 246(1–2):321–330
27. Zhang Y, Buchholz F, Muyrers JP et al (1998) A new logic for DNA engineering using recombination in Escherichia coli. Nat Genet 20 (2):123–128
28. Jiang Y, Chen B, Duan C et al (2015) Multigene editing in the Escherichia coli genome via the CRISPR-Cas9 system. Appl Environ Microbiol 81(7):2506–2514
29. Zhao J, Li Q, Sun T et al (2013) Engineering central metabolic modules of Escherichia coli for improving β-carotene production. Metabolic Eng 17:42–50
30. Amann E, Ochs B, Abel KJ (1988) Tightly regulated tac promoter vectors useful for the expression of unfused and fused proteins in Escherichia coli. Gene 69(2):301–315
31. Lutz R, Bujard H (1997) Independent and tight regulation of transcriptional units in Escherichia coli via the LacR/O, the TetR/O and AraC/I1-I2 regulatory elements. Nucleic Acids Res 25(6):1203–1210
32. Neidhardt FC, Bloch PL, Smith DF (1974) Culture medium for enterobacteria. J Bacteriol 119(3):736–747
33. Jiang H, Wong WH (2008) SeqMap: mapping massive amount of oligonucleotides to the genome. Bioinformatics 24(20):2395–2396
34. Magoč T, Salzberg SL (2011) FLASH: fast length adjustment of short reads to improve genome assemblies. Bioinformatics 27 (21):2957–2963
35. Liu SJ, Horlbeck MA, Cho SW et al (2017) CRISPRi-based genome-scale identification of functional long noncoding RNA loci in human cells. Science 355(6320):eaah7111
36. Storey JD, Tibshirani R (2003) Statistical significance for genomewide studies. Proc Natl Acad Sci U S A 100(16):9440–9445
37. Gerdes S, Scholle M, Campbell J et al (2003) Experimental determination and system level analysis of essential genes in Escherichia coli MG1655. J Bacteriol 185(19):5673–5684
38. Wetmore KM, Price MN, Waters RJ et al (2015) Rapid quantification of mutant fitness in diverse bacteria by sequencing randomly bar-coded transposons. MBio 6(3): e00306–e00315

Chapter 8

A CRISPR-Based Method for Constructing Conditional Mutations of Essential Genes in Cyanobacteria

Ju-Yuan Zhang, Tian-Cai Niu, Gui-Ming Lin, and Cheng-Cai Zhang

Abstract

Cyanobacteria, a group of diverse bacteria capable of oxygenic photosynthesis, are excellent models for investigating many important cellular processes, such as photosynthesis, nitrogen fixation, and prokaryotic cell differentiation. They also have great potential to become the next-generation cell factories for sustainable biosynthesis of valuable products. However, genetic manipulation in cyanobacteria is not as convenient as in other model bacteria. Particularly, handling essential genes in cyanobacteria has been difficult due to the lack of appropriate tools, limiting our understanding of many important cellular functions encoded by them. We recently develop a CRISPR-based method for constructing the conditional mutants of cyanobacterial essential genes by engineering the ribosome binding site to a theophylline-responsive riboswitch. Here, we provide the details of this method. The principle of this method could be used to construct conditional mutants in a wide range of bacterial species.

Key words Cyanobacteria, Essential gene, CRISPR, Conditional mutant

1 Introduction

Cyanobacteria, evolutionally related to plant chloroplasts, are a group of diverse bacteria capable of performing oxygenic photosynthesis. Due to their short generation times and amenability to genetic manipulation, cyanobacteria are important models for the investigation of photosynthesis [1]. Some filamentous cyanobacteria, such as *Anabaena* PCC 7120 and *Nostoc punctiforme* ATCC 29133, can undergo cell differentiation and nitrogen fixation, thus are also excellent models for studying the mechanisms of prokaryotic cell differentiation, nitrogen fixation, and carbon/nitrogen balance [2, 3]. In addition, cyanobacteria also show great potential in biotechnology and synthetic biology, because they can be engineered to synthesize pharmaceuticals, biofuels, and many other valuable bioproducts, using CO_2 as the sole carbon source and sunlight as the sole energy [4].

With the booming studies on cyanobacteria in fundamental biology and synthetic biology, the tools for cyanobacterial genetic manipulation are expanding rapidly [5]. Nevertheless, gene manipulations in cyanobacteria remain not as convenient as those in *E. coli*, *B. subtilis*, and many other heterotrophic model microorganisms. Particularly, manipulation of essential genes has been complicated or even impossible in most cyanobacterial species. The set of cyanobacterial essential genes consists of many genes predicted to participate in key metabolic pathways (including photosynthesis, DNA replication, cell division, etc) and many genes whose functions are completely unknown [6]. Usually, the cellular function of an essential gene can be characterized by investigating its conditional mutant, in which its expression can be strictly regulated by certain inducible promoter [7]. However, due to the lack of inducible systems for tight regulation and the inconvenience and low efficiency of genome modification with previously existing methods, only a handful conditional mutants of cyanobacterial essential genes have been reported until recently [8, 9]. The difficulty in studying essential genes genetically has impeded our understanding of many key cellular functions of cyanobacteria; tools and strategies for manipulation of essential genes in cyanobacteria are highly desired.

In recent years, CRISPR systems have emerged as revolutionary genetic tools for rapid, precise, marker-less, and efficient genome modification in wide range of organisms [10–12]. One of these systems, CRISPR-Cpf1, was also shown to greatly facilitate genome editing in several cyanobacterial species [13]. Recently, we have improved the CRISPR-Cpf1 system, achieving high editing efficiency and large genome region deletion in the cyanobacterium *Anabaena* PCC7120 [14]. Particularly, with this system we successfully created the conditional mutant of the essential gene *polA*, encoding the DNA polymerase I, by substituting its native ribosome binding site (RBS) with a theophylline-responsive riboswitch (Fig. 1). Theophylline-responsive riboswitches, which are synthetic RNA elements that can activate protein translation upon their binding to the small ligand theophylline, have been used for controlled gene expression in diverse organisms [15–17]. In contrast to those regulation systems functioning at the transcriptional level, theophylline riboswitches regulate gene expression at the translational level, thus do not interfere with the natural transcription regulation. In addition to *polA*, we have constructed the conditional mutants of many other essential cyanobacterial genes using the same method. Here, we provide a detailed procedure for making TRS-based conditional mutants in *Anabaena*. Note that the CRISPR-Cpf1 system used here also allows for the construction of the conditional mutant of an essential gene by replacing its native promoter with an inducible one. The ability to genetically manipulate essential genes could greatly enhance our understanding of key

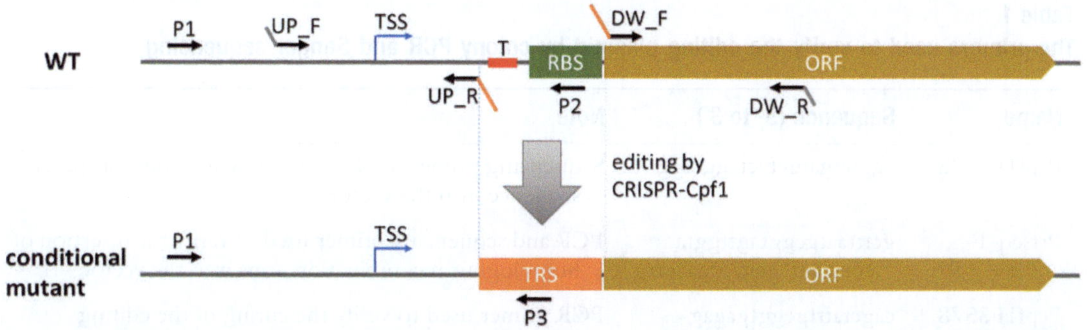

Fig. 1 The principle of making the conditional mutant of an essential gene by replacing the native ribosome binding site with a theophylline-responsive riboswitch, assisted by the CRISPR-Cpf1 genome editing system. The schematic shows the genotype of the wild type strain (WT) and the conditional mutant strain. ORF, the open reading frame of the gene of interest. RBS, the native ribosome binding site. TSS, the transcription start site of the gene (if multiple transcription start sites exist, it represents the one that is closest to the start codon). T, the target of Cpf1 selected for genome editing. TRS, the theophylline-responsive riboswitch. UP_F, UP_R, DW_F, DW_R are the primers for the amplification of the upstream and downstream homologous recombination arms used to construct the genome editing plasmid. P1, P2, and P3 are the primers for checking the genotypes of WT and the mutant by PCR. After genome editing by CRISPR-Cpf1, a small region compassing the gene's native RBS and the CRISPR target in the WT chromosome is replaced with the sequence of TRS, generating the conditional mutant strain in which the gene's translation is regulated by the concentration theophylline

cellular functions of cyanobacteria. Given that both the CRISPR-Cpf1 system and theophylline-responsive riboswitches can function in diverse bacteria, the method might be easily adapted for making conditional mutants in a broad range of bacterial species.

2 Materials

2.1 Strains, Plasmids, and Primers

1. Strains: *Anabaena* PCC 7120, *E. coli* DH5α, *E. coli* DH10B that contains the conjugation plasmid pRL443 (Addgene plasmid # 70261) and the helper plasmid pRL623 (Addgene plasmid # 58494).
2. Cyanobacterial genome editing vectors: pCpf1b (Addgene plasmid # 122187) or pCpf1b-Sp (Addgene plasmid # 122188).
3. The primers for sequencing and PCR verification (Table 1).

2.2 Media, Antibiotics, and Other Additives

1. Luria-Bertani (LB) medium (1 L):10 g Peptone, 5 g Yeast Extract, 5 g NaCl in 980 mL H_2O. Autoclave at 115 °C for 20 min.
2. SOC medium (1 L): 20 g Peptone, 5 g Yeast Extract, 4.8 g $MgSO_4$, 3.603 g glucose, 0.5 g NaCl, 0.186 g KCl. Sterilize the medium using 0.22-μm syringe filters.

Table 1
The primers used to verify the editing plasmid by colony PCR and Sanger sequencing

Name	Sequence (5′ to 3′)	Note
Pcpf1F3795	aggtctgatgctactaggtagg	Sequencing primer used to verify the insertion of spacer sequence into the vector
Prtseq_F Prtseq_R	gctttaatgcggtagttggtac tctagagtcgagcgcaacgttg	PCR and sequencing primer used to verify the insertion of homologous recombination arms into the vector
Pcpf1F3578 Pcpf1R3901	cagctatttgcggtgagag agttattcctattctgcacgaac	PCR primer used to verify the curing of the editing plasmid from the mutant cells by *sacB*-mediated counter-selection

3. BG11 medium (1 L): 1.5 g $NaNO_3$, 40 mg $K_2HPO_4 \cdot 3H_2O$, 75 mg $MgSO_4 \cdot 7H_2O$, 36 mg $CaCl_2 \cdot 2H_2O$, 6 mg citric acid, 6 mg ferric ammonium citrate, 1 mg EDTA (disodium magnesium salt), 20 mg Na_2CO_3, and 1 mL of 1000× trace-metal mix [18]. Autoclave at 121 °C for 30 min. BG11 agar is made by adding 2 g agar powder into 1 L of BG11 liquid before autoclave.
4. Trace-metal mix (1000×, 1 L): 2.86 g H_3BO_3, 1.81 g $MnCl_2 \cdot 4H_2O$, 0.222 g $ZnSO_4 \cdot 7H_2O$, 0.390 g $NaMoO_4 \cdot 5H_2O$, 0.079 g $CuSO_4 \cdot 5H_2O$, 0.0494 g $Co(NO_3)_2 \cdot 6H_2O$ [18].
5. Stock solutions of antibiotics and medium additives: 50 mg/mL kanamycin, 50 mg/mL neomycin, 100 mg/mL ampicillin, 25 mg/mL chloramphenicol (in ethanol), 1 M IPTG, 40 mg/mL X-gal (in dimethylformamide), 40 mM theophylline, 50% (w/v) sucrose. Sterilize these solutions (except for the stocks of chloramphenicol and X-gal) using 0.22-μm syringe filters.

2.3 Molecular Biology Reagents and Kits

1. Annealing Buffer: 10 mM Tris (pH 8.0), 50 mM NaCl, 1 mM EDTA.
2. DNA isolation and purification kits: plasmid MiniPrep kit, PCR purification kit, bacterial genomic DNA isolation kit.
3. Molecular biology enzymes: thermophilic DNA Polymerases, AarI (2 U/μL) (Thermo Scientific, Cat#: ER1581), BamHI (20 U/μL) (New England Biolabs, Cat#: R0144), BglII (20 U/μL) (New England Biolabs, Cat#: R3136), T4 DNA Ligase (5 U/μL) (Thermo Scientific, Cat#: EL0014).
4. Seamless DNA cloning kit.

2.4 Equipment, Glassware, and Disposable Materials

1. Equipment: handheld sonicator, shaking incubator with light, thermocycler.
2. Glassware: 200-mL baker, 100-mL flasks, 250-mL flasks.
3. Disposable materials: 2-mL Eppendorf tubes, Petri dishes (9 cm diameter), 0.2-mL tubes, 50-mL centrifuge tubes.

3 Methods

3.1 Design and Construction of the Genome Editing Plasmid

The workflow for design and construction of the genome editing plasmid is illustrated in Fig. 2.

1. Search for candidate CRISPR targets between the transcriptional start site (TSS) and the start codon of the gene to be conditionally mutated, on both the sense strand and the antisense strand (step 1 in Fig. 2; **Note 1**). Take the 22 bp DNA segments following a motif TTV (V represents for the nucleotide A, C, or G) within this region as the candidate targets (*see* **Note 2**).
2. Check the specificity of each CRISPR candidate target by searching a query sequence consisting of its PAM and the 18 bp nucleotides following the PAM against the genome of the target strain by BLASTN (*see* **Note 3**). A target should be discarded if the query sequence matches to more than one sites, perfectly or with just 1-base mismatch within the 18 bp nucleotides following the PAM (*see* **Note 4**). The query sequence should also be searched against the genome of *E. coli* and the sequence of the editing vector to make sure the absence of any matching site in them. If multiple candidate targets have passed the specificity check, choose the one that overlaps or is closest to the RBS sequence for the subsequent experiment (*see* **Note 5**).
3. According to the sequence of the chosen target, design two complementary oligonucleotides to generate the spacer sequence of the cognate crRNA (step 2 in Fig. 2). Make the forward oligonucleotide (Spacer_F) by adding "agat" to the sequence of target at the 5′-end and make the reverse one (Spacer_R) by adding "agac" to the reverse-complementary sequence of the target at the 5′-end. The two oligonucleotides will be annealed to produce a spacer duplex with cohesive ends compatible to the ends of the genome editing vector pCpf1b digested by AarI (*see* **Note 6**).
4. Design the primers for making the repair template. The repair template consists of three parts: the upstream homologous recombination arm (HR-UP), the downstream homologous recombination arm (HR-DW), and the sequence of theophylline-responsive riboswitch (TRS) between the two arms (*see* **Note 7**). Two pairs of primers (UP_F/UP_R and DW_F/DW_R) are required to amplify HR-UP and HR-DW from the chromosome (step 1 in Fig. 2). Their relative matching sites on the chromosome are shown in Figs. 1 and 2, and their detailed information is described in Table 2. Note that the sequence of TRS, a fragment of 60 bp, is split into two halves. The two halves are, respectively, attached to the 5′-ends of the

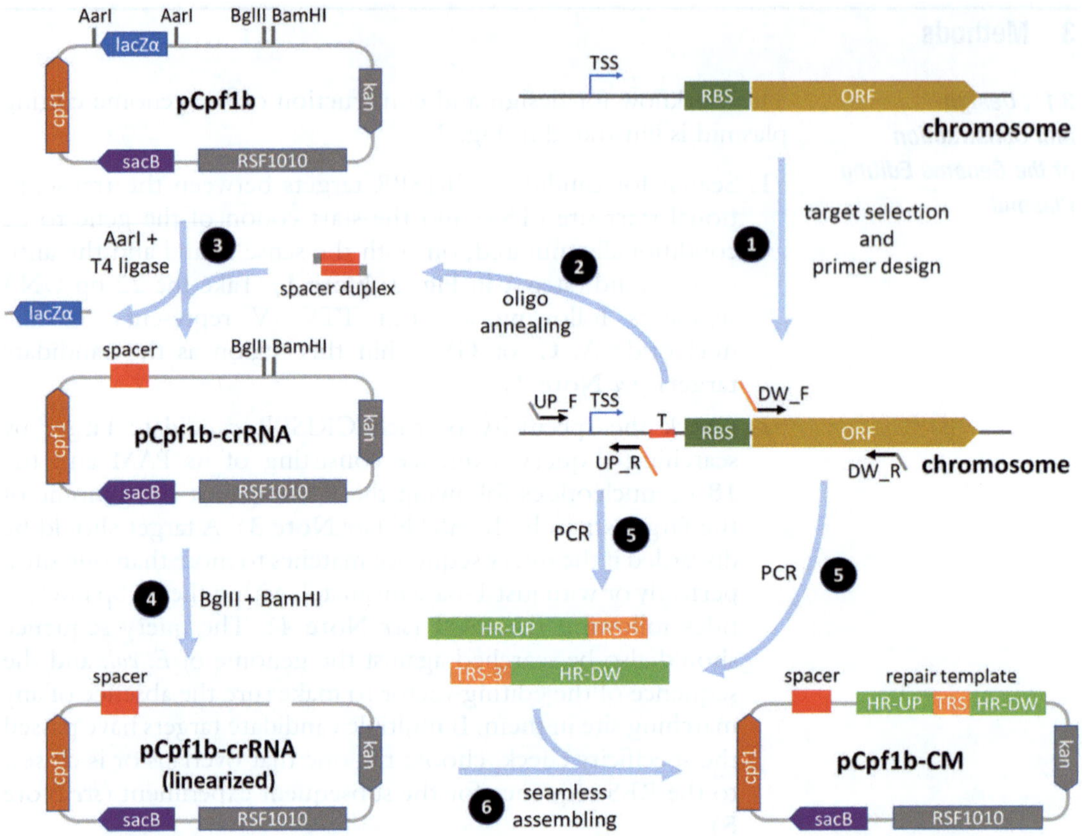

Fig. 2 The flowchart of constructing the genome editing plasmid for the creation of a conditional mutant. The plasmid construction procedure includes six major steps, and the detail of each step is described in the text. pCpf1b, the vector used for cyanobacterial genome editing. To make an editing plasmid, the spacer sequence that determines the specificity of the crRNA needs to be cloned between the two AarI sites, and the homologous repair template needs to be cloned at the BglII-BamHI site. pCpf1b-crRNA, an intermediate plasmid with the spacer having been cloned. pCpf1b-CM, the genome editing plasmid for the construction of the conditional mutant, with both the spacer and the repair template having been cloned. RSF1010, the replication region from the broad host range plasmid RSF1010; cpf1, the gene encoding the Cpf1 protein from *Francisella novicida*; sacB, the counter-selection gene conferring sucrose sensitivity; lacZα, the sequence encoding the α fragment of the β-galactosidase LacZ for blue/white screening during spacer cloning; kan, the kanamycin/neomycin resistance gene. HR-UP and HR-DW, the upstream and downstream of the homologous recombination arms for genome editing. ORF, the open reading frame of the gene of interest. RBS, the native the ribosome binding site. TSS, transcriptional start site of the gene. T, the CRISPR target selected for genome editing. TRS, the theophylline-responsive riboswitch. TRS-5′ and TRS-3′, the 5′-half and the 3′-half of the theophylline-responsive riboswitch sequence, respectively

primers UP_R and DW_F, so that the full-length TRS can be produced once HR-UP and HR-DW are cloned into the vector. The 5′-ends of these primers overlap the vector or with each other by ~20 bases, enabling the corresponding PCR fragments to be cloned into the vector via seamless cloning.

Table 2
The primers used to amplify the upstream and downstream homologous recombination arms

name	sequence (5′ to 3′)	Note
UP_F	tggcagaaattcgatatctaN(18–30)	The lowercase part overlaps the BamHI/BglII linearized vector at the BglII-end; the uppercase part should match to a site 500–1500 bp upstream the gene's start codon
UP_R	agggtgctgccaagggcatcaagacgatgctggtatcaccN(18–30)	The lowercase part is the 5′ half of the TRS sequence (reverse-complemented), and it overlaps the 5′-end of UP_R by 20 bp; the uppercase part should match to a site that is between the TSS and the CRISPR target
DW_F	gatgcccttggcagcaccctgctaaggaggtaacaacaagATGN(15–27)	The lowercase part is the 3′ half of the TRS sequence, and it overlaps the 5′-end of DW_R by 20 bp; the uppercase part should match the 5′-end of the gene's ORF, beginning with the gene's start codon
DW_R	cgcaacgttgttgccattgcN(18–30)	The lowercase part overlaps the BamHI/BglII linearized vector at the BamHI-end; the uppercase part should match to a site that is around 500–1500 bp downstream the gene's start codon

5. Prepare the spacer duplex. Dissolve the oligoribonucleotides of Spacer_F and Spacer_R with Annealing Buffer to the concentration of 10 μM. Mix 10 μL of each oligonucleotide in a 2-mL Eppendorf tube. Place the tube into a 200-mL glass baker that contains at least 100 mL hot water (>95 °C). Wait till the water is cooled down to room temperature. Annealing of the oligonucleotides will accomplish during the slow cooling process. Dilute the annealed oligonucleotides (i.e., the spacer duplex) to 100 nM by adding 980 μL water into the tube.
6. Clone the spacer duplex into the vector pCpf1b via the AarI sites (step 3 in Fig. 2). As the AarI sites are eliminated from the vector in the ligation product, so digestion and ligation can be carried out in one tube. Prepare the reaction mixture (Table 3) in a 0.2-mL PCR tube and incubate the mixture in a thermocycler with the program: 40 cycles of (37 °C for 2.5 min, 16 °C for 5 min) followed by 10 min incubation at 37 °C.
7. Transform the whole ligation mixture into 100 μL *E. coli* DH5α competent cells (*see* **Note 8**). Screen for the transformants in the LB agar plates containing 50 μg/mL kanamycin, 1 mM IPTG, and 40 μg/mL X-gal. Pick up 3 or more white

Table 3
The reaction mixture used to clone the spacer region into the pCpf1b vector

Component	Quantity
pCpf1b	185 ng (0.02 pmol)
Spacer duplex (100 nM)	2 μL (0.2 pmol)
T4 DNA ligase buffer (10×)	1 μL
Oligonucleotides (50×, provided with AarI)	0.4 μL
AarI (2 U/μL)	0.5 μL
T4 DNA ligase (5 U/μL)	0.5 μL
H_2O	Filled up to 10 μL

Table 4
The reaction mixture used to linearize the intermediate plasmid pCpf1b-crRNA

Component	Quantity
pCpf1b-crRNA	2 μg
NEBuffer 3.1 (10×)	2 μL
BamHI (20 U/μL)	0.5 μL
BglII (20 U/μL)	0.5 μL
H_2O	Filled up to 20 μL

colonies and grow each of them in 5 mL liquid LB medium supplemented with 50 μg/mL kanamycin for 12 h in a 37 °C shaking incubator (>200 rpm). Isolate the plasmids from the cultures using plasmid MiniPrep kit and verify the insertion of the spacer sequence by Sanger sequencing using the primer Pcpf1F3795. Name the plasmids have the correct insert as pCpf1b-crRNA and use one of them for subsequent cloning.

8. Linearize the plasmid pCpf1b-crRNA with BamHI and BglII (step 4 in Fig. 2). Prepare the reaction mixture (Table 4) and incubate it for 2 h at 37 °C. Purify the digestion product using PCR Purification kit.
9. Use one of the high-fidelity thermophilic DNA polymerases to amplify the homologous recombination arms HR-UP and HR_DW, using the primer pairs UP_F/UP_R and DW_F/DW_R, respectively (step 5 in Fig. 2). Purity the PCR fragments using PCR purification kit.
10. Assemble HR-UP, HR-DW, and the linearized plasmid pCpf1b-crRNA, using seamless cloning kit (step 6 in Fig. 2). Set up a 20-μL reaction mixture following the kit manual.

Transform 10 μL assembling mixture into 100 μL *E. coli* DH5α competent cells. Screen for transformants on LB plates supplemented with 50 μg/mL kanamycin.

11. Isolate the plasmids from 10 or more transformants as performed in **step** 7. Verify the plasmids by PCR using the primers Prtseq_F and Prtseq_R. If the assembling is successful, a fragment that has the summed size of HR-UP and HR-DW should be amplified; otherwise a 94 bp fragment should be amplified. Choose one of those plasmids verified by PCR for Sanger sequencing (*see* **Note 9**). If the sequencing result is correct, name the plasmid as pCpf1b-CM and use it for genome editing.

3.2 Conjugation

1. Grow 50 mL of fresh *Anabaena* culture in BG11 medium to the concentration around $OD_{700nm} = 0.5$.
2. Transform 10 ng of the editing plasmid pCpf1b-CM into 50 μL of the competent cells of the DH10B strain that contains the conjugation plasmid pRL443 and the helper plasmid pRL623. Spread 1/10 of the transformed cells onto an LB plate supplemented with 100 μg/mL ampicillin, 50 μg/mL kanamycin, and 25 μg/mL chloramphenicol. Put the plate at 37 °C overnight.
3. Pick up more than 5 colonies and inoculate them into a 250-mL flask that contains 50 mL LB medium supplemented with 100 μg/mL ampicillin, 50 μg/mL kanamycin, and 25 μg/mL chloramphenicol. Grow the culture until the OD600 is between 1.0 and 2.0.
4. Harvest 40 mL *E. coli* cells by centrifugation at 6000 × *g* for 5 min. Wash the cells with 20 mL LB medium. Resuspend the cell pellet in 1 mL LB medium.
5. Take 40 mL of *Anabaena* culture into a 50-mL tube and centrifuge at 6000 × *g* for 5 min, remove the supernatant. Resuspend the cells with 20 mL fresh BG11 supplemented with 2 mM theophylline and centrifuge again (*see* **Note 10**). Remove carefully most of the supernatant and leave about 1 mL of liquid in the tube. Resuspend the cells gently.
6. Make the conjugation mixture by gently mixing 0.2 mL *E. coli* cells and 0.2 mL *Anabaena* cells in a 2-mL Eppendorf tube. Close the tube and put it under the light condition of 30 μmol m^{-2} s^{-1} at 30 °C for 3 h.
7. Prepare three fresh BG11 agar plates by pouring 40 mL of melted BG11 agar that contains 2 mM theophylline (*see* **Note 11**). Spread 10 μL, 50 μL, and 250 μL of the conjugation mixture onto these plates, respectively. Dry the surface of the agar slightly before closing the lids.

8. Put the plates (lids on the top) under the light of 30 μmol m^{-2} s^{-1} at 30 °C for 24 h. Then add 0.5 mL of 4 mg/mL neomycin to the bottom of each plates (*see* **Note 12**). Dry the surface of the agar plate slightly before covering the lids.
9. Put the plates (upside down) at the light intensity of 30 μmol m^{-2} s^{-1} and the temperature of 30 °C. The agar surface will lose the green color within 1 week, as most *Anabaena* cells will die (*see* **Note 13**). Conjugants normally appear as individual green colonies within 2 weeks.

3.3 Genotype and Phenotype Verification

1. Once the conjugant colonies have the diameters between 1 and 2 mm, pick 5 or more of them and patch each of them onto fresh BG11 agar plates that contains 2 mM theophylline and 50 μg/mL neomycin. Put the plates (upside down) under the light of 30 μmol m^{-2} s^{-1} at 30 °C. The patches will become green for 1 week.
2. Take most cells from each patch into a 100-mL flask that contains 20 mL liquid BG11 medium supplemented with 2 mM theophylline and 50 μg/mL neomycin. Grow the cultures for more than 5 days until the culture becomes lime green. Take 10 mL of each culture to isolate the genome using bacterial genomic DNA isolation kit. Check the genotype of the cells from each culture by PCR with the primer pairs P1/P2 and P1/P3 (the relative positions of these primers are shown in Fig. 1), using the WT genomic DNA as the negative control (*see* **Note 14**). Use the cells that have the mutant genotype for subsequent phenotype analysis (*see* **Note 15**).
3. Check the phenotype of the mutant cells. Collect 10 mL of the cells, wash them with 10 mL BG11 medium, and then resuspend in 40 mL fresh BG11. Transfer the cells into two 100-mL flasks (20 mL each), and add 2 mM theophylline into one of the flasks. Put the flasks under the light condition of 30 μmol m^{-2} s^{-1} at 30 °C. If the gene is essential, and the conditional mutation is successful, the cells in the theophylline-containing flask should keep growing and the cells in the other flask should die within several days (*see* **Note 16**).

3.4 Curing the Mutant Cells of the Editing Plasmid

1. Cure the mutant cells of the editing plasmid by *sacB*-assisted counter-selection. Take 100 μL of the culture from the flask with addition of theophylline (grown in the previous step) and inoculate it into a 100-mL flask that contains 20 mL BG11 medium supplemented with 2 mM theophylline. Grow the culture under the light condition of 30 μmol m^{-2} s^{-1} at 30 °C, in a shaking incubator with a rotation speed of 200 rpm in a shaking incubator, until the cells reach a concentration close to OD700 = 1.0 (*see* **Note 17**). Take 1 mL of the

culture and breakdown the cell filaments into short segments by brief sonication with a handheld sonicator (*see* **Note 18**). Prepare three fresh BG11 plates supplemented with 2 mM theophylline and 5% sucrose. Spread 20 μL, 100 μL, and 500 μL of the sonicated cells onto these plates, respectively. Single colonies should appear within 10 days.

2. Pick up 3 or more colonies from the plate, patch them on agar plates, then grow in liquid medium, as performed in the **steps 1** and **2** in the Sect. 3.3 (here the agar and liquid media are supplemented with 2 mM theophylline only). Isolate the genomic DNA from the cells. Confirm the loss of the editing plasmid by PCR using the primer pairs Pcpf1F3578/Pcpf1R3901 (*see* **Note 19**). Use the mutant cells cured of the editing plasmid for subsequent analyses.

4 Notes

1. The CRISPR target needs to be eliminated together with RBS during genome editing, and it must locate between the TSS and the start codon; otherwise the transcription or the translation of the gene will be impaired by the editing. If the gene has multiple transcriptional start sites, the candidate targets should be searched between the start codon and the TSS that is closest to the start codon. If the gene's TSS information is unknown, we can simply search for candidate targets within the 40 bp upstream the start codon, because a gene's TSSs are outside this region in most cases.

2. Cpf1 in complex with the CRISPR RNA (crRNA) recognizes the target DNA sequence that is identical to the spacer region of the crRNA in sequence and has a protospacer adjacent motif (PAM) TTV located to its 5′-end. A spacer should match at least the first 18 base of the target to allow for specific target cleavage by Cpf1 [19]. However, the spacer should not be too long or contain too many G and C residues either, as crRNAs with longer spacer or high GC content are more likely to have a misfolding problem. By our experience, spacers with the length of 22 bp and the GC content between 30% and 70% work well most time.

3. As mentioned in **Note 2**, only the first 18 bp of the target determines if the target can be specific cleaved by Cpf1 associated with the cognate crRNA; therefore, this region is used for specificity check.

4. One mismatch within PAM will sufficiently render the target uncleavable; however, one mismatch within the region

following the PAM still allows for cleavage, although the cleavage efficiency might be low [19].

5. In some cases, no specific CRISPR targets can be found between TSS and the start codon of a gene. If so, we can use the CRISPR systems based on those Cpf1 variants that recognize alternative PAM sites [20, 21].
6. Two genome editing vectors, pCpf1b and pCpf1b-Sp, have been developed for making conditional mutants in cyanobacteria [14]. The two vectors differ only in antibiotic resistance, with pCpf1b conferring the kanamycin resistance while pCpf1b-Sp conferring the spectinomycin resistance. Here, we just demonstrate plasmid construction using the vector pCpf1b. The spacer sequence needs to be cloned into the vector via the AarI sites, which are separated by a lacZα fragment. AarI is one of the Type II-S restriction enzymes, which cut a double-stranded DNA beyond the recognition site. After AarI digestion, a linear vector with two non-compatible ends is generated (at the same time, the lacZα fragment is removed), allowing for the insertion of the spacer duplex by T4 DNA ligase.
7. Using long homologous recombination arms can improve recombination efficiency. However, as the size of the homologous recombination arms increases, the efficiency of PCR and cloning may decrease. By our experience, using homologous recombination arms sizing within 500 to 1500 bp balances well the recombination efficiency and the PCR/cloning efficiency.
8. Do the transformation as follows. Add 10 μL ligation mixture to the tube that contains 100 μL of competent cells, mix gently. Place the tube on ice for 30 min. Heat shock at 42 °C for 90 s. Place the tube on ice for 5 min. Add 1 mL SOC medium (pre-warmed at 37 °C) into the tube. Place the tube in a shaking incubator at 37 °C for 60 min. Collect the cells by centrifugation at 6000 × *g* for 3 min and resuspend in 200 μL LB medium. Spread the cells onto LB plates that contain appropriate selection reagents and put the plates in a 37 °C incubator until the transformants appear.
9. Make sure sufficient amount of sequencing reactions are performed so that the entire sequence of the homologous recombination arms can be covered.
10. The optimal concentration of theophylline required by TRS for activate translation is between 1 and 2 mM; theophylline at higher concentration may inhibit cell growth in cyanobacteria [17, 22]. We routinely use 2 mM theophylline.
11. To prepare the BG11 agar plates supplemented with additives (such as antibiotics, theophylline, and sucrose), melt the BG11 agar in the microwave oven, add the appropriate amount of the

additive stocks into the molten agar when it has cooled down to around 60 °C, mix well. Pour about 40 mL melted agar into each Petri dish (9 cm diameter).

12. The antibiotic will gradually diffuse through the agar from bottom to top and approach an equilibrium concentration of 50 μg/mL within days.
13. Most cells will die either because they have not successfully received the editing plasmid during conjugation thus will be killed by the antibiotic, or because they have not repaired the chromosomes that are cleaved by Cpf1 in time.
14. The primer P2 matches to the WT genome only and P3 matches to the mutant genome only, whereas P1 matches to both. Thus, if all the chromosome copies isolated from the cells have been successfully edited, PCR with P1/P2 should give no product while PCR with P1/P3 should give a product with the expected size. And if all the chromosome copies have not been modified, PCR with P1/P2 should produce a fragment with the expected size while PCR with P1/P3 should give no product. If only a portion of the chromosome copies have been edited (incomplete segregation), PCR with each primer pairs will give a product with the expected size.
15. If the cells from all the cultures contain only the WT genome copies, then the reason might be that the CRISPR target cannot be cleaved by Cpf1, or construction of the editing plasmid is problematic. If they all have both the mutant and the WT chromosome copies, then the reason might be that the CRISPR target cannot be cleaved by Cpf1 efficiently. To avoid the cleavage efficiency problem, we recommend to construct another editing plasmid which use a different CRISPR target. A good strategy is to always use two editing plasmids that targeting different sites to improve editing efficiency. We call this "two-spacer" strategy and have shown that it can greatly enhance the chance of successful editing [14].
16. If cells in both flasks can grow, there might be two reasons. First, the TRS does not function properly due to the folding problem, which is likely caused by the influence of its flanking sequences. This can be known by examining the protein expression by Western blot when the antibodies against the gene's product are available. This problem could be solved by redesigning the primer UP_R by adjusting its chromosomal matching site, and using it to construct another editing plasmid. Second, the mutated gene is non-essential. This can also be tested by a Western blot assay. If the gene is indeed non-essential, it should be able to be deleted from the chromosome.

17. Make sure cells have been grown under the optimal growth condition in antibiotic-free medium for more than 5 generations, as the loss of plasmid occurs more frequently when cell grows fast.
18. Check the cell filaments under microscope after each sonication pulse, stop sonicating if the most of filaments consist of less than 10 cells. Over-sonication may greatly reduce the cell viability. This sonication step is only necessary for filamentous strains.
19. Both primers match to the ORF of the *cpf1* gene present on the editing plasmid. If the plasmid is not in the genomic DNA, no product can be amplified; otherwise a 324 bp fragment can be amplified. A control PCR reaction using the editing plasmid pCpf1b-CM as the template should be included.

Acknowledgments

This work was supported by the Featured Institute Service Project from the Institute of Hydrobiology, the Chinese Academy of Sciences (Grant No.: Y85Z061601), the National Natural Science Foundation of China (Grant No.: 32070037), and the Key Research Program of Frontier Sciences of the Chinese Academy of Sciences (Grant No.: QYZDJ-SSW-SMC016).

References

1. Lea-Smith DJ, Bombelli P, Vasudevan R, Howe CJ (2016) Photosynthetic, respiratory and extracellular electron transport pathways in cyanobacteria. Biochim Biophys Acta 1857:247–255
2. Muro-Pastor AM, Hess WR (2012) Heterocyst differentiation: from single mutants to global approaches. Trends Microbiol 20:548–557
3. Zhang CC, Zhou CZ, Burnap RL, Peng L (2018) Carbon/nitrogen metabolic balance: lessons from cyanobacteria. Trends Plant Sci 23:1116–1130
4. Lau NS, Matsui M, Abdullah AA (2015) Cyanobacteria: photoautotrophic microbial factories for the sustainable synthesis of industrial products. Biomed Res Int 2015:754934
5. Santos-Merino M, Singh AK, Ducat DC (2019) New applications of synthetic biology tools for cyanobacterial metabolic engineering. Front Bioeng Biotechnol 7:33
6. Rubin BE, Wetmore KM, Price MN, Diamond S, Shultzaberger RK, Lowe LC, Curtin G, Arkin AP, Deutschbauer A, Golden SS (2015) The essential gene set of a photosynthetic organism. Proc Natl Acad Sci U S A 112: E6634–E6643
7. Texier P, Coddeville M, Bordes P, Genevaux P (2018) A useful gene cassette for conditional knock-down of essential genes by targeted promoter replacement in mycobacteria. BioTechniques 65:159–162
8. Tous C, Vega-Palas MA, Vioque A (2001) Conditional expression of RNase P in the cyanobacterium *Synechocystis* sp. PCC6803 allows detection of precursor RNAs. Insight in the in vivo maturation pathway of transfer and other stable RNAs. J Biol Chem 276:29059–29066
9. Ramos-León F, Mariscal V, Frías JE, Flores E, Herrero A (2015) Divisome-dependent subcellular localization of cell-cell joining protein SepJ in the filamentous cyanobacterium *Anabaena*. Mol Microbiol 96:566–580
10. Liu Y, Tao W, Wen S, Li Z, Yang A, Deng Z, Sun Y (2015) In vitro CRISPR/Cas9 system for efficient targeted DNA editing. MBio 6: e01714–e01715

11. Bayat H, Modarressi MH, Rahimpour A (2018) The conspicuity of CRISPR-Cpf1 system as a significant breakthrough in genome editing. Curr Microbiol 75:107–115
12. Hille F, Richter H, Wong SP, Bratovič M, Ressel S, Charpentier E (2018) The biology of CRISPR-Cas: backward and forward. Cell 172:1239–1259
13. Ungerer J, Pakrasi HB (2016) Cpf1 is a versatile tool for CRISPR genome editing across diverse species of cyanobacteria. Sci Rep 6:39681
14. Niu TC, Lin GM, Xie LR, Wang ZQ, Xing WY, Zhang JY, Zhang CC (2019) Expanding the potential of CRISPR-Cpf1-based genome editing technology in the cyanobacterium *Anabaena* PCC 7120. ACS Synth Biol 8:170–180
15. Suess B, Fink B, Berens C, Stentz R, Hillen W (2004) A theophylline responsive riboswitch based on helix slipping controls gene expression in vivo. Nucleic Acids Res 32:1610–1614
16. Topp S, Reynoso CM, Seeliger JC, Goldlust IS, Desai SK, Murat D, Shen A, Puri AW, Komeili A, Bertozzi CR, Scott JR, Gallivan JP (2010) Synthetic riboswitches that induce gene expression in diverse bacterial species. Appl Environ Microbiol 76:7881–7884
17. Nakahira Y, Ogawa A, Asano H, Oyama T, Tozawa Y (2013) Theophylline-dependent riboswitch as a novel genetic tool for strict regulation of protein expression in cyanobacterium *Synechococcus elongatus* PCC 7942. Plant Cell Physiol 54:1724–1735
18. Stanier RY, Kunisawa R, Mandel M, Cohen-Bazire G (1971) Purification and properties of unicellular blue-green algae (order Chroococcales). Bacteriol Rev 35:171–205
19. Zetsche B, Gootenberg JS, Abudayyeh OO, Slaymaker IM, Makarova KS, Essletzbichler P, Volz SE, Joung J, van der Oost J, Regev A, Koonin EV, Zhang F (2015) Cpf1 is a single RNA-guided endonuclease of a class 2 CRISPR-Cas system. Cell 163:759–771
20. Nishimasu H, Yamano T, Gao L, Zhang F, Ishitani R, Nureki O (2017) Structural basis for the altered PAM recognition by engineered CRISPR-Cpf1. Mol Cell 67:139–147.e2
21. Zetsche B, Strecker J, Abudayyeh OO, Gootenberg JS, Scott DA, Zhang F (2019) A survey of genome editing activity for 16 cas12a orthologs. Keio J Med 69:59–65
22. Ma AT, Schmidt CM, Golden JW (2014) Regulation of gene expression in diverse cyanobacterial species by using theophylline-responsive riboswitches. Appl Environ Microbiol 80:6704–6713

Chapter 9

A Method of Transposon Insertion Sequencing in Comprehensively Identifying *Vibrio vulnificus* Genes Required for Growth in Human Serum

Miguel Carda-Diéguez and Carmen Amaro

Abstract

One of the most powerful approaches to detect the loci that enable a pathogen to cause disease is the creation of a high-density transposon mutant library by transposon insertion sequencing (TIS) and the screening of the library using an adequate in vivo and/or ex vivo model of the disease. Here we describe the procedure for detection of the putative loci required for a septicemic pathogen to cause sepsis in humans by using TIS plus an ex vivo model of septicaemia: to grow the pathogen in fresh and inactivated human serum. We selected *V. vulnificus* because it is a highly invasive pathogen capable of spreading from an infection site to the bloodstream, causing sepsis and death in less than 24 h. To survive and proliferate in blood (or host serum), the pathogen requires mechanisms to overcome the innate immune defenses and metabolic limitations of this host niche. Initially, genes under-represented for insertions can be used to estimate the *V. vulnificus* essential gene set. Analysis of the relative abundance of insertion mutants in the library after exposure to serum would detect which genes are essential for the pathogen to overcome the diverse limitations imposed by serum.

Key words TnSeq, Transposon insertion sequencing, *Vibrio vulnificus*, Human serum, Capsule, YJ016

1 Introduction

1.1 Transposon Insertion Sequencing

Transposon insertion sequencing (TIS also known as TnSeq, InSeq, TraDIS, or HITS) is an approach that combines the transposon mutagenesis with next-generation sequencing [1]. Over the last decade, TIS has been improved and has allowed scientists to sequence multiple transposon mutants simultaneously, thus allowing genome-wide analyses. Consequently, this technique has been applied for the analyses of the genetic requirements to grow in different conditions in a series of bacterial species [2–11]. In particular, Waldor and collaborators (https://waldorlab.bwh.harvard.edu/) have used this technique in various occasions to study the

Ren Zhang (ed.), *Essential Genes and Genomes: Methods and Protocols*, Methods in Molecular Biology, vol. 2377, https://doi.org/10.1007/978-1-0716-1720-5_9,

virulence and metabolism of *Vibrio* species [12–15] and even developed a pipeline for its bioinformatics analysis [16].

TIS is based on the growth of a transposon library (that ideally covers the whole genome) in a challenge medium or situation where the bacterium will require the use of specific functions that normally does not use in a control medium. In other words, if a gene required for the growth in the host is mutated by the insertion of the transposon, the mutated clone will not be able to grow or it will grow slower than in the control medium. Thanks to the NGS technology, we can incubate the whole transposon library in our selected challenge medium and, in parallel, in a culture medium and sequence all the resultant clones after the growth and compare them.

Obviously, it is not feasible to sequence the whole genome of those clones. Therefore, basic genetic approaches are used to sequence only the transposon and the immediate region of the genome where it is inserted, this is, the gene that is mutated. In conclusion, using the sequenced reads the transposon is located, quantified, and compared between challenge and control media. According to differences in the abundance of the reads in the genes, those are classified as "neutral" or "not essential" (no difference), "under-represented" (gene under-represented in challenged media in the whole gene), and regional (reads under-represented in the challenged media only in a part of the gene).

When TIS is used as a whole-genome approach to unravel the relevant virulence genes for a pathogen to cause a disease, the key is to find a good challenge medium. This medium should simulate the key step of the natural disease in an in vivo or ex vivo model for bacterial growth.

1.2 Microorganism Model: The Septicemic Bacterium *Vibrio vulnificus*

We have used *V. vulnificus* as a microorganism model for this chapter. *Vibrio vulnificus* is an aquatic, Gram-negative bacterium capable of causing a range of pathologies (vibriosis) upon infecting fish or human hosts [17, 18]. Among all the *Vibrio* species, *V. vulnificus* is the only one recognized as a true zoonotic agent [19]. Two distinct types of human vibriosis, severe skin lesions and gastroenteritis, result from *V. vulnificus* infection. Skin lesions develop following exposure of wounds to seawater or marine animals colonized by *V. vulnificus*, whereas gastroenteritis results from ingestion of seafood contaminated by the pathogen [18]. Both skin and enteric infections can result in septicemia and death by sepsis, particularly in immunocompromised individuals and those with high blood iron levels associated with chronic liver diseases, who exhibit 80-fold higher risk of *V. vulnificus* septicemia than healthy individuals [17, 20]. The rapidity with which the pathogen spreads from the infection site to the blood and the high mortality (50–90%) associated with septicemia are two of the properties that make this organism particularly frightening.

A recent study performed with a representative zoonotic strain demonstrated the correlation between ability to grow in host serum and virulence [21]. Therefore, human serum should be selected as challenge medium for the study.

Serum is the soluble fraction of the blood after clotting [22]. Serum does not contain either blood cells (leukocytes) or platelets but contains different components of the humoral (complement system) and nutritional immunity (the iron-chelator transferrin and other nutritional constraints). The complement system consists of around 20 small proteins, liberated in the blood as pro-proteases. Once the first element is activated, this acts as a protease that activates the following element, and this event is produced successively. The final result of this cascade of activations is the opsonization and/or destruction of the pathogen as well as the activation of inflammation. Complement is thermolabile and can be inactivated by heating 30 min at 56 °C. Transferrin and the rest of nutritional constraints are still active after heating and, consequently, are thermostable. The specific objective of the chapter is to explain the methodology necessary to discover all the genes from the pathogen involved in resistance to humoral innate immunity in blood and discriminate those involved in resistance to complement from the rest. For that, the bacterium can be grown in human serum (HS), heat-inactivated human serum (hiHS), and a control medium such as LB (Luria-Bertani broth) to identify the essential genes to cause septicemia.

2 Materials

All media must be adjusted and at pH 7 and, together with all materials, correctly sterilized except those from commercial kits or ethanol and isopropanol.

1. Strains and plasmids (Table 1).
2. Commercial agitator.
3. Centrifuge.
4. Luria-Bertani (LB) broth and agar.
5. NaCl.
6. Glycerol.
7. Antibiotics: kanamycin and streptomycin.
8. Phosphate buffer saline (PBS) is used with 1% of NaCl.
9. A spectrophotometer.
10. Basic material for electrophoresis and agarose gels.
11. Human serum (HS) from a commercial source. In order to inactivate the complement and obtain heat-inactivated human

Table 1
Species, strains and plasmids used in the study

Specie, strain, plasmid	Characteristics	Isolation/ Source	Country	Year of isolation
Vibrio vulnificus				
YJ016 (YJ)	Lineage 1 (formerly Biotype 1); Serovar non-E, non-A, non-O	Human blood	Taiwan	1993
CECT529^T (ATCC27562)	Lineage 2 (formerly Biotype 1); Serovar non-E, non-A, non-O	Human blood	USA	1977
CECT5168 (CDC-7184)	Lineage 1 (formerly Biotype 1); Serovar non-E, non-A, non-O	Human blood	USA	1977
CECT4999	Lineage 2, pv *piscis* (formerly Biotype 2), Serovar E	Diseased eel	Spain	1999
CECT5769	Lineage 2, pv piscis (formerly Biotype 2), Serovar A	Diseased eel	Spain	2004
12	Lineage 3 (formerly Biotype 3), Serovar O	Clinical	Israel	2002
11,028	Lineage 3 (formerly Biotype 3), Serovar O	Clinical	Israel	1997
YJSm^r	YJ016 streptomycin resistant	This study		
Escherichia coli				
DH5α	Cloning strain F$^-$ Φ80*lac*ZΔM15 Δ(*lac*ZYA-*arg*F) U169 *rec*A1 *end*A1 *hsd*R17(r$_k^-$, m$_k^+$) *pho*A *sup*E44*thi*-1 *gyr*A96 *rel*A1 λ$^-$	Invitrogen		
SM10 λ*pir*	Cloning strain and transposon donor *thi thr leu tonA lacy supE recA::RP4–2-Tc::Mu λpir Kan^r*	[30]	USA	1983
Nova blue	Cloning strain and protein expression			
Plasmids				
pGEM-T easy	T/A cloning vector, Amp^r	Promega		
pDM4	Suicide vector, deletion mutant construction	[31]		
pYB742	Transposon vector	Waldor		
pMMB207	Deletion mutant complementation	Addgene		
pET30a	Expression vector (his-tag)			

serum (hiHS), the HS can be incubated 30 min at 56 °C in a water bath.

12. 0.45 μm-filters (commercial source)
13. 24.5 × 24.5 cm^2 plates.
14. A commercial kit for genomic DNA extraction.
15. Commercial kits for PCR.
16. Ethanol 70 and 100%.
17. Isopropanol.
18. T4 DNA ligase.
19. Sonicator.
20. Nuclease-free water.
21. Thermocycler.
22. DNA polymerase.
23. High-Fidelity DNA Polymerase.
24. Agarose (1 and 2%).
25. DNA blunting Kit.
26. PCR Purification Kit.
27. Restriction enzymes.
28. All the programs used for bioinformatics analyses are free and available: BLAST, Protein homology/analogy recognition engine V 2.0 (Phyre^2), MATLAB and ARTIST (Waldor lab website: http://waldorlab.bwh.harvard.edu/resources/).

3 Methods

3.1 Strain Selection

All the strains (Table 1) can be routinely grown in LB + 0.5% NaCl with shaking (200 rpm on an commercial agitator) or on LB-agar (LBA) at 37 °C for 18–24 h. Antibiotic concentrations to be used in the experiments for genetic modification are as follows: streptomycin (Sm) (500 μg/ml) and kanamycin (Km) (150 μg/ml). *E. coli* SM10 *lambda pir* carrying the Himar1 suicide transposon (Tn) vector pYB742 should be used for Tn mutagenesis [23]. All strains can be kept in LB+ 20% glycerol at −80 °C. The selected strain must be first tested for its ability to grow in HS at 37 °C for 6 h (Fig. 1). In the study of Carda-Diéguez et al. (2018), the selected strain was YJ016 because it was the most efficient in growth in HS [15]. The protocol to test ability to grow in HS is as follows:

1. Inoculate the strains in different tubes with 5 ml of LB and incubate at 37 °C overnight.

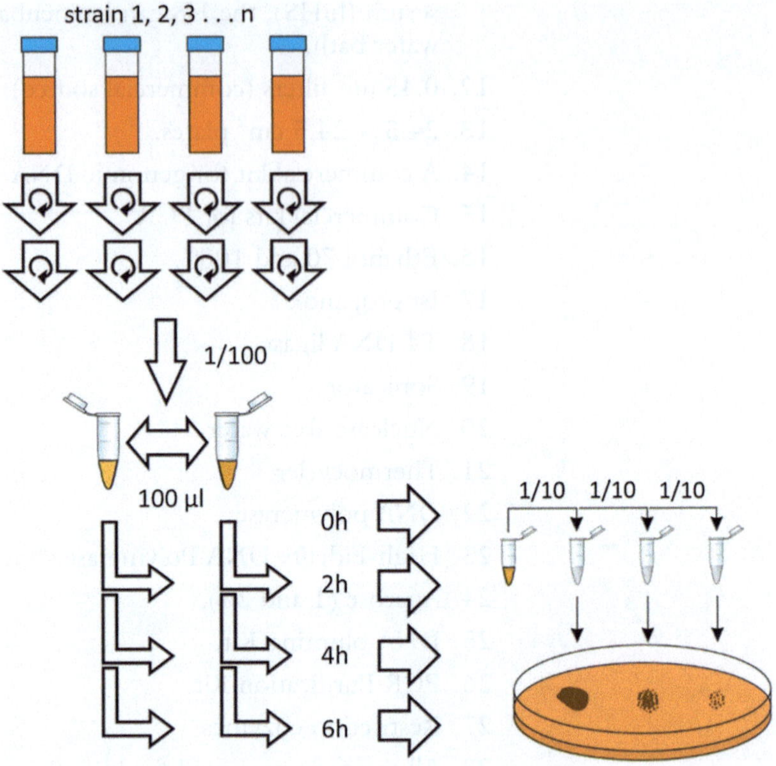

Fig. 1 Selection of the better strain for human serum growth. Grow the selected strains overnight. Eliminate the medium by centrifugation and inoculate the strains in human serum and heat-inactivated human serum. Perform plate counting after serial ten-fold dilutions

2. Centrifuge 1 ml of each bacterial culture at exponential phase and wash the pellet twice with PBS. Resuspend with 1 ml of fresh LB. At this point, the concentration is approximately 10^{8-9} cfu/ml. Use the spectrophotometer and/or perform plate counting to check the concentration.
3. Dilute the culture a 100 times and add 100 µl ($\approx 10^{5-6}$ cfu) in tubes of 5 ml containing 0.9 ml of HS or hiHS. Replicate this process at least three times. Incubate tubes at 37 °C with shaking.
4. Plate on LBA plates at 0, 2, 4, and 6 h post-inoculation (human complement starts to inactivate spontaneously after 6–8 h of incubation at 37 °C; unpublished result) in order to monitor bacterial growth. Spot 10 µl of serial ten-fold dilutions in PBS on LBA plates (drop plate count). Perform plate counting by triplicate and select the strain with higher counts.

3.2 Library Preparation (Fig. 2)

1. To obtain a spontaneous streptomycin resistant mutant (in Carda-Diéguez et al (2018) [15]: YJSm^r), plate 10^{10} cfu

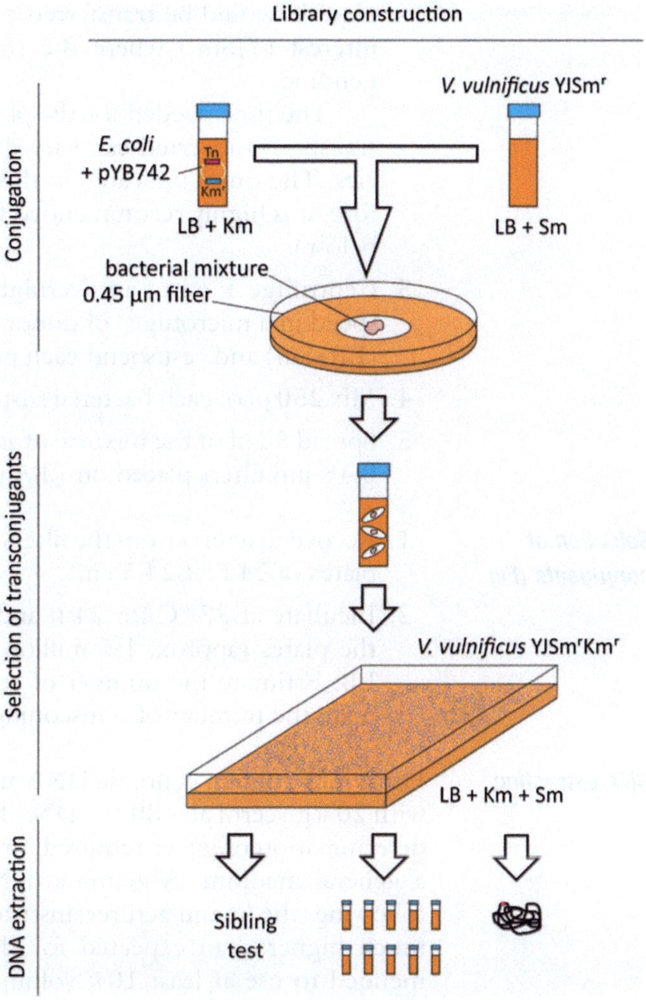

Fig. 2 Library construction. Obtain a spontaneous Streptomycin resistant strain (YJSmr) and mix it with the transposon donor strain (*E. coli*) in a 0.45 μm filter to promote the conjugation. Recover the filters and spread into a 24.5 × 24.5 cm^2 plate. After 24 h at 37 °C, count and recover the colonies and proceed with Sibling test, DNA extraction, and freeze the rest

of the selected strain (exponential phase) on LBA-Sm (streptomycin, 500 μg/ml) and select 3–5 colonies. Check that the mutants and the wild-type's growth curves in LB are statistically similar.

2. Use the YJSm^r for the transposon mutagenesis. The Mariner-based Himar1 Tn (which inserts into TA sites) and a previously published protocol [16, 24] can be used to construct two independent libraries. For that, the strain *E. coli* SM10 *lambda pir* carrying the Himar1 suicide Tn vector pYB742 (Km resistance) (Table 1) can be used as donor strain. This

plasmid would be transferred by conjugation to the species of interest (YJSm^r) where the transposon would insert in the genome.

The time needed for the plasmid to be conjugated and the transposon inserted can vary depending on the bacterial species. The times indicated in this chapter could change. Therefore, it is highly recommended to perform a "Sibling test" (see below).

3. Centrifuge 1 ml of an overnight culture (5 min at maximum speed in a microfuge) of donor (LB+Km) and recipient strains (LB+Sm) and resuspend each pellet with 250 μl of fresh LB.
4. Mix 250 μl of each bacterial suspension with 250 μl of fresh LB.
5. Spread 50 μl of the mixture of *E. coli* and *V. vulnificus* in several 0.45-μm filters placed on LBA plates and incubate for 6 h.

3.3 Selection of Transconjugants (Fig. 2)

1. Recover bacteria from the filters and spread onto LBA+Sm+Km plates of 24.5 × 24.5 cm^2.
2. Incubate at 37 °C for 24 h and then scrap bacteria grown on the plates (approx. 1.5 million colonies) with 12 ml of fresh LB. Estimate the number of colonies grown in order to estimate the number of transconjugants.

3.4 DNA Extraction

Use 3 ml to obtain genomic DNA and keep the rest (around 20 ml) with 20% glycerol at −80 °C. DNA from this library will be used for determination of genes required for *V. vulnificus* growth in vitro in a general medium. A genomic DNA extraction kit can be used following the manufacturer instructions. Since the volumes are much higher than expected for the extraction kit, it is recommended to use at least 10× volumes of all reagents for this lysis. At the final DNA precipitation step, after isopropanol addition, a blunted glass Pasteur pipette is needed to pull the DNA out of the isopropanol solution. The DNA pellet must be successively washed in 70% and 100% ethanol, spooling between each wash into a new tube. Dry out remaining ethanol at room temperature for 30 min. Finally, dry DNA pellets must be finally resuspended in 1.5–2.5 ml nuclease-free water at 65 °C for at least 1 h, with frequent mixing.

3.5 Sibling Test

If the time of incubation during the conjugation step has been too long, this can promote the over-representation of the transposon in sites where it is more plausible for the transposon to insert. Usually, 10–20% of sibling is accepted. In order of assessing how many unique insertion mutants are detected, 10–20 colonies must be selected and its transposon insertion points sequenced. For that:

1. Select 10–20 colonies from the library and extract the genomic DNA from each colony, separately.
2. Digest 2 μg of genomic DNA with a restriction enzyme (i.e., *AluI*) in a final volume of 20 μl. Incubate overnight at 37 °C.

Table 2
Primers used for Sibling test

Primer	Sequence
AB1	5′-GAGAGGGAAGAGAGCAGECAAGGAATGGAAGCTGTCTGTCGCAGGAGGGAAG −3′
AB2	5′-GACTCTCCCTTCTCGAATCGTAACCGTTCGTACGA GAATCGCTGTCCTCTCCTTC -3′
UV	5′- CGAATCGTAACCGTTCGTACGAGAATCGCT -3′

3. Vectorette PCR. Prepare the anchor bubble primers by mixing 10 μl of AB1 and AB2 primers with 230 μl of water/elution buffer (Table 2) and incubate at 65 °C for 20 min. Add 5 μl of 100 mM $MgCl_2$ and let it cool at room temperature and then place it on ice.
4. Ligate 20 μl of the digested genomic DNA with 1 μl of the bubble primers overnight at 16 °C with a T4 DNA ligase.
5. Amplify the region where the transposon is inserted using a primer that matches the transposon (5′-CCAACCTTCAAAT GATTCCC and 5′- GGTACTATATAAAAATAATATGCATT TAATACTAGCG) and another that aligns to the bubble primers (UV primer, Table 2). The PCR consisted in two processes: the first is 10 cycles of 94 °C 0.5 min, 69 °C (−2 °C per cycle) 0.5 min, and 72 °C 1 min, and the second one is 25 cycles of 94 °C 0.5 min, 59 °C 0.5 min, and 72 °C 1 min. The PCR starts with 15 min at 95 °C and ends with 10 min at 72 °C.
6. Finally, analyze 10 μl of the PCR reaction on an agarose (1.5%) gel. A single strong band should be obtained. If that is the case, clean the remaining PCR product and sequence it using Sanger. Annotate the resultant reads using a simple approach, i.e., BLAST. If the percentage of colonies with the same mutated gene is bigger than 20, decrease the incubation time and repeat the process.

3.6 Library Selection (Fig. 3)

1. Inoculate the obtained Tn-library in LB (control) and, in parallel, in HS (tested condition 1) and hiHS (tested condition 2) at a ratio 1:30 vol/vol (approximately 10^8 cfu/ml) and incubate at 37 °C for 2 h.
2. Recover cells by centrifugation, resuspend in 3 ml of LB, and immediately spread on LBA+Sm+Km plates (24.5 × 24.5 cm^2, Corning). Incubate at 37 °C overnight.
3. Recover the grown bacteria by scraping with approximately 12 ml of LB. Use 3 ml to extract genomic DNA as explained before. Keep the remaining bacterial suspension with 20% glycerol at −80 °C.

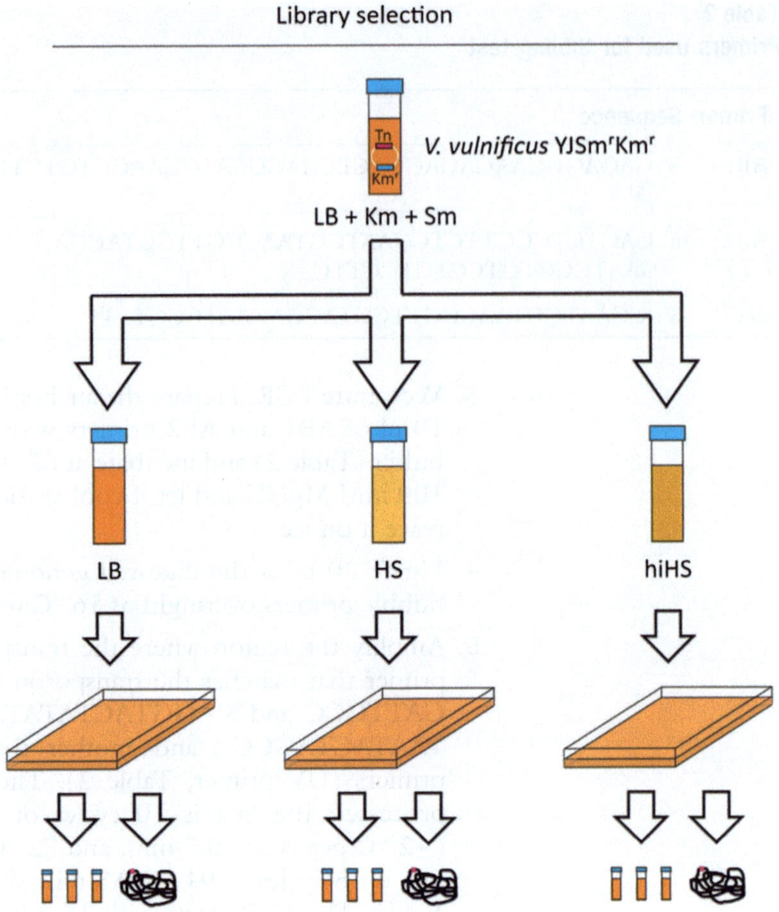

Fig. 3 Library selection. Inoculate the obtained library into the control medium (LB), human serum (HS), and heat-inactivated human serum (hiHS) and incubate at 37 °C for 2 h. After that, spread into a 24.5 × 24.5 cm^2 plate and recover after 24 h at 37 °C. Extract the DNA and freeze the rest at −80 °C

3.7 Sequencing (Fig. 4)

Samples must be kept on ice during this part of the protocol.

1. Sonication. Fragment 5–10 μg of genomic DNA into pieces of ≈350 bp by sonication, then repair DNA ends with a commercial kit for blunting (*see* **Note 1**).
2. Transposon amplification. The Tn junctions must be amplified by PCR using random primers and the kanamycin resistance gene primers (the same used in the sibling test). Clean PCR product and proceed with next step.
3. Attach adaptor sequences from Illumina or your preferred sequencer and purify the final PCR amplified products on a 2% agarose gel to isolate 200–500 bp fragments.
4. Subject the cleaned PCR product to qPCR using primers against the Illumina P5 and P7 hybridization sequence in

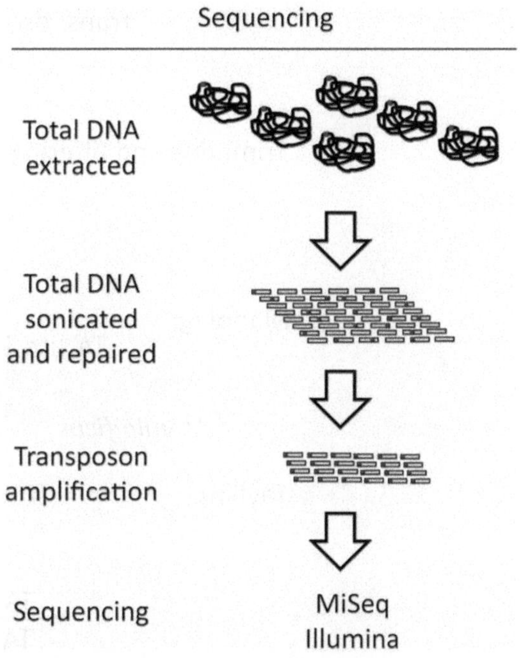

Fig. 4 Sequencing. Total DNA extracted is fragmented by sonication and repaired before DNA is amplified using random and specific-transposon primers. Illumina adaptors are attached to cleaned PCR products and quantified by qPCR and sequenced

order to quantify the amount of reads per sample. Pool the samples according to how much of each sample you want to sequence and sequence each Tn-library on the preferred sequencing platform. In Carda-Diéguez et al (2018), MiSeq (Illumina, San Diego, CA) was the selected platform [15, 16, 24]. These last two steps can be done by professional sequencing services.

3.8 Analysis of Transposon Insertion Sites. Under-Represented Genes Detection (Fig. 5)

1. Trimming and filtering. Sequencing output will putatively result in reads that contain the end of the transposon, the flanking genomic region mutated and, if the genomic fragment is too short, the 3′ adaptor sequence. In order to map the reads to the *Vibrio* genome, both, the transposon and the adaptor, must be removed. After trimming the Mariner-based transposon, check that all your reads start or end with a TA (reverse strand). Before to map the transposon-free reads, filter them by length. We recommend to use a minimum length of 15 nucleotides.
2. Mapping. Map reads to the *V. vulnificus* YJ016 genome, which is composed of two chromosomes (chromosome 1, NC_005139.1; and 2, NC_005140.1) and one plasmid (pYJ016, NC_005128.1) [25]. Discard the reads that do not

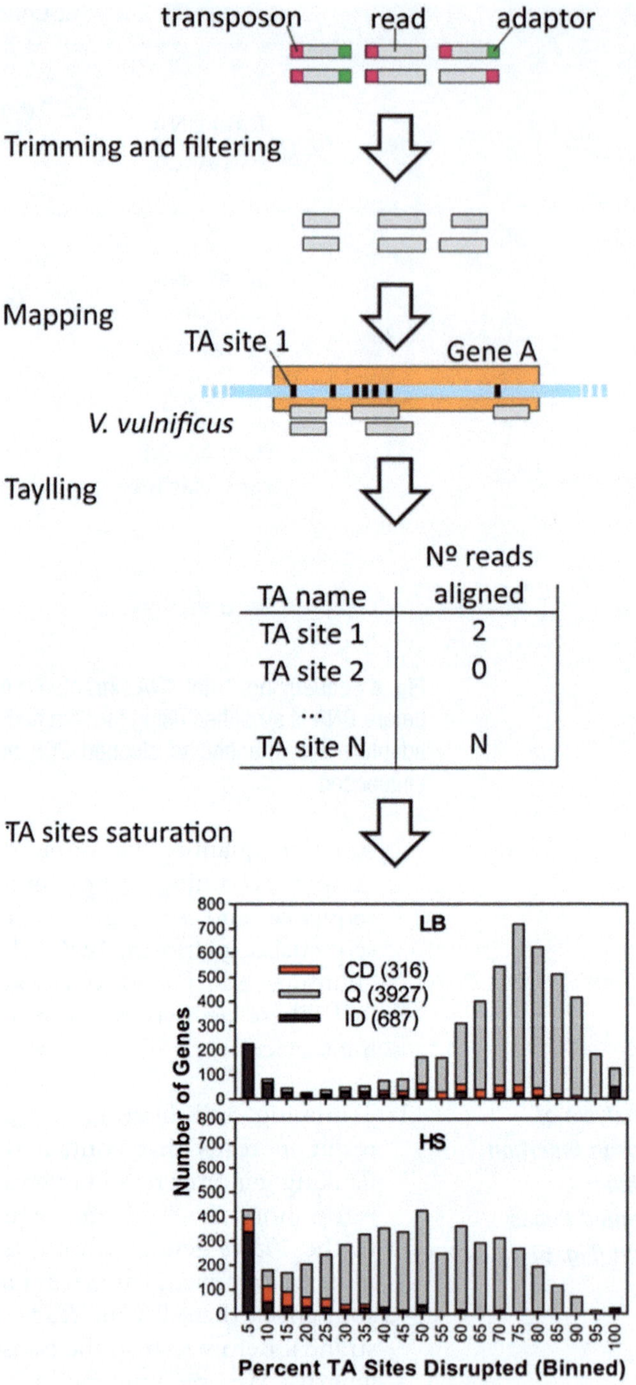

Fig. 5 Analysis of transposon insertion sites. Remove the transposon and adaptor sequences from the reads before these are mapped against the genome of the selected strain. Count the reads that aligned to each TA site, discard the rest, and calculate the saturation of the TA sites

align to any TA site in the genome and randomly distribute the reads mapping to multiple TA sites. Bowtie can be used to map the reads (.SAM file) [26]. Use the "--very-sensitive" option in the command line to discard low identity matches.

3. Create a compatible gene list. Find in online databases and download the corresponding GTF (gene transfer format) or GFF (general feature format) files of the mentioned chromosomes and plasmids.
4. Find the TA sites in the genome. Since Mariner-based transposon only inserts in TA sites in the genome, find each TA site and ignore those reads that suggests an insertion elsewhere. Use the manually created script "TAfinder.py" which looks for TA sites in the genome and creates a two-column file with the name (the script calls it a "misc_feature") and the position of the T for the TA site on the positive strand.
5. Tallying the read counts for every TA site. The number of reads at each TA site must be tallied. The *Python* script "Readcount_-SAM.py" will calculate the number of reads mapped to each TA site using the SAM file, the gene list file, and the TA sites file.
6. Library saturation (Fig. 6). Before to proceed with library comparison, you must assure that these libraries are complex enough. In order to test the saturation of the library, the percentage of unique TA sites present in the libraries must be calculated. Again, a manually created Python script can be used as follows: "TApercent.py." This script will generate a table with the number of TA sites with at least 1 read mapped and the average number of reads mapped to a TA site. Usually, the saturation will be smaller when the library is grown in a more demanding media in comparison to the control media since the number of genes essential will be higher.

   To assess the library saturation at gene level, another manually created script should be used as follows: "TAcounts_gene.py." This will generate a table with the number of TA sites disrupted in each gene, the fraction of TA sites hit per gene, the number of reads in the locus, and the average number of reads per TA. Using Excel command "=frequency," the number of genes with high (60–100%) (non-essential or neutral) and low (0–20%)(under-represented) percentage of TA sites disrupted must be calculated. Usually, a bimodal distribution with a significant higher number of genes non-essential and a small left-shift in the challenged libraries (increase in the number of under-represented genes) is found.
7. Visualization of transposon insertion. DNA plotter [27] and/or Artemis [28] can be used to map and visualize the reads across the genome. In both cases, the tally file needs to

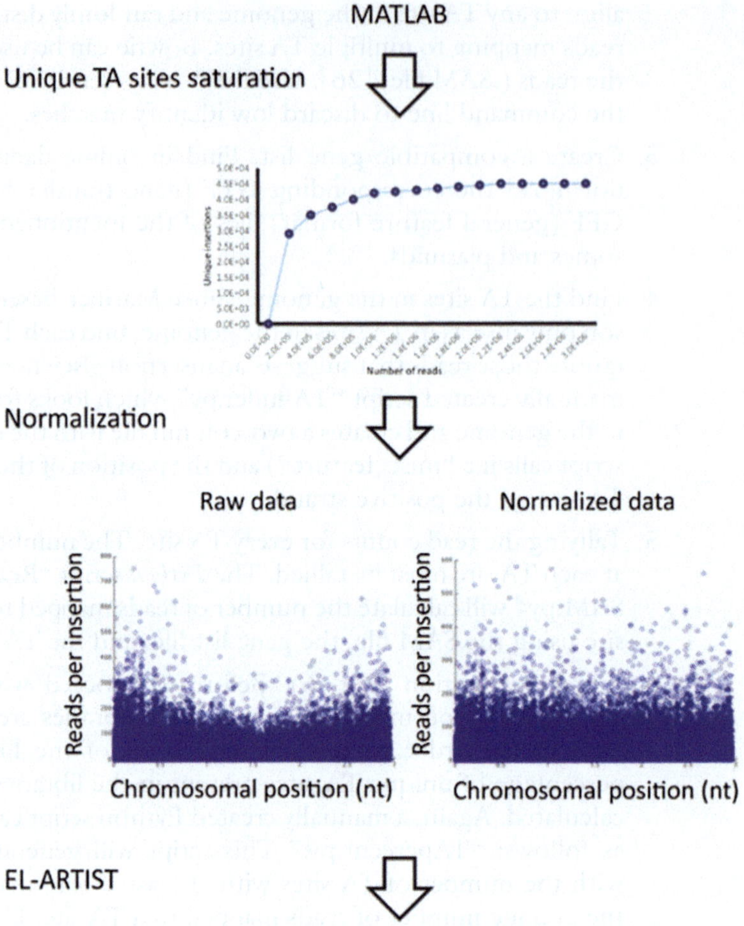

Gene	classified
gene 1	neutral
gene 2	under-represented
...	
gene N	regional

Libraries comparison: MWU

Gene	%MWU significant test	significant	average p-value	SD p-value	average Fold change	SD Fold change
gene 1	90	yes	10^{-5}	10^{-6}	15	2
gene 2	60	no	10^{-2}	10^{-3}	4	1
...						
gene N	100	yes	10^{-7}	10^{-8}	0.007	0.002

Fig. 6 Gene classification and libraries comparison. Calculate the unique TA sites saturation and normalize the read counts across the genome to avoid positional bias. Use EL-ARTIST software to classify genes into "neutral," "under-represented," and "regional" depending on the number of TA sites detected. Finally, use Mann-Whitney test in order to compare the TA abundances between libraries

be modified. "DNAplotter.py" and "Artemize.py" can be used to modify the file for DNA plotter and Artemis, respectively.

8. Transposon abundance (Fig. 6). MATLAB standalone command line computational program should be used. Custom Matlab scripts and functions to analyze the TIS data must be used.

   First of all, calculate the sequencing saturation. The script "checkseqdepth" can be used to plot the number of unique insertions found according to the number of reads mapped. The amount of reads needed to saturate the library should be observed.

9. Normalize the read counts across the genome. *V. vulnificus*, as other fast-growing organisms, may have positional biases in the read count data. Multiple origin firings and incomplete rounds of replication can create a gradient of transposon reads near the origin toward the terminus. In order to solve this bias, the data need to be normalized so the read counts truly reflect the fitness of transposon mutants across genome and not their DNA replication state.

   With "scatter" script in Matlab, any replication bias should be detected. This would be reflected as a "V" shape in the raw data caused by the higher abundance of reads at the beginning and end of the chromosome (close to the origin of replication) than at the middle (terminus).

   The normalization can be done using "normalized_input_-reads" Matlab script which will take large windows (100,000 nucleotides is recommended) across the genome and will average the reads within each window and derive a local scaling factor compared to the average reads of the whole chromosome. The reads in each window will be multiplied by the scaling factor in the corresponding window and therefore normalize the local positional bias.

10. Unique names and indices. Files with a list of unique names and indices for *V. vulnificus* genome must be created. "uniquenames, uniqueindices" is the name of the Matlab script that will give us these lists.

11. EL-ARTIST: essential locus analysis. Then, EL-ARTIST software [16] can be used to analyze the two chromosomes and the plasmid pYJ016, independently. Obtain chromosome 1 classifications using hidden Markov model analysis following usual values for sliding window (SW) training (10 TA sites; $P < 0.005$) while use 15 TA sites for chromosome 2 classifications. For chromosome 1 and plasmid, genetic loci with fewer than 10 TA sites should be retroactively excluded from classification. In chromosome 2, at least 15 TA sites should be used.

The Matlab script is "Slidingwindow" and requires the definition of 3 variables: number of independent simulations (1000 is recommended), sliding window sizes (see above), and the maximum p-value (0.005 should be used) (*see* **Note 2**). As a result, 4 output matrices will be obtained as follows:

(a) Bootstats. The first row shows the number of reads in the bootstrapped window, while the second row the proportion of TA sites without reads.

(b) Essentialregions is a classification of every TA site in neutral (0) or under-represented (1) by sliding window analysis.

(c) Essentiallist is the same than Essentialregions but with every annotated feature (e.g., gene and intergenic elements). Specifically, the feature is only called under-represented (1) if all windows in feature were called essential in all simulations.

(d) Essentialpvals is a list of each TA site and its significance value in 1000 simulations for being under-represented in reads.

12. The essential regions obtained with the previous sliding window are used to train a hidden Markov model (HMM) to refine the essential regions. "essentialregions" is the script used for this purpose.
13. Since PCR amplification during library preparation may introduce some variation in the exact number of reads detected, appropriate cutoffs for discretization of the reads in each insertion must be created. Boxplot will be used ("boxplot" script) to judge how insertions will be grouped according to the number of reads. Usually, the box in the boxplot appears compact: many values are zero (likely essential regions), and the median is usually 1 read. The median, 25th, 75th percentile, and the outlier boundary values must be used as cutoffs to discretize reads. These values will be introduced in the script "discretize5."
14. The output from the previous script will be used to calculate the HMM's transition and emission probabilities using the Baum-Welch algorithm. Use the script "hmmestimate."
15. Train the HMM using the transition probabilities generated above and predict the essentiality of every TA site in the genome. Use the "hmmconverge" script.
16. Assign the final HMM prediction to each TA site ("convergencematrixessentiality" script) and create a final table with the classification of the genes (Fig. 6). Considering the TA positions that could have been disrupted and the TA positions covered in the Tn-library, classify genomic loci as "under-

represented"/"regional" when the number of sequenced transposons in the loci is significantly low, or "neutral" (1 in the table), when reads are mapped in practically the total sequence of the gene. Further, two categories can be differentiated within essential genes on the basis of the location of reads, "under-represented" (2 in the table) and "regional" (3 in the table) when the reads are located in the full sequence or in a part of the gene, respectively.

17. Libraries comparison. Because of the differences in TA % disrupted between Tn-libraries, use Mann-Whitney test in order to compare the TA abundances. Minimum 90% Mann-Whitney U (MWU) test values, p-value<0.00001, and fold change (FC) bigger than 5 are recommended for the analysis. The FC represents the number of times a gene is more represented in one Tn-library vs. other.
18. Simulation-based normalization for bottleneck effects. Usually, the challenged media represents an experimental bottleneck. Therefore, the comparison of control and challenged media libraries needs to simulate the bottleneck effect in the input/control library. In order to simulate the genetic drift and sampling error in the experiment, use "simulateequalsaturation" script (*see* **Note 3**).
19. Mann-Whitney U test. The script "runmwuallboots" will compare the abundance of each gene in the two libraries to be compared. It will perform as much test as previous simulations were performed and will give us an output matrix with the p-values for all tests.
20. Calculate the average p-value and reproducibility for each gene using "MWUsummarystats" script. The output table will have the p-value and fold change for each locus.

3.9 Bioinformatic Analyses

Genes listed as under-represented for survival/growth in HS must be subjected to additional bioinformatic analyses; in particular, genes annotated as hypothetical proteins can be re-annotated using HHpred or BLASTP against GenBank. Moreover, domains can be analyzed using the Conserved Domain Database (CDD) on the NCBI website. The Protein homology/analogy recognition engine V 2.0 (Phyre^2) pipeline can be used for analysis of amino acid sequences [29].

4 Notes

1. The number, magnitude, and duration of pulses during sonication will depend on the sonicator. It is recommended to perform a test in order to check the sizes of the resultant

fragments. For that, we recommend to run 3 pulses of 10 s while sonicate different DNA aliquots from the library with different magnitude and run the sonicated DNA in a 1% agarose gel. Keep the magnitude that fragment the DNA into 350 bp length pieces.

2. EL-ARTIST. Depending on the transposon saturation level and sequencing depth of the library, different parameters (i.e., p-value or window size) should be used. The analysis should be repeated using different cutoffs until regions are called essential and non-essential. In general, less saturated libraries benefit from using larger window sizes.
3. Mann-Whitney U test. The average read count ratio from all insertions within each locus between the experimental and control dataset across 100 simulations is calculated (Experiment reads/Controlsim reads). To avoid mathematical errors, some adjustments are made. When reads in the Controlsim dataset are detected for a locus but not in the Experiment dataset, arbitrarily set the Experiment read to 1. Similarly, if there are reads in Experiment, but not Controlsim, set Controlsim reads to 1. Finally, if there are no reads in either the Experiment or Controlsim library, the ratio reported is "0."

Acknowledgments

The authors are grateful to Dr. Mattew Waldor for hosting Miguel Carda-Dieguez in his laboratory and teaching him to perform the experimentation and bioinformatic analysis related to transposon insertion sequencing (TIS). The authors also thank the Ministry of Science and Innovation of Spain (Project AGL2017—87723) for funding the stay of Miguel Carda in the Waldor's lab and the Generalitat Valenciana (Project AICO/2018/123) for funding a part of the research that have allowed us to write this chapter.

References

1. van Opijnen T, Bodi KL, Camilli A (2009) Tn-seq: high-throughput parallel sequencing for fitness and genetic interaction studies in microorganisms. Nat Methods 6:767–772. https://doi.org/10.1038/nmeth.1377
2. Gao B, Vorwerk H, Huber C et al (2017) Metabolic and fitness determinants for in vitro growth and intestinal colonization of the bacterial pathogen campylobacter jejuni. PLoS Biol 15:e2001390. https://doi.org/10.1371/journal.pbio.2001390
3. Cole BJ, Feltcher ME, Waters RJ et al (2017) Genome-wide identification of bacterial plant colonization genes. PLoS Biol 15:e2002860. https://doi.org/10.1371/journal.pbio.2002860
4. de Moraes MH, Desai P, Porwollik S et al (2017) Salmonella Persistence in Tomatoes Requires a Distinct Set of Metabolic Functions Identified by Transposon Insertion Sequencing. Appl Environ Microbiol 83. https://doi.org/10.1128/AEM.03028-16
5. McCarthy AJ, Stabler RA, Taylor PW (2018) Genome-wide identification by transposon insertion sequencing of Escherichia coli K1 genes essential for in vitro growth,

gastrointestinal colonizing capacity, and survival in serum. J Bacteriol 200:e00698-17. https://doi.org/10.1128/JB.00698-17

6. Armbruster CE, Forsyth-DeOrnellas V, Johnson AO et al (2017) Genome-wide transposon mutagenesis of Proteus mirabilis: essential genes, fitness factors for catheter-associated urinary tract infection, and the impact of polymicrobial infection on fitness requirements. PLoS Pathog 13:e1006434. https://doi.org/10.1371/journal.ppat.1006434
7. Capel E, Barnier J-P, Zomer AL et al (2017) Peripheral blood vessels are a niche for blood-borne meningococci. Virulence 8:1808–1819. https://doi.org/10.1080/21505594.2017.1391446
8. Zhu L, Charbonneau ARL, Waller AS et al (2017) Novel Genes Required for the Fitness of Streptococcus pyogenes in Human Saliva. mSphere 2. https://doi.org/10.1128/mSphereDirect.00460-17
9. George AS, Cox CE, Desai P et al (2017) Interactions of Salmonella enterica Serovar typhimurium and Pectobacterium carotovorum within a tomato soft rot. Appl Environ Microbiol 84:e01913-17. https://doi.org/10.1128/AEM.01913-17
10. White KM, Matthews MK, Hughes RC et al (2018) A metagenome-wide association study and arrayed mutant library confirm Acetobacter lipopolysaccharide genes are necessary for association with *Drosophila melanogaster*. Genomes Genet 8:1119–1127. https://doi.org/10.1534/g3.117.300530
11. Price MN, Wetmore KM, Waters RJ et al (2018) Mutant phenotypes for thousands of bacterial genes of unknown function. Nature 557:503–509. https://doi.org/10.1038/s41586-018-0124-0
12. Wang Q, Millet YA, Chao MC et al (2015) A genome wide screen reveals that Vibrio cholerae phosphoenolpyruvate phosphotransferase system (PTS) modulates virulence gene expression. Infect Immun 83:IAI.00411–15. https://doi.org/10.1128/IAI.00411-15
13. Chao MC, Abel S, Davis BM, Waldor MK (2016) The design and analysis of transposon insertion sequencing experiments. Nat Rev Microbiol 14:119–128. https://doi.org/10.1038/nrmicro.2015.7
14. Hubbard TP, Chao MC, Abel S et al (2016) Genetic analysis of *Vibrio parahaemolyticus* intestinal colonization. Proc Natl Acad Sci U S A 113(22):6283–6288. https://doi.org/10.1073/pnas.1601718113
15. Carda-Diéguez M, Silva-Hernández FX, Hubbard TP et al (2018) Comprehensive identification of Vibrio vulnificus genes required for growth in human serum. Virulence 9:981–993. https://doi.org/10.1080/21505594.2018.1455464
16. Pritchard JR, Chao MC, Abel S et al (2014) ARTIST: high-resolution genome-wide assessment of fitness using transposon-insertion sequencing. PLoS Genet 10:e1004782. https://doi.org/10.1371/journal.pgen.1004782
17. Oliver JD (2015) The biology of Vibrio vulnificus. Microbiol Spectr 3. https://doi.org/10.1128/microbiolspec.VE-0001-2014
18. Amaro C, Sanjuán E, Fouz B et al (2015) The fish pathogen Vibrio vulnificus biotype 2: epidemiology, phylogeny, and virulence factors involved in warm-water Vibriosis. Microbiol Spectr 3. https://doi.org/10.1128/microbiolspec.VE-0005-2014
19. Amaro C, Biosca EG, Fouz B et al (1995) Evidence that water transmits Vibrio vulnificus biotype 2 infections to eels. Appl Environ Microbiol 61:1133–1137
20. Crim SM, Griffin PM, Tauxe R, et al (2015) Preliminary incidence and trends of infection with pathogens transmitted commonly through food — foodborne diseases active surveillance network, 10 U.S. sites, 2006–2014. Atlanta
21. Hernández-Cabanyero C, Lee C, Tolosa-Enguis V et al (2019) Adaptation to host in Vibrio vulnificus , a zoonotic pathogen that causes septicemia in fish and humans. Environ Microbiol 21:3118–3139. https://doi.org/10.1111/1462-2920.14714
22. Owen J, Punt J, Stranford S, Jones P (2018) Kuby Immunology, 8th edn. W.H. Freeman, New York
23. Yamaichi Y, Chao MC, Sasabe J et al (2015) High-resolution genetic analysis of the requirements for horizontal transmission of the ESBL plasmid from Escherichia coli O104:H4. Nucleic Acids Res 43:348–360. https://doi.org/10.1093/nar/gku1262
24. Chao MC, Pritchard JR, Zhang YJ et al (2013) High-resolution definition of the vibrio cholerae essential gene set with hidden Markov model-based analyses of transposon-insertion sequencing data. Nucleic Acids Res 41:9033–9048. https://doi.org/10.1093/nar/gkt654
25. Chen C-Y (2003) Comparative genome analysis of Vibrio vulnificus, a marine pathogen. Genome Res 13:2577–2587. https://doi.org/10.1101/gr.1295503.7
26. Langmead B, Salzberg SL (2012) Fast gapped-read alignment with bowtie 2. Nat Methods

9:357–359. https://doi.org/10.1038/nmeth.1923

27. Carver T, Thomson N, Bleasby A et al (2009) DNAPlotter: circular and linear interactive genome visualization. Bioinformatics 25:119–120. https://doi.org/10.1093/bioinformatics/btn578
28. Carver TJ, Rutherford KM, Berriman M et al (2005) ACT: the Artemis comparison tool. Bioinformatics 21:3422–3423. https://doi.org/10.1093/bioinformatics/bti553
29. Kelley LA, Mezulis S, Yates CM et al (2015) The Phyre2 web portal for protein modeling, prediction and analysis. Nat Protoc 10:845–858. https://doi.org/10.1038/nprot.2015.053
30. Simon R, Priefer U, Pühler A (1983) A broad host range mobilization system for in vivo genetic engineering: transposon mutagenesis in Gram negative bacteria. Bio/Technology 1:784–791. https://doi.org/10.1038/nbt1183-784
31. Milton DL, O'Toole R, Horstedt P, Wolf-Watz H (1996) Flagellin A is essential for the virulence of *Vibrio anguillarum*. J Bacteriol. 178:1310–1319

Chapter 10

The Use of Tn-Seq and the FiTnEss Analysis to Define the Core Essential Genome of *Pseudomonas aeruginosa*

Bradley E. Poulsen, Anne E. Clatworthy, and Deborah T. Hung

Abstract

Transposon-insertion sequencing (Tn-Seq) allows for identification of bacterial genes and pathways essential for growth under a given condition. A transposon mutant is created by the stable and random integration of a transposable element into a genome of interest, followed by a period of outgrowth and selection for relative fitness on one or more growth media. By pooling hundreds of thousands of mutants, sequencing the transposon-genomic DNA junctions, and mapping sequencing reads to the genome, one can identify an abundance of reads in nonessential insertion regions and the absence of reads in essential regions and thus identify which genes are essential for a given growth condition. By performing this method iteratively across multiple strains and growth conditions, one can define a core essential genome for a species. Here, we describe this methodology in detail and its application for the species *Pseudomonas aeruginosa*, from generating mutants to the analysis of nonessential versus essential genes using the freely available software "*FiTnEss*".

Key words Tn-Seq, Essential gene analysis, *Pseudomonas aeruginosa*, Transposon mutagenesis, Infection-relevant media

1 Introduction

Determining which genes and biological pathways are essential for a bacterial species in a given environment is important for understanding the biology of the bacteria's lifestyle and, more practically, what genes are required for bacterial survival, which could ultimately serve as targets for antibiotics or other therapeutics. Transposon insertion sequencing (Tn-Seq) is a valuable tool used to determine how each gene functions in the overall fitness of a bacterium [1–5]. It is performed by first mutagenizing a bacterial strain using transposon mutagenesis, which results in the insertion of a transposable element into multiple random sites in the genome, often disabling a gene where it inserts. A large random pool of transposon mutants, covering all possible insertion sites, is constructed and grown for a period of time on a given medium.

Ren Zhang (ed.), *Essential Genes and Genomes: Methods and Protocols*, Methods in Molecular Biology, vol. 2377, https://doi.org/10.1007/978-1-0716-1720-5_10,

The transposon insertion regions are then amplified and sequenced, and essential and nonessential genes are defined based on the number of sequencing reads obtained for each gene. Nonessential genes are identified by the presence of a high number of sequencing reads within the gene, while essential genes are defined by having no reads or extremely low read counts within the gene. In practice, however, categorizing some genes as essential or nonessential is not obvious as the ability of each mutant in a pool to grow on a given medium is a continuum rather than a binary phenomenon of full growth or no growth. To meet this challenge, multiple statistical software packages have been developed to infer the relative fitness of mutants growing in a pool on a given medium and thus classify genes as essential or nonessential [1, 6–8], although the complexity of statistical modeling and the stringency in determining essential genes of each method differ.

While the use of Tn-Seq to probe the essential genes of a strain on one or more media has provided useful insights into many species [9–12], we have recently shown that in the case of the important pathogen *P. aeruginosa*, ~4 strains should be used for Tn-Seq experiments to more comprehensively define the core essential genome of a species [13]. We used 2 commonly used lab-based media (LB and M9 minimal) and 3 infection-relevant media (serum, sputum, and urine) to define a core set of essential genes and to expand our knowledge of conditionally essential genes. This paradigm can be applied to other bacterial species to aid our quest for new antibiotics and for more advanced comparisons between species.

Here, we present in detail the methodology we employed to define the core essential genome of *P. aeruginosa* as highlighted in Fig. 1. In Subheading 3.1, we describe the creation of saturated transposon libraries of many diverse strains, which are stored before a selection process on one of the five distinct media. In Subheading 3.2, we report the molecular biology steps to create next-generation sequencing libraries from purified genomic DNA isolated from selected transposon libraries. Finally, in Subheading 3.3, we outline the steps required to perform the data analysis required to classify essential vs nonessential genes using our statistical software *FiTnEss* (Finding Tn-Seq Essentials), which uses minimal assumptions and has good predictive power [13]. This detailed description should facilitate the implementation of this approach for other bacterial species.

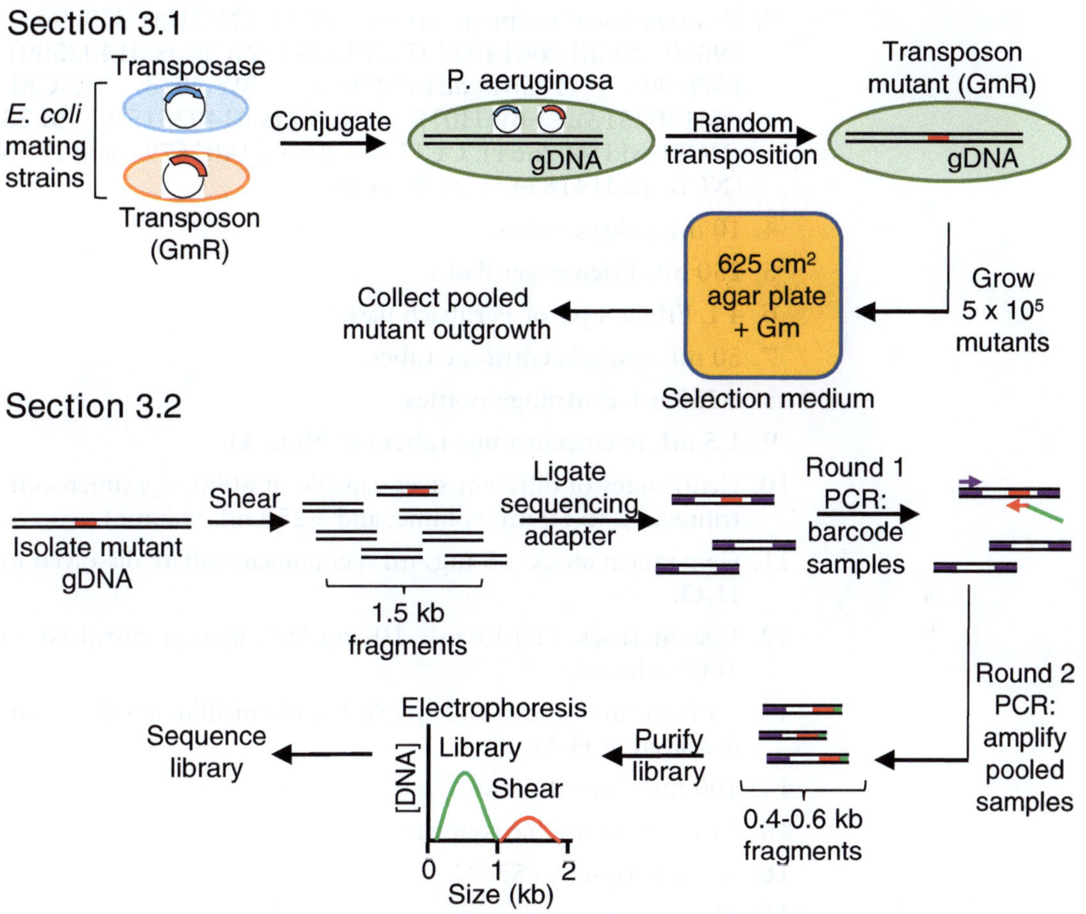

Fig. 1 Methodology for creating Tn-seq libraries. Subheading 3.1, top. Two mating competent strains carrying either a transposon plasmid containing a gentamicin resistance cassette or a transposase plasmid are conjugated in a tri-partite mating to the *P. aeruginosa* recipient strain, whereupon a transposon mutant is created after transposition occurs at a random location in the recipient genome. A saturated library of mutants is plated on an agar selection medium supplemented with gentamicin and grown for a period of time before the pooled mutants are collected and stored. Subheading 3.2, bottom. Genomic DNA from selected mutant libraries are independently isolated and sheared to 1.5 kb. The sheared DNA is ligated with an Illumina sequencing adapter, and a unique barcode identifier is introduced during a PCR reaction using a primer internal to the transposon and a primer at the adapter end. Barcoded samples are pooled, and the ~500 bp library is amplified before size selection during purification

2 Materials

2.1 Plasmids, Strains, Antibiotics, and Agar Selection Media

1. Transposon donor strain: *E. coli* SM10 conjugation strain (KanR) harboring pMAR plasmid (himar1 transposon; GentR; oriT) (*see* **Note 1**).
2. Transposase donor: *E. coli* SM10 conjugation strain (KanR) harboring pC9 plasmid (pC9 transposase; AmpR; oriT) (*see* **Note 2**).

3. *P. aeruginosa* recipient strains: PA14 (NCBI:txid208963); 19660 (NCBI:txid1402497); X13273 (NCBI:txid1402500); BWH005 (NCBI:txid1402505); BWH013 (NCBI:txid1402513); BWH015 (NCBI:txid1402515); BL23 (NCBI:txid1402564); CF77 (NCBI:txid1402579); and PS75 (NCBI:txid1418247) (*see* **Note 3**).
4. 10 mL culture tubes.
5. 250 mL Erlenmeyer flasks.
6. 4 L Erlenmeyer or Fernbach flasks.
7. 50 mL conical centrifuge tubes.
8. >250 mL centrifuge bottles.
9. 1.5 mL microcentrifuge tubes (*see* **Note 4**).
10. Centrifuges of different sizes capable of 5000 × *g* (microcentrifuge, up to 50 mL volume, and >250 mL volume).
11. Gentamicin stock: 15 mg/mL gentamicin sulfate dissolved in H_2O.
12. Irgasan stock (Triclosan): 10 mg/mL irgasan dissolved in 100% ethanol.
13. Carbenicillin stock: 100 mg/mL carbenicillin disodium salt dissolved in H_2O.
14. 100 mm Petri dishes.
15. 245 mm square bioassay dishes.
16. Water bath set to 55 °C.
17. Microwave.
18. 37 °C and 42 °C incubators (shaking and stationary).
19. Spectrophotometer (600 nm).
20. Disposable cell scrapers.
21. Phosphate buffered saline (PBS).
22. Glycerol: 80% v/v in H_2O. Sterilize by 0.2 μm filtration.
23. 100% ethanol.
24. Liquid nitrogen or ethanol/dry ice bath.
25. Agar (concentrated): 5% w/v agar in H_2O. Prepurchase sterile or sterilize by autoclave.
26. LB liquid growth medium: 5 g of yeast extract, 10 g of tryptone, and 5 g of (86 mM) NaCl. Dissolve in 1 L of H_2O. Adjust pH to 7.0. Sterilize by autoclave or by 0.2 μm filtration.
27. Mating plates: LB medium containing 1.5% w/v agar, sterilize by autoclave. Sterilely add 30 mL to 100 mm Petri dishes. Can be stored <30 days at 4 °C.
28. Counting plates: LB medium containing 1.5% w/v agar, sterilize by autoclave. Add gentamycin to 30 μg/mL final

concentration and irgasan to 5 μg/mL final concentration. Sterilely add 30 mL to 100 mm Petri dishes. Can be stored <30 days at 4 °C.

29. Concentrated LB medium: 5 g of yeast extract, 10 g of tryptone, and 5 g of NaCl. Dissolve in 700 mL of H_2O. Adjust pH to 7.0. Sterilized by autoclave or by 0.2 μm filtration. Can be stored <6 months at room temperature.
30. Concentrated M9 minimal medium: 42 mmol Na_2HPO_4, 24 mmol KH_2PO_4, 8.5 mmol NaCl, 19 mmol NH_4Cl, 2 mmol $MgCl_2$, 100 μmol $CaCl_2$, and 2 g of D-glucose. Dissolve in 700 mL of H_2O. Adjust pH to 7.0. Sterilized by 0.2 μm filtration. Can be stored <6 months at room temperature.
31. Concentrated synthetic cystic fibrosis medium (SCFM): 1.3 mmol NaH_2PO_4, 1.25 mmol Na_2HPO_4, 348 μmol KNO_3, 271 μmol K_2SO_4, 2.28 mmol NH_4Cl, 14.9 mmol KCl, 51.8 mmol NaCl, 10 mmol MOPS, 827 μmol L-aspartate, 1.072 mmol L-threonine, 1.446 mmol L-serine, 1.549 mmol L-glutamate, 1.661 mmol L-proline, 1.203 mmol L-glycine, 1.78 mmol L-alanine, 160 μmol L-cysteine HCl, 1.117 mmol L-valine, 633 μmol L-methionine, 1.12 mmol L-isoleucine, 1.609 mmol L-leucine, 802 μmol L-tyrosine, 530 μmol L-phenylalanine, 676 μmol L-ornithine HCl, 2.128 mmol L-lysine HCl, 519 μmol L-histidine, 13 μmol L-tryptophan, 306 μmol L-arginine HCl, 1.754 mmol $CaCl_2$, 606 μmol $MgCl_2{\cdot}7H_2O$, and 3.6 μmol $FeSO_4{\cdot}7H_2O$. Dissolve in 700 mL of H_2O. Adjust pH to 6.8. Sterilize by 0.2 μm filtration. Can be stored <30 days at 4 °C.
32. Fetal bovine serum medium: Fetal bovine serum, sterilized by filtration from purchased source (*see* **Note 5**).
33. Urine medium: Pool several sources and sterilize by 0.2 μm filtration. Store at −80 °C until ready for use (*see* **Note 6**).

2.2 Illumina Library Preparation and Sequencing

1. DNeasy genomic DNA extraction kit with provided buffer reagents: ATL, AL, Proteinase K, RNase A, AW1, AW2, and AE (Qiagen).
2. Nanodrop spectrophotometer or other DNA quantification method.
3. DNA shearing sonicator instrument (Covaris E303).
4. Sonicator sample tubes (Covaris AFA slip cap).
5. NEBNext End Repair Module (NEB).
6. NEBNext dA-Tailing Module (NEB).
7. NEBNext Quick Ligation Module (NEB).
8. NEBNext Q5 HotStart HiFi PCR Master Mix (NEB).
9. Thermocycler.

10. PCR tubes or plates.
11. Magnetic AMPure XP Beads (Beckman Coulter).
12. Wash buffer (80% ethanol v/v in water, freshly prepared).
13. Elution buffer (10 mM Tris–HCl, pH 8 in ultrapure water).
14. Magnetic microcentrifuge tube stand.
15. Magnetic PCR tube stand.
16. Top adapter: TTCCCTACACGACGCTCTTCCGATCT, synthesize with a phosphorothioate bond at 3′ end CT; dissolve to 100 μM in elution buffer.
17. Bottom adapter: GATCGGAAGAGCGTCGTGTAGGGAA, 5′ phosphorylated; dissolve to 100 μM in elution buffer.
18. Barcode PCR Primer Forward: CAAGCAGAAGACGGCATACGAGAT -**INDEX** - AGACCGGGGACTTATCATCCAACCTGT, HPLC purified; dissolve to 10 μM in elution buffer (*see* **Note 7**).
19. Barcode PCR Primer Reverse: AATGATACGGCGACCACCGAGATCTACACTCTTTCCCTACACGACGCTCTTCCGATCT, HPLC purified; dissolve to 10 μM in elution buffer.
20. Amplification primer Forward: AATGATACGGCGACCACCGA; dissolve to 10 μM in elution buffer.
21. Amplification primer Reverse: CAAGCAGAAGACGGCATACGA; dissolve to 10 μM in elution buffer.
22. TapeStation electrophoresis detection instrument (Agilent).
23. D5000 high-sensitivity tape (Agilent).
24. Illumina sequencing platform.
25. Custom Index Read Primer: TAACAGGTTGGATGATAAGTCCCCGGTCT, dissolve to 100 μM in elution buffer.
26. Custom Read 2 Primer: AGACCGGGGACTTATCATCCAACCTGTTA, dissolve to 100 μM in elution buffer.

2.3 Analysis Requirements

1. A UNIX-based computer or server.
2. Genomic sequences (.fasta file) and gene annotations (.gff file).
3. Microsoft Excel or similar.
4. Text editor such as BBEdit.
5. Python.
6. Bowtie (http://bowtie-bio.sourceforge.net/index.shtml).
7. R (https://cran.r-project.org).
8. Rstudio (https://rstudio.com).
9. *FiTnEss* and genome processing scripts (https://github.com/broadinstitute/FiTnEss).

3 Methods

3.1 Generating Transposon Mutant Libraries

The following protocol is intended to create transposon libraries for 1–2 *P. aeruginosa* strains, with enough independent mutants to select on 5 different media. For each strain and medium, at least 2 independent transposon mutant libraries should be created.

3.1.1 Day 1

1. Prepare mating and counting agar plates.
2. Inoculate transposase donor strain into 2 culture tubes containing 5 mL of LB medium supplemented with 100 μg/mL carbenicillin. Shake 250 RPM at 37 °C for 16 h.
3. Inoculate transposon donor strain into 2 culture tubes containing 5 mL of LB medium supplemented with 15 μg/mL gentamicin. Shake 250 RPM at 37 °C for 16 h.
4. Inoculate *P. aeruginosa* recipient into a 250 mL Erlenmeyer flask containing 15 mL of LB medium. Shake 250 RPM at 37 °C for 16 h.

3.1.2 Day 2

1. Subculture transposase donor strain (1:100) by adding 2 × 5 mL cultures to 1 L of LB medium supplemented with 100 μg/mL carbenicillin in a 4 L Erlenmeyer or Fernbach flask. Shake 250 RPM at 37 °C for 3.5 h.
2. Subculture transposon donor strain (1:100) by adding 2 × 5 mL cultures to 1 L of LB medium supplemented with 15 μg/mL gentamicin in a 4 L Erlenmeyer or Fernbach flask. Shake 250 RPM at 37 °C for 3.5 h.
3. Subculture *P. aeruginosa* recipient (1:4) by adding 60 mL of LB to the culture. Shake 250 RPM at 42 °C for 4 h.
4. After 3.5 h of donor strain growth, measure OD to ensure OD600 nm ~ 0.4–0.6.
5. Centrifuge each donor strain at 5000 × *g* for 10 min at 20 °C. Decant supernatants.
6. Resuspend pellets in 1 mL of LB.
7. For each donor strain, measure OD at 600 nm of a 1:200 dilution (5 μL diluted in 995 μL of LB).
8. Measure OD at 600 nm of a 1:2 dilution of the recipient strain.
9. Mix 1.2×10^{11} CFU transposon donor, 1.2×10^{11} CFU transposase donor, and 6×10^{10} CFU *P. aeruginosa* recipient. Use the formula: Volume = CFU/(OD600 nm × dilution factor × 10^9).
10. Spin in a 50 mL falcon tube at 5000 × *g* for 10 min at 20 °C and decant supernatant.
11. Resuspend pellet in 1 mL of LB. Bring volume up to 2.8 mL.

12. To ~4 dry mating plates, pipette cell mating mixture in 30 μL spots (*see* **Note 8**).
13. Allow spots to soak in on bench for 5 min and then incubate at 37 °C for 90 min with the plate upright (*see* **Note 9**).
14. Scrape cell mating mixture into 6 mL of PBS (assume 0.5 mL cell volume), add 6.5 mL of 80% glycerol, and mix well with gentle vortexing.
15. Aliquot 1.25 mL of cell mating mixture to 10 Eppendorf tubes and 100 μL to one Eppendorf tube used for counting mating efficiency.
16. Flash freeze all aliquots in a dry ice ethanol bath or liquid nitrogen and store at −80 °C.
17. Thaw the 100 μL counting tube. Create a 1:10 dilution of the counting tube by adding 900 μL of PBS. Sequentially create 1:100, 1:1000, and 1:10000 serial dilutions in PBS by using 100 μL of each previous dilution.
18. Plate 100 μL of each dilution to a counting plate and incubate 37 °C overnight.

3.1.3 Day 3

1. Count colonies on one of the counting plates from a dilution that has ~30–300 colonies.
2. Calculate the successful transposon mutants per μL of the frozen mating mixture glycerol stock using the formula: mutants/μL = (# colonies × dilution factor)/volume plated μL.
3. Prepare concentrated media.
4. Thaw FBS and urine media stored at −80 °C.
5. Prepare selection agar plates as follows: warm 700 mL of FBS, urine, and concentrated LB, M9 minimal, and SCFM media to 55 °C for 20 min; slowly melt 5% agar using a microwave and add 300 mL of melted agar to each prewarmed medium (1.5% final concentration); add 2 mL of gentamycin and 0.5 mL of irgasan for 30 μg/mL and 5 μg/mL final concentrations, respectively; mix well by swirling to prevent bubbles and return to water bath for 5 min to allow media to settle; sterilely pour 250 mL of each agar medium into a 245 mm square bioassay dish and allow to solidify on the bench.

3.1.4 Day 4

1. Dry selection plates in a laminar flow hood for a total of 1 h, rotating plates after 30 min to ensure even drying.
2. Thaw the required volume of mating pools frozen in glycerol for 500,000 transposon mutants.
3. Transfer thawed aliquots to 50 mL conical centrifuge tube and increase volume to 40 mL with PBS.

4. Centrifuge 5000 × *g* for 10 min at 20 °C. Decant supernatant.
5. Resuspend pellet in 2 mL of PBS.
6. Plate total volume to selection plates and incubate at 37 °C for 24–48 h (*see* **Note 10**).

3.1.5 Days 5–6

1. After incubation of each selected transposon mutant library, scrape cells into a 50 mL centrifuge tube containing 5 mL of PBS and resuspend the cells.
2. Measure the OD600 nm (dilution is likely required) and determine the volume required for 2×10^9 CFU using the formula: volume $= 2 \times 10^9/(\text{OD600 nm} \times \text{dilution factor} \times 10^9)$. This is the recommended CFU for the Qiagen DNeasy purification column.
3. Aliquot this volume to 4 microcentrifuge tubes and collect cells by centrifugation at 5000 × *g* for 10 min.
4. Store cell pellets at −80 °C until DNA extraction step.
5. To remaining mutant library, add and equal volume of 80% glycerol, flash freeze, and store at −80 °C for future use, if desired.

3.2 Preparing DNA Libraries for Illumina Sequencing

1. For each sample, thaw a single pellet and isolate genomic DNA using the DNeasy kit as follows (*see* **Note 11**).
   (a) To each pellet of 2×10^9 CFU, add 180 μL of ATL and resuspend;
   (b) Add 20 μL of Proteinase K and mix by vortexing;
   (c) Incubate 1 h at 56 °C with the occasional vortex;
   (d) Add 4 μL of RNase A, vortexed and incubated on the benchtop for 2 min;
   (e) Add 200 μL of Buffer AL and vortex;
   (f) Add 200 μL of 100% ethanol and vortex;
   (g) Add mixture to column and spin at 6000 × *g* for 1 min;
   (h) Transfer column to new collection tube, add 500 μL of Buffer AW1, and spin at 6000 × *g* for 1 min;
   (i) Transfer column to new collection tube, add 500 μL of Buffer AW2, and spin at 16,000 × *g* for 3 min;
   (j) Carefully transfer column to low-binding 1.5 mL microcentrifuge tube, add 150 μL of elution buffer AE, rest 1 min, and spin 10,000 × *g* for 1 min.
2. Quantify purified genomic DNA using a nanodrop spectrophotometer.
3. Dilute purified genomic DNA to 38.5 ng/μL in a total volume of 130 μL (5 μg) using AE buffer. If concentration is lower than 38.5 ng/μL, isolate genomic DNA from an additional pellet.

4. Transfer 130 μL of diluted DNA to a Covaris AFA slip cap tube and shear to a final size of 1.5 kb using a Covaris E303 sonicator using the following settings (*see* **Note 12**):
   (a) Peak Incident Power (W): 140.
   (b) Duty Factor: 2%.
   (c) Cycles/burst: 200.
   (d) Treatment time: 15 s per sample.
   (e) Temperature (°C): 5–10.
   (f) Water Level: 6.
   (g) Intensifier: Yes.
5. Retain 1 μL for Tapestation.
6. End repair fragmented DNA using the NEBNext End Repair Module as follows (*see* **Note 13**):
   (a) Mix the following components in a sterile microfuge tube:
   - 127.5 μL of sheared DNA.
   - 15 μL of NEBNext End Repair Reaction Buffer (10×).
   - 7.5 μL of NEBNext End Repair Enzyme Mix.
   - 150 μL Total volume.

   (b) Split into two 75 μL aliquots in PCR tubes and incubate in a thermal cycler for 30 min at 20 °C.
7. Cleanup using AMPure XP Beads as follows:
   (a) Vortex AMPure XP beads to resuspend;
   (b) Add 240 μL (1.6×) of resuspended AMPure XP Beads to a microcentrifuge tube. Add 150 μL of sample and mix by pipetting up and down ten times;
   (c) Incubate for 5 min on the benchtop;
   (d) Place tube on a magnetic stand for 5 min and carefully remove and discard the supernatant.
   (e) Add >500 μL of wash buffer to the tube while in the magnetic stand, ensuring beads are covered, and incubate on the benchtop for 30 s.
   (f) Carefully remove and discard the wash buffer and repeat wash step.
   (g) Air dry beads for 5 min while the tube is on the magnetic stand with the lid open. Do not over dry the beads.
   (h) Remove the tube from the magnet. Elute the DNA target from the beads by adding 45 μL of elution buffer and mix beads well by pipetting up and down and incubate for 1 min at room temperature.
   (i) Put the tube in the magnetic stand until the solution is clear (2–5 min). Without disturbing the bead pellet,

carefully transfer 43 μL of the supernatant to a fresh, sterile PCR tube. Remove 1 μL and retain for Tapestation.

8. dA-tail end repaired DNA using the NEBNext dA-Tailing Module as follows:
   (a) Mix the following components in a sterile PCR tube from above:
   - 42 μL End Repaired, Blunt DNA.
   - 5 μL NEBNext dA-Tailing Reaction Buffer (10×).
   - 3 μL Klenow Fragment (3′→5′ exo$^-$).
   - 50 μL total volume.

   (b) Incubate in a thermal cycler for 30 min at 37 °C.
9. Cleanup using AMPure XP Beads as follows:
   (a) Vortex AMPure XP beads to resuspend;
   (b) Add 90 μL (1.8×) of resuspended AMPure XP Beads to the reaction and mix by pipetting up and down ten times;
   (c) Incubate for 5 min on the benchtop;
   (d) Place PCR tube/plate on a magnetic PCR stand for 5 min and carefully remove and discard the supernatant;
   (e) Add 200 μL of wash buffer to the tube while in the magnetic stand, ensuring beads are covered, and incubate on the benchtop for 30 s.
   (f) Carefully remove and discard the wash buffer and repeat wash step.
   (g) Air dry beads for 5 min while the tube is on the magnetic stand with the lid open. Do not over dry the beads.
   (h) Remove the tube from the magnet. Elute the DNA target from the beads by adding 28 μL of elution buffer and mix beads well by pipetting up and down and incubate for 1 min at room temperature.
   (i) Put the tube in the magnetic stand until the solution is clear (2–5 min). Without disturbing the bead pellet, carefully transfer 26 μL of the supernatant to a fresh, sterile PCR tube. Remove 1 μL and retain for Tapestation.
10. Create custom Illumina adapter for ligation as follows:
   (a) In PCR tube, add:
   - 10 μL 100 μM Top adapter.
   - 10 μL 100 μM Bottom adapter.

   (b) In a thermocycler, heat at 95 °C for 2 min and decrease temperature 1 °C every 10 s to 20 °C (*see* **Note 14**);
   (c) Add 80 μL elution buffer;

(d) Aliquot 10 μL single-use volumes in PCR tubes (10 μM) and freeze at −80 °C;

11. Thaw a single custom Illumina adapter tube for each sample on ice.
12. Ligate custom Illumina adapter to dA-tailed DNA as follows:
    (a) Mix the following components in a sterile microfuge tube:
    - 25 μL dA-Tailed DNA.
    - 10 μL Quick Ligation Reaction Buffer (5×).
    - 10 μL Custom Illumina Adapter (10 μM).
    - 5 μL Quick T4 DNA Ligase.
    - 50 μL Total volume.

    (b) Incubate in a thermal cycler for 15 min at 20 °C.
13. Cleanup using AMPure XP Beads as follows:
    (a) Vortex AMPure XP beads to resuspend;
    (b) Add 60 μL (1.2×) of resuspended AMPure XP Beads to the reaction and mix by pipetting up and down ten times;
    (c) Incubate for 5 min on the benchtop;
    (d) Place PCR tube/plate on a magnetic PCR stand for 5 min and carefully remove and discard the supernatant;
    (e) Add 200 μL of wash buffer to the tube while in the magnetic stand, ensuring beads are covered, and incubate on the benchtop for 30 s;
    (f) Carefully remove and discard the wash buffer and repeat wash step;
    (g) Air dry beads for 5 min while the tube is on the magnetic stand with the lid open. Do not over dry the beads;
    (h) Remove the tube from the magnet. Elute the DNA target from the beads by adding 33 μL of elution buffer and mix beads well by pipetting up and down and incubate for 1 min at room temperature;
    (i) Put the tube in the magnetic stand until the solution is clear (2–5 min). Without disturbing the bead pellet, carefully transfer 31 μL of the supernatant to a fresh, sterile PCR tube. Remove 1 μL and retain for Tapestation.
14. PCR of adapter-transposon region using custom barcoded primers as follows:
    (a) Mix the following components in sterile strip tubes:
    - 30 μL adapter-ligated DNA Fragments.
    - 10 μL Barcode PCR Primer Forward.
    - 10 μL Barcode PCR Primer Reverse.
    - 50 μL NEBNext Q5 Hot Start HiFi PCR Master Mix.
    - 100 μL Total volume.

- Split 25 μL volumes into 4 PCR tubes/wells of plate for each sample.

(b) PCR cycling conditions (4 cycles):
- Initial Denaturation: 98 °C, 30 s.
- Denaturation: 98 °C, 10 s.
- Annealing & Extension: 65 °C, 75 s.
- Final Extension: 65 °C, 5 min.
- Hold: 4 °C, ∞.

15. Pool all 100 μL PCR reactions together into an appropriately sized tube (one pool per sequencing library; *see* **Note 15**).
16. Size-selected cleanup of barcoded PCR product using Ampure XP Beads as follows:
    (a) Vortex AMPure XP beads to resuspend;
    (b) Add 50 μL (0.5×) of resuspended AMPure XP Beads to the reaction for each 100 μL PCR reaction volume (sample). Mix by pipetting up and down ten times;
    (c) Incubate for 5 min on the benchtop;
    (d) Split sample into multiple microcentrifuge tubes for larger pools, if necessary;
    (e) Place tube(s) on a magnetic PCR stand for 5 min and carefully transfer the supernatant(s) to 70 μL per # samples (1.2× final) of fresh AMPureXP Beads in a new tube and mix by pipetting up and down ten times;
    (f) Incubate for 5 min on the benchtop;
    (g) Split sample into multiple microcentrifuge tubes for larger pools, if necessary;
    (h) Place tube(s) on a magnetic stand for 5 min and carefully remove and discard the supernatant;
    (i) Add 800 μL of wash buffer to the tube while in the magnetic stand, ensuring beads are covered, and incubate on the benchtop for 30 s;
    (j) Carefully remove and discard the wash buffer and repeat wash step;
    (k) Air dry beads for 5 min while the tube is on the magnetic stand with the lid open. Do not over dry the beads.
    (l) Remove the tube from the magnet. Elute the size-selected DNA by adding 23 μL of elution buffer per # samples. If pooling >6 samples, elute with 160 μL.
    (m) Mix well by pipetting up and down and incubate for 1 min at room temperature.

(n) Place the tube in the magnetic stand until the solution is clear (2–5 min). Transfer 21 μL per # samples of the supernatant (or 151 μL for >6 samples) to a new tube;

(o) Remove 1–2 μL and retain for Tapestation.

17. Amplify the size selected barcoded product by PCR amplification as follows:

(a) Mix the following components in a sterile PCR tube/well and adjust volumes depending on size of library:

- 20 μL Purified size-selected barcoded PCR product.
- 2.5 μL Amplification primer Forward.
- 2.5 μL Amplification primer Reverse.
- 25 μL NEBNext Q5 Hot Start HiFi PCR Master Mix.
- 50 μL Total volume.
- Aliquot PCR reaction into 25 μL per PCR tube/well.

(b) PCR cycling conditions (**14 cycles**):

- Initial Denaturation: 98 °C, 30 s.
- Denaturation: 98 °C, 10 s.
- Annealing: 65 °C, 20 s.
- Extension: 72 °C, 30 s.
- Final Extension: 72 °C, 5 min.
- Hold: 4 °C, ∞.

18. Pool all PCR reactions.

19. Cleanup of Illumina library using AMPure XP Beads as follows (the volumes are for a 50 μL PCR reaction):

(a) The following volume values are for 1 sample (50 μL). Increase bead and elution volumes as required;

(b) Vortex AMPure XP Beads to resuspend;

(c) Add 27.5 μL (0.55×) of resuspended AMPure XP Beads to the PCR reaction. Mix well by pipetting up and down at least ten times;

(d) Incubate for 5 min at room temperature;

(e) Place PCR tube/PCR plate on an appropriate magnetic stand for 5 min and carefully remove the supernatant and place in new PCR tube/well containing 32.5 μL (1.2× final) of fresh AMPure XP Beads. Mix well by pipetting up and down at least ten times;

(f) Incubate for 5 min at room temperature;

(g) Place PCR tube/PCR plate on an appropriate magnetic stand for 5 min and carefully remove and discard the supernatant;

(h) Add 200 μL of wash buffer to the tube/PCR plate while in the magnetic stand. Incubate at room temperature for 30 s;

(i) Carefully remove and discard the wash buffer and repeat wash step;

(j) Air dry beads for 5 min while the tube is on the magnetic stand with the lid open. Do not over dry the beads.

(k) Remove the tube/plate from the magnet. Elute the purified library from the beads by adding 25 μL of elution buffer;

(l) Mix well by pipetting up and down and incubate for 1 min at room temperature;

(m) Place the tube/PCR plate in the magnetic stand until the solution is clear (2–5 min). Transfer 22 μL of supernatant to a new tube/well;

(n) Retain 2 μL for the TapeStation. Use 1 μL as undiluted, and 1 μL to create 1:10 and 1:100 dilutions in elution buffer for the Tapestation.

20. Run samples on the TapeStation system using a D5000 high-sensitivity DNA tape.
21. Calculate the concentration of the purified library using the TapeStation software, measuring the area under the curve of the library between the 400 and 600 bp region (*see* **Note 16**).
22. Sequence library using an Illumina Sequencer such as MiSeq (for 2 pooled samples) or NextSeq (for up to 18 pooled samples) (*see* **Note 15**). When submitting samples, be sure to note and provide the custom index read primer (6 bp reads for provided indices) and the custom read 2 primer (50 bp reads). The standard Illumina read 1 primer is used for 25 bp reads. Provide the index sequences and request demultiplexed . fastq files.

3.3 Data Analysis and Identifying Essential and Nonessential Genes

The following protocol prepares a genome of interest for mapping Tn-Seq reads and the removal of insertion sites that are indistinguishable and may create bias in classification. Second, the protocol outlines the procedure for using our statistical model *FiTnEss* to classify essential vs nonessential genes, resulting in a p-value output and classification.

3.3.1 Preparing your Computer and Files for Analysis

1. Download and save/install the necessary software from the materials list.
2. On your computer, create a main folder for this work.
3. This main folder should contain 2 files:

   (a) execute_TAsite_preprocessing.sh,
   (b) execute_mapping_tallying.sh,
4. This main folder should contain 4 subfolders:
   (a) genome_info (empty),
   (b) scripts (containing 50bp_TA_sequence_extractor.py, TAFinder.py, homology_finder.py, and nonpermissive_-TAFinder.py, tally.py),
   (c) TAsite_info (empty).
   (d) sample_data (empty),
5. To the "genome_info" folder, add "your_genome_name.fasta" and "your_genome_name_gff.txt" files for the genome of interest.
6. Create "your_genome_name_reverse_complement.fasta" and add to the "genome_info" folder (*see* **Note 17**).
7. Create a "genelist.txt" file using the .gff file and a program such as Microsoft Excel. Save to the genome_info folder. This file should be in tab-delimited text format, where each row is a gene extracted from the annotations file for your genome. Column 1 is the chromosome #, column 2 is Gene_ID, columns 3–5 are blank, column 6 is the chromosomal gene start locus, and column 7 is the chromosomal gene stop locus.
8. To the "sample_data" folder, add your "sample_name.fastq" files obtained from Illumina sequencing (Read 2).
9. To the "sample_data" folder, add a copy of the "your_genome_name.fasta" file.
10. Using a text editor, open "execute_TAsite_preprocessing.sh" file and modify to reflect "your_genome_name.fasta" as indicated on line #2 and save.
11. Using a text editor, open the "tally.py" file and modify to reflect the length of your genome as indicated on line #3 and save.
12. Using a text editor, open the "execute_mapping_tallying.sh" file and modify to reflect your_genome_name, your_sample_-names, as indicated, and save.

3.3.2 Identifying and Filtering TA Insertion Sites from the Genomic Sequence

1. On your computer, open the command line ('Terminal' on a Mac).
2. Navigate to the main folder created for this work.
3. Run the "execute_TAsite_preprocessing.sh" file by typing:

```
sh execute_TAsite_preprocessing.sh
```

4. Output files are automatically written to the folder "TAsite_info".

3.3.3 Mapping Sample Data to the Genome and Tallying Mapped Reads

1. From the command line, navigate to the main folder created for this work.

   Run the "execute_mapping_tallying.sh" file by typing:

```
sh execute_mapping_tallying.sh
```

2. Mapped .sam files and tally output files are automatically written to the sample_data folder.

3.3.4 Using FiTnEss *to Identify Essential and Nonessential Genes*

1. On your computer, open RStudio to run *FiTnEss*.
2. For each sample, run *FiTnEss* by typing and modifying your_-sample_name and your_genome_name where required:

```
FiTnEss_Run(""your_sample_name",
 "/your/folder/location/sample_data/your_sample_name_tally.
txt",
 "/your/folder/location/TAsite_info/nonpermissive_TA_sites.
txt",
 "/your/folder/location/TAsite_info/homologous_TA_sites.txt",
 "/your/folder/location/genome_info/your_genome_name_gff.
txt",
 "/your/folder/location//sample_data/your_sample_name_re-
sults.xlsx",
 repeat_time = 3)
```

3. Open the "your_sample_name_results.xlsx" file in the sample_data folder. Each gene has an analysis output with an adjusted p-value and essentiality classification (*see* **Note 18**).

4 Notes

1. Himar1 transposons integrate at the dinucleotide sequence "TA". Other himar1 transposons can be used for this protocol, but the primer sequences used may need to be modified depending on the sequence of the transposon.
2. The C9 transposase may be used in other bacterial species, but the upstream region driving expression may require modification for the species of interest.
3. Genetically diverse strains should be used for comparison across a species. Not all strains allow for efficient transposition, and thus, small-scale screening of efficiency may be necessary when using other strains.

4. Low DNA binding tubes should be used for DNA manipulation steps.
5. When comparing strains, ensure that the same lot # of FBS is used due to lot variability. Contrarily, when comparing replicates, the use of a different lot # may be informative for identifying genes that are consistently essential.
6. Create a large pool and aliquot into single use bottles for freezing.
7. Indices can be user defined, but are usually 6–8 bp. A set of 18 that have been successfully used are ATCACG, CGATGT, TTAGGC, TGACCA, ACAGTG, GCCAAT, CAGATC, ACTTGA, GATCAG, TAGCTT, GGCTAC, CTTGTA, AGTCAA, AGTTCC, ATGTCA, CCGTCC, GTCCGC, and GTGAAA.
8. For quicker plating, use a multichannel pipette with tips placed every second barrel to plates that have been dried in a laminar hood for 1 h. Additionally, rectangular style plates may be used to reduce the number of plates to 2.
9. Allowing the mating to occur for too long a period may result in sibling mutants.
10. Cells on some minimal media such as M9 minimal and Urine grow slowly, and therefore, a 48-h incubation time may be necessary to obtain enough cells. Placing plates in a large Tupperware container prevent the agar from over drying and cracking over these long incubation periods.
11. Other isolation kits may be used from suppliers such as NEB, Zymo research, etc.
12. Other shearing methods or machines may be used to achieve the target size of 1.5 kb.
13. Alternatively, DNA Polymerase I Large (Klenow) Fragment and T4 polynucleotide kinase can be used individually and sequentially to blunt and 5′ phosphorylate DNA, respectively.
14. Alternatively, primers can be mixed and heated to 95 °C and allow to cool slowly on the bench for 10 min before freezing.
15. The desired number of reads per sample is empirically 10,000,000. For an MiSeq sequencing technology, 2 samples can be sequenced together in a pool. Significantly more samples can be pooled for sequencing on a NextSeq, HiSeq, or NovaSeq. We have successfully pooled 18 samples on a NextSeq and 2 samples for an MiSeq.
16. A small amount of sheared template may still be present, but this does not affect sequencing quality.

17. The reverse complement .fasta file can be generated with any free online reverse complement software such as http://reverse-complement.com.
18. Two *p*-values are presented. P-fwer is based on a family-wise error rate adjustment and is regarded as maximally stringent classification. P-fdr is based on a false discovery rate adjustment that is slightly less stringent.

Acknowledgements

We would like to thank Nicole Chumbler for critical reading of this manuscript. This work was supported by a generous gift from Anita and Josh Bekenstein, NIH Grant R33AI098705 (to D.T.H.), and a Cystic Fibrosis Canada Fellowship (to B.E.P.).

References

1. Chao MC, Abel S, Davis BM et al (2016) The design and analysis of transposon insertion sequencing experiments. Nat Rev Microbiol 14(2):119–128. https://doi.org/10.1038/nrmicro.2015.7
2. Gawronski JD, Wong SM, Giannoukos G et al (2009) Tracking insertion mutants within libraries by deep sequencing and a genome-wide screen for Haemophilus genes required in the lung. Proc Natl Acad Sci U S A 106 (38):16422–16427. https://doi.org/10.1073/pnas.0906627106
3. Goodman AL, McNulty NP, Zhao Y et al (2009) Identifying genetic determinants needed to establish a human gut symbiont in its habitat. Cell Host Microbe 6(3):279–289. https://doi.org/10.1016/j.chom.2009.08.003
4. Langridge GC, Phan MD, Turner DJ et al (2009) Simultaneous assay of every salmonella Typhi gene using one million transposon mutants. Genome Res 19(12):2308–2316. https://doi.org/10.1101/gr.097097.109
5. van Opijnen T, Bodi KL, Camilli A (2009) Tn-seq: high-throughput parallel sequencing for fitness and genetic interaction studies in microorganisms. Nat Methods 6 (10):767–772. https://doi.org/10.1038/nmeth.1377
6. DeJesus MA, Ambadipudi C, Baker R et al (2015) TRANSIT--A software tool for Himar1 TnSeq analysis. PLoS Comput Biol 11(10):e1004401. https://doi.org/10.1371/journal.pcbi.1004401
7. Pritchard JR, Chao MC, Abel S et al (2014) ARTIST: high-resolution genome-wide assessment of fitness using transposon-insertion sequencing. PLoS Genet 10(11): e1004782. https://doi.org/10.1371/journal.pgen.1004782
8. Zomer A, Burghout P, Bootsma HJ et al (2012) ESSENTIALS: software for rapid analysis of high throughput transposon insertion sequencing data. PLoS One 7(8):e43012. https://doi.org/10.1371/journal.pone.0043012
9. DeJesus MA, Gerrick ER, Xu W et al (2017) Comprehensive essentiality analysis of the Mycobacterium tuberculosis genome via saturating transposon mutagenesis. MBio 8(1): e02133-16. https://doi.org/10.1128/mBio.02133-16
10. Goodall ECA, Robinson A, Johnston IG et al (2018) The essential genome of *Escherichia coli* K-12. MBio 9(1). https://doi.org/10.1128/mBio.02096-17
11. Lee SA, Gallagher LA, Thongdee M et al (2015) General and condition-specific essential functions of *Pseudomonas aeruginosa*. Proc Natl Acad Sci U S A 112(16):5189–5194. https://doi.org/10.1073/pnas.1422186112
12. Turner KH, Wessel AK, Palmer GC et al (2015) Essential genome of *Pseudomonas aeruginosa* in cystic fibrosis sputum. Proc Natl Acad Sci U S A 112(13):4110–4115. https://doi.org/10.1073/pnas.1419677112
13. Poulsen BE, Yang R, Clatworthy AE et al (2019) Defining the core essential genome of *Pseudomonas aeruginosa*. Proc Natl Acad Sci U S A 116(20):10072–10080. https://doi.org/10.1073/pnas.1900570116

Chapter 11

Analysis of *Escherichia coli* K1 Virulence Genes by Transposon-Directed Sequencing

Alex J. McCarthy and Peter W. Taylor

Abstract

Transposon-directed insertion site sequencing (TraDIS) combines random transposon mutagenesis and massively parallel sequencing to shed light on bacterial gene function on a genome-wide scale and in a high-throughput manner. The technique has proven to be successful in the determination of the fitness contribution of every gene under specific conditions both in vitro and in vivo. In this contribution, we describe the procedure used for the identification of *Escherichia coli* K1 genes essential for in vitro growth, survival in pooled human serum and gastrointestinal colonisation in a rodent model of neonatal invasive infection. TraDIS has broad application for systems-level analysis of a wide range of pathogenic, commensular and saprophytic bacteria.

Key words Transposon-directed insertion site sequencing, Essential genes, *Escherichia coli*, Systems microbiology, Bacterial pathogenesis, Bacterial fitness

1 Introduction

The development of massively parallel sequencing in combination with more established transposon mutagenesis techniques has enabled genome-wide enhanced genetic screening of large mutant pools and for the first time provides a high-throughput means of assigning function to multiply disrupted genes in a single experiment. A number of variations of transposon sequencing were described simultaneously and independently in 2009 and variously referred to as transposon insertion site sequencing (Tn-seq) [1], insertion sequencing (INSeq) [2], high-throughput insertion tracking by deep sequencing (HITS) [3] and transposon-directed insertion site sequencing (TraDIS; Fig. 1) [4]. All rely on the creation of a pool of insertion mutants in which every gene has been disrupted at multiple sites followed by determination of the sites of transposon insertion through sequencing of transposon junctions within chromosomal DNA before (input pool) and after (output pool) the application of a specific condition. The large

Ren Zhang (ed.), *Essential Genes and Genomes: Methods and Protocols*, Methods in Molecular Biology, vol. 2377,
https://doi.org/10.1007/978-1-0716-1720-5_11,

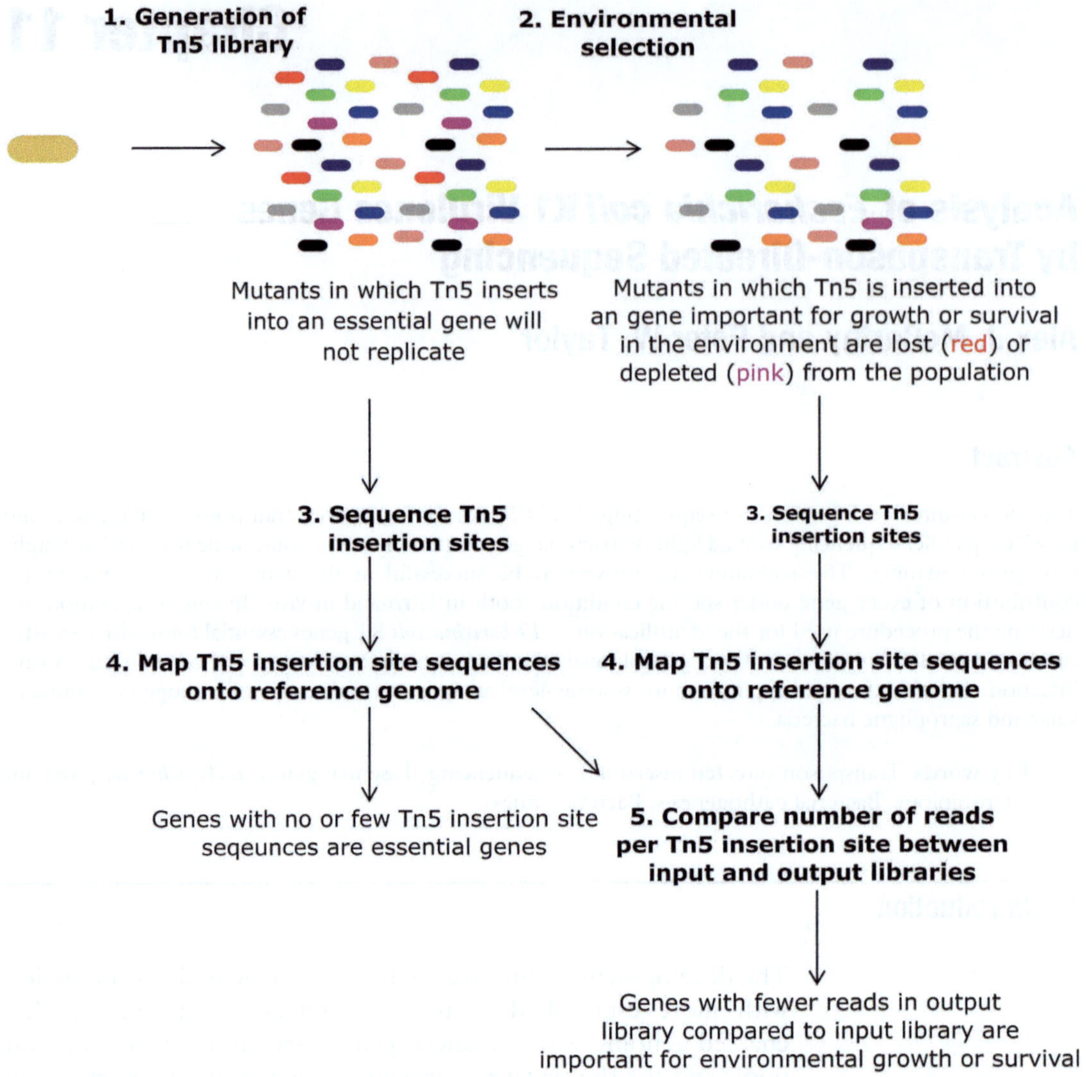

Fig. 1 Overview of the TraDIS procedure

majority of studies have employed either random mutagenesis using transposon Tn*5* or mariner transposon derivatives that target thymine-adenine dinucleotides [5]. Mutants carrying genes essential or required under a specific condition will be outcompeted by other mutants and will be under-represented in the output pool. Although initially used to examine the relationship between different in vitro growth conditions and gene essentiality on a genome-wide scale, transposon sequencing has provided insights into the role of key genes required for fitness of a wide range of bacterial pathogens during various stages of the infection process in laboratory animal models [5–7].

We have a longstanding interest in the capacity of *Escherichia coli* strains expressing the α-2,8-linked polysialic acid capsular K1 polysaccharide to cause potentially lethal sepsis and meningitis in new born children and to function as a benign but frequent component of the gut microbiota in non-neonatal infants and adults [8–10]. These bacteria are acquired by the neonate at birth and colonise the small intestine prior to invasion of the blood compartment and central nervous system. In order to identify genes essential for gastrointestinal colonisation and survival in blood or serum, prerequisites for experimental invasive infection in neonatal rat pups, we employed a TraDIS library and constituent sub-libraries of ~775,000 Tn*5* transposon mutants of the highly virulent *E. coli* K1 strain A192PP [11]. The processing of bacterial DNA is illustrated in Fig. 2. This study not only enabled the identification of a number of bacterial genes that had not previously been implicated in invasive *E. coli* K1 disease but also highlighted one of a number of confounding factors that must be borne in mind during analyses of gene essentiality [12]: the presence of a bottleneck during bacterial translocation from gut to blood and from the blood circulation to cerebrospinal compartment invalidated TraDIS analysis at these physical barriers by restricting bacterial population diversity, thereby eliminating genotypes from the translocated gene pool in a stochastic, rather than gene-dependent manner [11]. Here, we describe the generation of a TraDIS library and its use as a tool to determine gene essentiality for *E. coli* K1 pathogenesis; the procedures are sufficiently robust for analytical use to determine gene function for a broad range of bacteria in a wide variety of scenarios.

2 Materials

Unless otherwise stated, we use ultrapure water when referring to water. All buffers described below are filter-sterilized (≤0.45 μm) after preparation.

2.1 Preparation of Electrocompetent E. coli

1. *E. coli* strain cultured on an appropriate agar plate.
2. Sterile loop.
3. Sterile culture vessels.
4. Luria-Bertani (LB) broth.
5. Optical density metre for tubes.
6. 10% glycerol: Mix 100 mL of glycerol with 900 mL of dH_2O.
7. Sterile 1.5 mL microcentrifuge tubes.

2.2 Transformation of E. coli with EZ-Tn5 Transposome

1. Electrocompetent *E. coli*.
2. 2 mM electroporation cuvettes.
3. MicroPulser Electroporator or equivalent.

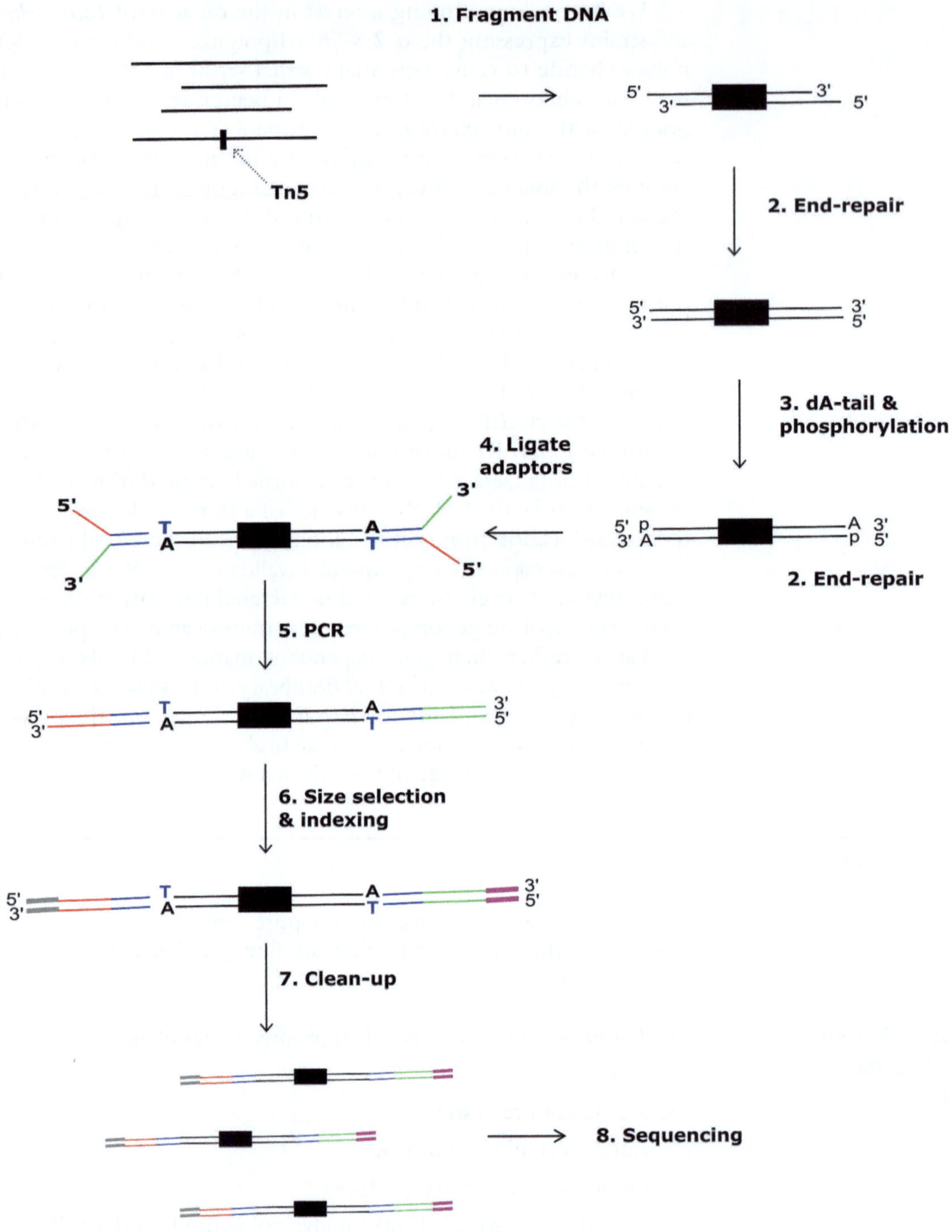

Fig. 2 Processing of bacterial DNA for TraDIS

4. Sterile 1.5 mL microcentrifuge tube.
5. EZ-Tn5™ <KAN-2>Tnp Transposome™ Kit (Epicentre).
6. SOC medium: 0.5% yeast extract, 2% tryptone, 10 mM NaCl, 2.5 mM KCl, 10 mM $MgCl_2$, 10 mM $MgSO_4$ and 20 mM glucose. Add glucose to autoclaved and cooled solution containing the remaining ingredients. Filter sterilise the final solution with a 0.2 μm filter.
7. Mueller-Hinton agar plates supplemented with 50 μg/mL kanamycin.
8. Sterile spreader.

2.3 Harvesting and Pooling *E. coli* Tn5 Mutants

1. Sterile 15 mL centrifuge tube.
2. 20% glycerol in PBS: Mix 200 mL of glycerol with 800 mL of PBS.
3. Cell scrapers.

2.4 Counting Colony-Forming Units in *E. coli* Tn5 Pool

1. PBS (1×).
2. Mueller-Hinton agar plates supplemented with 50 μg/mL kanamycin.

2.5 Extracting DNA from *E. coli* Tn5 Library

1. DNA extraction kit (e.g. PureElute Bacterial Genomic DNA preparation kit; EdgeBiosystems, UK).
2. Sterile dH_2O.
3. PCR purification kit (e.g. QIAgen MinElute PCR purification kit).
4. 1.2% agarose gel, containing ethidium bromide or alternative DNA dye.
5. NanoDrop SpectroPhotometer.

2.6 Fragmentation of DNA

1. Sterile dH_2O.
2. Sonication device (e.g. Bioruptor Sonication Device).
3. PCR purification kit (e.g. QIAgen MinElute PCR purification kit).
4. 1.2% agarose gel containing DNA visualisation dye (e.g. Ethidium bromide)
5. TBE buffer (1×): 89 mM Tris-borate, 2 mM EDTA at pH 8.3.
6. DNA ladder (e.g. 100 bp DNA ladder, NEB).
7. DNA gel loading dye (6×).
8. Imaging System (e.g. ChemiDoc imaging system).

2.7 Repairing the Ends of Fragmented DNA

1. Sterile dH_2O.
2. NEBNext DNA library prep reagent kit for Illumina.
3. PCR purification kit (e.g. QIAgen MinElute PCR purification kit).

2.8 Addition of A-Tail to Repaired DNA Fragments

1. Sterile dH_2O.
2. NEBNext DNA library prep reagent kit for Illumina.
3. PCR purification kit (e.g. QIAgen MinElute PCR purification kit).
4. Agilent High Sensitivity DNA analysis kit and Bioanalyzer (Agilent).

2.9 Ligation of Adapters to DNA Fragments

1. Sterile dH_2O.
2. NEBNext DNA library prep reagent kit for Illumina.
3. PCR purification kit (e.g. QIAgen MinElute PCR purification kit).
4. Agilent 1000 DNA kit and Bioanalyzer (Agilent).
5. 100 mM adapter Ind_Ad_T (desalted and with phosphorothioate): diluted in dH_2O
   ACACTCTTTCCCTACACGACGCTCTTCCGATC*T
6. 100 mM adapter Ind_Ad_B (desalted): diluted in dH_2O.
   p GATCGGAAGAGCGGTTCAGCAGGAATGCCGA GACCGATCTC.

2.10 PCR Amplification of Tn Insertion Sites

1. HotStarTaq polymerase mix.
2. 100 mM adapt FO (desalted): diluted in dH_2O.
   CGTCGGCAGCGTCAGATGTGTATAAGAGA-CAGCGGGGATCCTCTAGAGTCGACCTGC.
   100 mM adapt RO (desalted): diluted in dH_2O.
   GTCTCGTGGGCTCGGAGATGTGTATAAGAGACA GACACTCTTTCCCTACACGACGCTCTTCCGATC
3. Sterile dH_2O.
4. Sterile PCR tubes.

2.11 Purification of PCR Amplicons

1. UV transilluminator box.
2. 1% agarose gel containing DNA visualisation dye (e.g. Ethidium bromide).
3. TBE buffer (1×): 89 mM Tris-borate, 2 mM EDTA at pH 8.3.
4. DNA ladder (e.g. 100 bp DNA ladder, NEB).
5. DNA gel loading dye (6×): Dissolve 25 mg of bromophenol blue, 25 mg of xylene cyanol FF and 1.5 g of Ficoll 400 in 10 mL of Milli-Q dH_2O.
6. DNA gel extraction kit (e.g. QIAquick Gel purification kit).
7. Agilent High Sensitivity DNA analysis kit and Bioanalyzer (Agilent).

2.12 Sample Indexing

1. Kapa HotStart ReadyMix (2×) (Kapa Biosystems).
2. Nextera XT 24 indexes (Illumina).
3. Sterile PCR tubes.

2.13 Clean-up Size Indexed Fragments

1. AMPure XP beads (Beckman Coulter).
2. Sterile 96-well plate.
3. 96-well plate seal.
4. 96-well plate shaker.
5. 100% ethanol.
6. Sterile dH_2O.
7. Tris buffer: 10 mM Tris in water, pH 8.5.
8. Agilent High Sensitivity DNA analysis kit and Bioanalyzer (Agilent).

2.14 Sequencing on MiSeq

1. MiSeq Reagent Kit v2 (300-cycle).

3 Methods

3.1 Preparation of Electrocompetent *E. coli*

1. Pick a single colony of *E. coli* strain from an appropriate agar plate using a sterile loop.
2. Inoculate 10 mL of Luria-Bertani (LB) broth in a sterile culture vessel.
3. Incubate at overnight at 37 °C with shaking.
4. Add 0.5 mL of the *E. coli* culture to 50 mL of LB broth in a sterile culture vessel.
5. Incubate the *E. coli* culture at 37 °C.
6. At regular intervals, measure the absorbance at OD_{600}.
7. When absorbance at $OD_{600} = 0.4$, place the *E. coli* culture on ice for 20 min (*see* **Note 1**).
8. Centrifuge the *E. coli* culture for 15 min at 4 °C and 4000 × *g*.
9. Remove supernatant.
10. Resuspend the *E. coli* pellet in 25 mL of ice-cold sterile 10% glycerol.
11. Centrifuge the *E. coli* culture for 15 min at 4 °C and 4000 × *g*.
12. Remove supernatant.
13. Resuspend the *E. coli* pellet in 10 mL of ice-cold sterile 10% glycerol.
14. Centrifuge the *E. coli* culture for 15 min at 4 °C and 4000 × *g*.
15. Remove supernatant.

16. Resuspend the *E. coli* pellet in 1 mL of ice-cold sterile 10% glycerol.
17. Aliquot 200 μL of the electrocompetent *E. coli* into 5× sterile 1.5 mL microcentrifuge tubes (*see* **Note 2**).
18. Use immediately or store aliquots at −80 °C.

3.2 Transformation of E. coli with EZ-Tn5 Transposome

1. Place 1 vial of electrocompetent *E. coli* on ice for 15 min (*see* **Note 3**).
2. Place 2 mM electroporation cuvette and 1.5 mL microcentrifuge tube on ice for 15 min (*see* **Note 4**).
3. Pipette 60 μL of electrocompetent *E. coli* cells into a sterile 1.5 mL microcentrifuge tube and immediately place on ice.
4. Add 0.2 μL of EZ-Tn5 transposome to the electrocompetent *E. coli* cells.
5. Mix by gently pipetting up and down.
6. Transfer the mixture to a 2 mM electroporation cuvette.
7. Place electroporation cuvette into a MicroPulser Electroporator or equivalent and run electroporation using Ec2 (2.5 Kv, 25 μF, 200 Ω) settings.
8. Add 1 mL of SOC medium to the electroporation cuvette.
9. Mix by gently pipetting up and down.
10. Transfer the *E. coli* cells to a sterile 1.5 mL microcentrifuge tube.
11. Incubate at 37 °C for 60 min.
12. Pipette 100 μL of *E. coli* culture onto Mueller-Hinton agar plates supplemented with 50 μg/mL kanamycin.
13. Spread the culture across the plate using a sterile spreader.
14. Repeat **steps 11** and **12** until all *E. coli* culture is spread on a plate.
15. Incubate the plates at 37 °C overnight.

3.3 Harvesting and Pooling E. coli Tn5 Mutants

1. Pipette 3 mL of sterile 20% glycerol in PBS to a 15 mL centrifuge tube.
2. Count the number of kanamycin-resistant *E. coli* colonies on a plate.
3. Using a cell scraper, collect all *E. coli* colonies from a plate and resuspend in the 20% glycerol in PBS. Use a vortex to assist the resuspension of colonies in 20% glycerol in PBS.
4. Repeat **steps 2** and **3**, until 3000–5000 *E. coli* colonies are pooled into one tube.
5. Note the number of colonies in the minipool and store at −80 °C until required.

6. To form a large *E. coli* Tn*5* pool, defrost minipools of *E. coli* Tn*5* on ice and mix 100 μL of each minipool in a 15 mL centrifuge tube.

3.4 Counting Colony-Forming Units in E. coli Tn5 Pool

1. Defrost pool of *E. coli* Tn*5* on ice.
2. Mix 20 μL of *E. coli* Tn*5* pool with 180 μL of sterile PBS.
3. Repeat **step 2** for 10 serial dilutions.
4. Pipette 50 μL of each serial dilution onto Mueller-Hinton agar plates supplemented with 50 μg/mL kanamycin.
5. Incubate the plates at 37 °C overnight.
6. Count colony-forming units (CFU) per mL.

3.5 Extracting DNA from E. coli Tn5 Library

1. Extract DNA from an aliquot of the *E. coli* Tn5 library using a selection of standard DNA extraction kits.
2. Elute DNA into sterile dH_2O.
3. Measure the concentration of purified DNA using NanoDrop SpectroPhotometer.

3.6 Fragmentation of DNA

1. Dilute DNA to 10 μg/mL in sterile dH_2O, to a total volume of 200 μL.
2. Transfer DNA sample to sterile 1.5 mL microcentrifuge tube.
3. Load microcentrifuge tube into sonication device.
4. Add ice and water to the sonication device vessel to cover the microcentrifuge tubes but not the microcentrifuge tube holder.
5. Sonicate on medium density for 15 min, set at 30 s on/30 s off.
6. Repeat **steps 4** and **5**.
7. Purify fragmented DNA using a selection of standard PCR purification kits.
8. Elute the purified fragmented DNA into 80 μL of dH_2O.
9. Load a 1.2% agarose gel containing a DNA visualisation dye into a gel electrophoresis tank and cover with TBE buffer.
10. Add 6 μL of an appropriate DNA ladder into the first well of the agarose gel.
11. Mix 5 μL of purified fragmented DNA with 1 μL of DNA gel loading dye.
12. Load the sample into the well of a 1.2% agarose gel.
13. Run the gel at 110 V for 30 min.
14. Image the gel using imaging system.
15. If fragmentation has been successful, the DNA should appear as a smear with a mean band size of around 500 bp.

3.7 Repairing the Ends of Fragmented DNA

1. To repair DNA ends, use the NEBNext DNA library prep reagent kit for Illumina.
2. For each 75 μL sample of purified fragmented DNA, add:
   10.00 μL of phosphorylation buffer.
   4.00 μL of dNTP mix.
   1.50 μL of Klenlow DNA polymerase.
   5.00 μL of T4 PNK.
   5.00 μL of T4 DNA polymerase.
3. Gently pipette up and down to mix reagents.
4. Incubate the microcentrifuge tubes at room temperature for 30 min.
5. Purify DNA fragments using a selection of standard PCR purification kits.
6. Elute the purified fragmented DNA into 35 μL of sterile dH_2O.

3.8 Addition of A-Tail to Repaired DNA Fragments

1. To polyadenylate the ends of repaired DNA, use the NEBNext DNA library prep reagent kit for Illumina.
2. For each 35 μL sample of purified repaired DNA fragments, add:
   5.00 μL of NEB2 buffer.
   10.00 μL of dATP.
   3.00 μL of Klenow fragment (exo−).
3. Gently pipette up and down to mix reagents.
4. Incubate the microcentrifuge tube at 37 °C for 1 h (best performed in water bath).
5. Purify DNA fragments using a selection of standard PCR purification kits.
6. Elute the purified fragmented DNA into 22 μL of sterile dH_2O.
7. Measure concentration of DNA sample and mean fragment size using Agilent 1000 DNA kit and BioAnalyzer following standard protocol.

3.9 Ligation of Adapters to DNA Fragments

1. For each sample, mix the following reagents in a microcentrifuge tube:
   0.40 μL of Adapter Ind_Ad_T.
   0.40 μL of Adapter Ind_Ad_T.
   0.40 μL of 10X annealing buffer.
   2.80 μL of dH_2O.
2. Gently pipette up and down to mix reagents.

3. Incubate the microcentrifuge tube as follows in a PCR machine:
   98 °C for 2 min.
   70 °C for 5 min.
   55 °C for 5 min.
4. Incubate the microcentrifuge tube on ice for 15 min.
5. Calculate the amount of adapters required per sample using the following equation:

$$(\text{Adaptor excess} \times \text{DNA concentration} \times \text{sample DNA volume} \times 25)/650 \times \text{mean DNA size} = (10 \times ? \times 17.5\ \mu\text{L} \times 25)/650 \times ?$$

6. Calculate the volume of 2× buffer required per sample using the following equation:

$$(\text{sample DNA volume} + \text{adapter volume} + \text{ligase volume}) = (17.5\ \mu\text{L} + ? + 5\ \mu\text{L})$$

7. For each sample, mix the following in a sterile 1.5 mL microcentrifuge tube:
   17.50 μL of fragmented DNA sample.
   ? μL annealed adapters (volume calculated in **step 4**).
   ? μL of 2× ligation buffer (NEB kit; volume calculated in **step 5**).
   5.00 μL of ligase.
8. Gently pipette up and down to mix reagents.
9. Incubate microcentrifuge tubes at room temperature for 15 min.
10. Purify DNA fragments using a selection of standard PCR purification kits.
11. Elute the purified fragmented DNA into 35 μL of sterile dH_2O.

3.10 PCR Amplification of Tn Insertion Sites

1. In order to minimise amplification bias, PCR is best performed on small volumes of the ligated material in 10 parallel reactions.
2. Mix a PCR master mix in a microcentrifuge tube as follows:
   150.0 μL of HotStarTaq polymerase mix.
   3.60 μL of F primer (Adapt RO at concentration 100 μM).
   3.60 μL of R primer (Tn FO at concentration 100 μM).
   118.80 μL of sterile dH_2O.
3. Gently pipette up and down to mix reagents.
4. Aliquot 24.00 μL of reaction mix into 11 sterile PCR tubes.

5. Add 1.00 μL of the sample to each of 10 PCR tubes.
6. Add 1.00 μL of sterile dH_2O to control PCR tube.
7. Run the PCR on the following program:

   94 °C 15 min.

   94 °C 45 s.

   60 °C 45 s × 22 cycles.

   72 °C 45 s.

   72 °C 10 min.

   10 °C continuous.

3.11 Purification of PCR Amplicons

1. Prepare a deep 1% agarose gel using a wide well comb containing DNA visualisation dye (*see* **Note 5**).
2. Load the 1% agarose gel containing a DNA visualisation dye into a gel electrophoresis tank and cover with 1 × TBE buffer.
3. Add 6 μL of an appropriate DNA ladder into the first well of the agarose gel.
4. Mix each 25 μL of PCR product with 5 μL of DNA gel loading Dye 6×.
5. Load as much sample into the wells of a 1% agarose gel (*see* **Note 5**).
6. Repeat **step 5** until all PCR products are loaded into wells.
7. Run the gel at 70 V for 90 min.
8. Image the gel on a UV transilluminator.
9. Using a sterile scalpel blade, cut out the band of gel from each lane that comprises size 150 to 700 bp.
10. Purify DNA fragments using a selection of standard PCR gel extraction purification kits.
11. Elute the purified fragmented DNA into 25 μL of sterile dH_2O.
12. Measure concentration of DNA sample and mean fragment size using Agilent 1000 DNA kit and BioAnalyzer following standard protocol.

3.12 Sample Indexing

1. In a sterile PCR tube, mix the following:

   5.00 μL of DNA sample.

   5.00 μL of Nextera index primer 1 (N7XX).

   5.00 μL of Nextera index primer 2 (S5XX).

   25.00 μL of 2× Kapa Hifi HotStart Ready Mix.

   10.00 μL of sterile dH_2O.
2. Gently mix by pipetting up and down.

3. Incubate the PCR tube as follows:

   1 cycle: 95 °C for 3 min.

   8 cycles: 95 °C for 30 s, 55 °C for 30 s and 72 °C for 30 s.

   1 cycle: 72 °C for 5 min.

4. Store samples on ice.

3.13 Clean-up Size Indexed Fragments

1. Place AMPure XP beads at room temperature for 30 min.
2. Shake the AMPure XP beads for 1 min.
3. Add 56 μL of AMPure XP bead mixture to a well of sterile 96-well plate.
4. Add 50 μL of indexing PCR product.
5. Gently pipette up and down ten times to mix the sample with AMPure XP beads.
6. Incubate the 96-well plate at room temperature for 5 min.
7. Cover the plate with a 96-well plate seal.
8. Place the 96-well plate on a shaker for 2 min at 1800 rpm.
9. Incubate the 96-well plate at room temperature for 5 min.
10. Place the 96-well plate on a 96-well plate magnetic stand for 2 min.
11. Using a pipette, carefully remove supernatant from wells without disturbing the AMPure XP beads on the bottom of the well.
12. Mix 4 mL of 100% ethanol with 1 mL of sterile dH_2O in a 15 mL centrifuge tube.
13. Add 200 μL of fresh 80% ethanol to the well.
14. Incubate the 96-well plate at room temperature for 1 min.
15. Using a pipette, carefully remove supernatant from wells without disturbing the AMPure XP beads on the bottom of the well.
16. Repeat **steps 13–15**.
17. Incubate the 96-well plate at room temperature for 10 min.
18. Add 27.5 μL of 10 mM Tris pH 8.5 to the well.
19. Remove the 96-well plate from the 96-well plate magnetic stand.
20. Gently pipette up and down ten times to mix the Tris buffer with AMPure XP beads.
21. Incubate the 96-well plate at room temperature for 2 min.
22. Place the 96-well plate on the 96-well magnetic stand for 2 min.

23. Using a pipette, carefully transfer the supernatant from wells to a sterile 1.5 mL microcentrifuge tube without disturbing the AMPure XP beads on the bottom of the well.
24. Measure concentration of DNA sample and mean fragment size using Agilent 1000 DNA kit and BioAnalyzer following standard protocol.

3.14 Sequencing of Amplicons with MiSeq and Processing of Sequence Reads

1. Sequence amplicons using MiSeq platform following standard protocol.
2. Use bioinformatic tools to
   (a) perform quality assessment of sequence reads,
   (b) to identify reads containing the Tn5,
   (c) to trim Tn5 sequence off reads and identify Tn5 insertion site sequences,
   (d) To map Tn5 insertion site sequences onto reference genome.

4 Notes

1. *E. coli* should be kept chilled at all subsequent stages in order to maximise electro-competency. All reagents and plasticware should be chilled at 4 °C or on ice. Centrifugation steps should be performed at 4 °C.
2. It is beneficial to place the 1.5 mL microcentrifuge tubes on ice prior to aliquotting.
3. In our experience, there is enhanced transformation of electro-competent *E. coli* that are freshly prepared compared to those previously stored at −80 °C.
4. Electroporation cuvettes can also be stored at −20 °C.
5. We find that it is more efficient for downstream purification to maximise the amount of PCR product loaded into a single well of a 1% agarose gel. Therefore, 1% agarose gels should be prepared that are deep and using a comb that generates wide wells. This permits purification of DNA from a smaller quantity of agarose gel.

Acknowledgments

This work was supported by research grant MR/K018396/1 from the Medical Research Council.

References

1. van Opijnen T, Bodi KL, Camilli A (2009) Tn-seq: high-throughput parallel sequencing for fitness and genetic interaction studies in microorganisms. Nat Methods 6:767–772
2. Goodman AL, McNulty NP, Zhao Y, Leip D, Mitra RD, Lozupone CA et al (2009) Identifying genetic determinants needed to establish a human gut symbiont in its habitat. Cell Host Microbe 6:279–289
3. Gawronski JD, Wong SM, Giannoukos G, Ward DV, Akerley BJ (2009) Tracking insertion mutants within libraries by deep sequencing and a genome-wide screen for *Haemophilus* genes required in the lung. Proc Natl Acad Sci U S A 106:16422–16427
4. Langridge GC, Phan MD, Turner DJ, Perkins TT, Parts L, Haase J et al (2009) Simultaneous assay of every *Salmonella* Typhi gene using one million transposon mutants. Genome Res 19:2308–2316
5. Chao MC, Abel S, Davis BM, Waldor MK (2016) The design and analysis of transposon insertion sequencing experiments. Nat Rev Microbiol 14:119–128
6. Barquist L, Boinett CJ, Cain AK (2013) Approaches to querying bacterial genomes with transposon-insertion sequencing. RNA Biol 10:1161–1169
7. Kwon YM, Ricke SC, Mandal RK (2016) Transposon sequencing: methods and expanding applications. Appl Microbiol Biotechnol 100:31–43
8. Mushtaq N, Redpath MB, Luzio JP, Taylor PW (2004) Prevention and cure of systemic *Escherichia coli* K1 infection by modification of the bacterial phenotype. Antimicrob Agents Chemother 48:1503–1508
9. Birchenough GM, Dalgakiran F, Witcomb LA, Johansson ME, McCarthy AJ, Hansson GC et al (2017) Postnatal development of the small intestinal mucosa drives age-dependent, regio-selective susceptibility to *Escherichia coli* K1 infection. Sci Rep 7:83
10. McCarthy AJ, Birchenough GMH, Taylor PW (2019) Loss of trefoil factor 2 Sensitises rat pups to systemic infection with the neonatal pathogen *Escherichia coli* K1. Infect Immun 87:e00878–e00818
11. McCarthy AJ, Stabler RA, Taylor PW (2018) Genome-wide identification by transposon insertion sequencing of *Escherichia coli* K1 genes essential for *in vitro* growth, gastrointestinal colonising capacity, and survival in serum. J Bacteriol 200:e00698–e00617
12. van Opijnen T, Camilli A (2013) Transposon insertion sequencing: a new tool for systems-level analysis of microorganisms. Nat Rev Microbiol 11:435–442

Chapter 12

Using Genome Scale Mutant Libraries to Identify Essential Genes

Kevin S. Myers, Piyush Behari Lal, Daniel R. Noguera, and Timothy J. Donohue

Abstract

Identification of essential genes is key to understanding the required processes and gene products of organisms under one or more conditions. Transposon sequencing (Tn-seq) has been used to predict essential genes or ones that conditionally impact fitness in a wide variety of organisms. Here, we describe the generation of genome-scale mutant libraries and the analysis of Tn-seq data to identify essential genes from cultures grown in a single condition as well as those that are conditionally important by analyzing the behavior of these mutant libraries in different growth environments. While we illustrate the approach using data derived from Tn-seq analysis of the α-proteobacteria *Rhodobacter sphaeroides* and *Zymomonas mobilis*, the protocols and systems we describe should be generally applicable to a variety of organisms.

Key words Essential genes, Conditional fitness, Transposon, Sequencing, Computational biology, Gene disruption, Genomics, Bacteria, Microbial lifestyles

1 Introduction

Advances in high-throughput nucleic acid sequencing have revolutionized our ability to analyze the content and potential products of genomes [1]. However, this analysis has also brought into stark relief the lack of high-throughput tools to annotate gene functions. Among the critical pieces of information that are often lacking from genome sequencing alone are the contribution of genes to growth or fitness in a particular environment. Indeed, even in well-studied bacterial systems, about 30% of the genes are of unknown function and their role or essentiality in a given condition cannot always be easily predicted using only bioinformatic tools [1, 2]. While individual and systematic deletion of individual genes is possible and has been done for some organisms [3–5], this method is typically far too time consuming, costly, or otherwise impractical to be performed for all species of interest. In recent years, genome-scale methods have been developed to increase the ability to determine

Ren Zhang (ed.), *Essential Genes and Genomes: Methods and Protocols*, Methods in Molecular Biology, vol. 2377, https://doi.org/10.1007/978-1-0716-1720-5_12,

the function and essentiality of genes. Globally referred to as transposon insertion sequencing (termed Tn-seq in this article), these methods make use of transposons that insert generally randomly one time per genome, with the assumption that, in such transposon-derived mutant libraries, the transposon insertion prevents gene product function, and thus, essential genes will lack insertions under conditions when they are required for growth or fitness [6]. Thus, by examining the frequency of transposon insertions across a genome when a mutant library is grown in either a single growth condition to identify genes essential for growth or two growth conditions for conditional essentiality, Tn-seq can be used to aid in the functional analysis of genomes. While Tn-seq has now been deployed to determine either essential genes or their contribution to fitness or growth in a wide set of bacteria, it has also been applied in yeast, protozoans, mice, and human cells [7–13].

There are two main Tn-seq methods that have been used to date. One method (termed Mariner in this article) uses a variant of the Himar I Mariner transposon that inserts preferentially at TA dinucleotides throughout the genome [14, 15]. The other method (termed Tn5 in this article) uses a derivative of the Tn5 transposon, which can insert generally randomly throughout the genome [16, 17]. Both transposons have been used in a variety of bacterial species. While Mariner-based libraries generally have fewer total insertions, the saturation percentage can more easily be measured since there are a measurable number of TA dinucleotides in an organism's genome. However, in organisms with high GC content—and thus a low percentage of TA dinucleotides—it might be more difficult to use Mariner-based libraries to achieve the number of unique insertions required for optimal essential gene identification. Tn5-based mutagenesis can result in a larger number of unique insertions in the mutant library, and there are several software packages to aid in the identification of essential genes [6]. Furthermore, both ends of the Tn5 transposon can be sequenced, allowing the user to determine directionality of the transposon and thus estimate possible polar effects on genes downstream.

Here, we describe a Tn5-based library method used for the analysis of essential and conditionally essential genes in the industrially important α-proteobacteria *Rhodobacter sphaeroides* [18] and *Zymomonas mobilis* (manuscript in preparation). However, this protocol can be adapted for use with virtually any bacterial species that is genetically tractable, has a genome sequence available, and is sensitive to kanamycin (or alternative drug resistance genes present on other Tn5 derivatives [19]). We will describe steps required to generate a large number of strains with independent Tn5 insertions, to propagate the resulting mutant library, purify the DNA, generate the sequencing libraries needed to analyze the position of transposon insertions, and sequence the amplified nucleic acid.

Finally, we describe some available computational tools for analysis of Tn-seq data and provide experience-based suggestions for public domain programs that may fit the analytic needs of each particular experiment. Many of the analysis programs that we describe are likely useful for transposon sequencing in eukaryotic organisms as well. Using Tn-seq will provide the investigator candidate essential and conditionally essential genes, improve the knowledge of the metabolic and regulatory networks of bacterial species, and generate new testable hypothesis for future experiments.

2 Materials

2.1 Growing Bacteria

Both rich and minimal media for construction of the Tn-seq mutant library and propagation of the bacterial strains. The initial mutant library (transposon insertion library) was propagated on a rich medium (LB) in order to recover as many strains with individual insertions as possible. If one chooses to propagate the initial mutant library on a defined, minimal medium, many additional genes (those involved in amino acid, nucleotide synthesis, etc.) will be scored as essential for growth.

1. For *R. sphaeroides*, the following media were used for library creation and strain propagation: Minimal Media: Sistrom's Minimal Medium (SMM) [20] supplemented with 25 μg/mL kanamycin; Rich Media: Luria Bertani (LB) medium [21] supplemented with 50 μg/mL kanamycin and 0.1 mM *meso*-2,6-diaminopimelic acid (DAP) (*see* **Note 1**); Storage Media: LB medium with glycerol (25% vol/vol final concentration).
2. For *Z. mobilis*, the following media were used for library creation and strain propagation. Minimal Media: Glucose-minimal media [22] supplemented with 160 μg/mL kanamycin; Rich Media: ZRMG [23] supplemented with 160 μg/mL kanamycin and 0.15 mM *meso*-2,6-diaminopimelic acid (DAP) (*see* **Note 1**); Storage Media: ZRMG with glycerol (25% vol/vol final concentration).
3. *E. coli* WM6439 containing plasmid pRL27 [24] (Fig. 1a). The pRL27 plasmid is the transposon delivery vector designed by and obtained from the study by Larsen et al. [24]. This is a suicide vector and thus ensures a single transposon insertion event per genome.
4. Bacterial strain of interest (i.e., *R. sphaeroides* 2.4.1 or *Z. mobilis*) (*see* **Note 2**).
5. Phosphate-buffered saline (PBS) solution: 137 mM NaCl, 2.7 mM KCl, 10 mM Na_2HPO_4, and 1.8 mM KH_2PO_4, in water.
6. Nunc Cell Scraper.

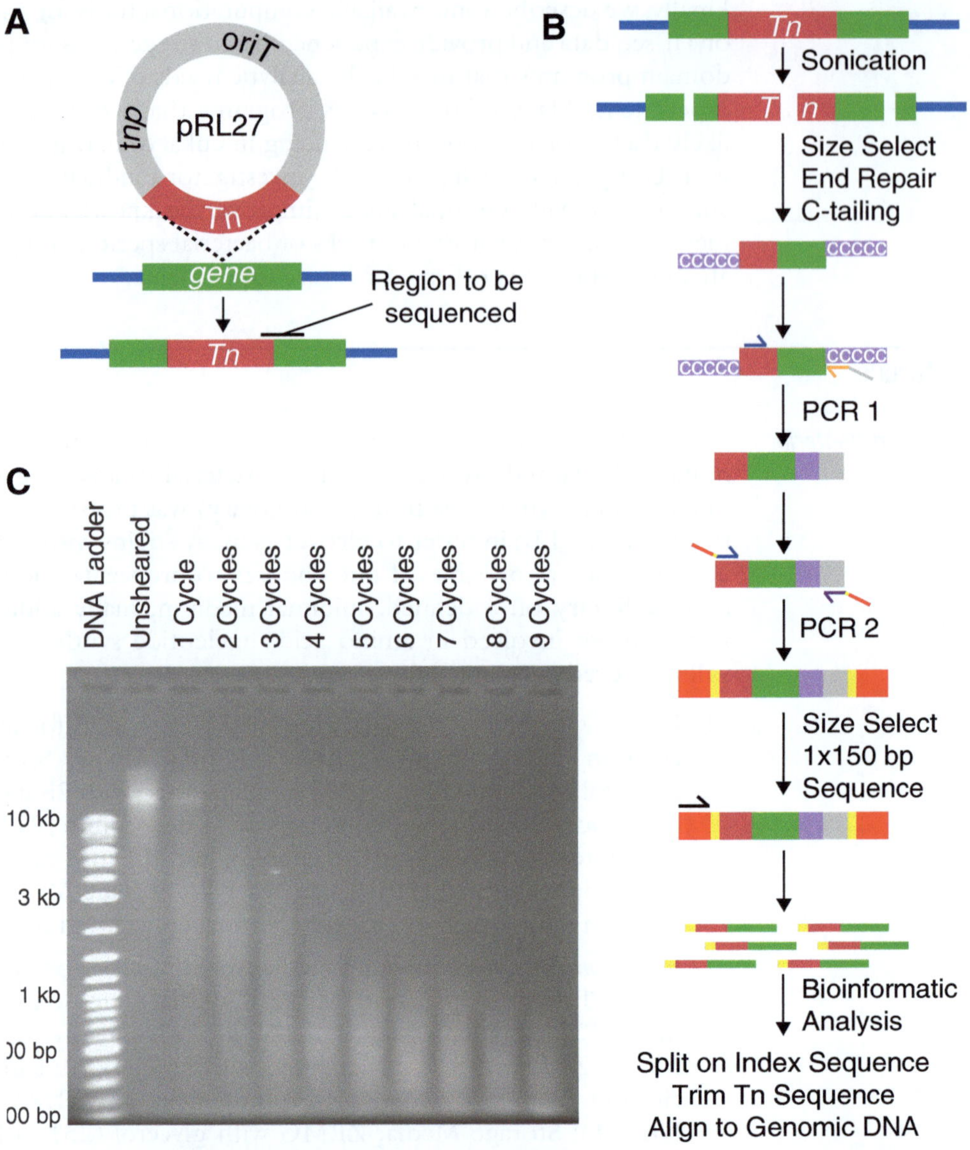

Fig. 1 (**a**) Map of Tn-5 transposon plasmid pRL27 [24] with the transposon (Tn, red), the mobilization element (*oriT*), and the Tn-5 transposase gene (*tnp*) indicated [24, 72]. The transposon contains a kanamycin resistance gene to confer antibiotic resistance upon insertion into the genome. (**b**) General overview of Tn-5 protocol. After the transposon insertion library is created and grown on the desired growth condition, the DNA is isolated, sheared with sonication, size selected, end repaired, and C-tailed (purple). PCR 1 is performed to amplify the transposon-genomic DNA boundary using the C-tail (orange primer) followed by PCR 2 to produce fragments for Illumina sequencing. Note the unique indexing barcodes (yellow) downstream of the Illumina sequencing adapter (bright red). After size selection, the material is sequenced, and the sequencing reads are aligned to the genome to identify the location of transposon insertion. (**c**) A 1% agarose gel image showing that at least four cycles of sonication are required to obtain fragments around 300–500 bp in length

2.2 Tn-seq Sample Preparation

1. DNeasy Blood and Tissue kit.
2. Misonix S4000 sonicator with cup horn 431C.
3. QIAquick PCR Purification Kit.
4. NEBNext End Repair Enzyme Mix.
5. AxyPrep Mag PCR Clean-up Kit.
6. MinElute PCR Purification Kit.
7. Elution buffer (10 mM Tris–HCl, pH 8.0–8.5) or Qiagen EB buffer.
8. Terminal Deoxynucleotidyl Transferase.
9. dCTP/ddCTP solution: 9.5 mM dCTP, 0.5 mM ddCTP, and Water to 20 μL.
10. KAPA SYBR Fast qPCR Master Mix (*see* **Note 3**).
11. KAPA HotStart HiFi Polymerase ReadyMix (*see* **Note 3**).
12. KAPA Illumina Library Quantification Kit.
13. Tn_Binding_Amp primer from preferred source (Standard desalting) that binds to the transposon sequence for amplification: 5′GCAAGCTTCAGGGTTGAGATGTGTATAAG3′.
14. olj376 primer from preferred source (Standard desalting) that binds to the 3′ C-tail for amplification: 5′GTGACTGGAGTTCAGACGTGTGCTCTTCCGATCTGGGGGG GGGGGGGGG3′.
15. Tn_Binding_Illumina_Adapter primer from preferred source (PAGE purified) that binds to the amplified region of the transposon and contains the Illumina sequencing adapter at the 5′ end and an experiment specific index barcode (**NNNNNNNN**). The 8 Ns should be replaced with a unique barcode specific to each experiment. This allows for multiplexing multiple experimental samples on the same sequencing run. Barcode sequences can be found at the Illumina website (www.illumina.com) (*see* **Note 4**). 5′AATGATACGGCGACCACCGAGATCTACACTCTTTCCCTACACG-ACGCTCTTCCGATCT **NNNNNNNN** GCAAGCTTCAGGGTT-GAGATGTGTATAAG 3′.
16. Index_Illumina_Adapter primer from preferred source (HPLC purified) that binds to the region amplified by olj376 and contains both the Illumina sequencing adapter and an experiment specific index barcode (**NNNNNN**). The 6 Ns should be replaced with a unique barcode specific to each experiment. This allows for multiplexing multiple experimental samples on the same sequencing run. Barcode sequences can be found at the Illumina website (www.illumina.com) (*see* **Note 4**). 5′CAAGCAGAAGACGGCATACGAGAT**NNNNNN**GTGACTGGAG-TTCAGACGTGTGCTCTTCCGATCT 3′.

2.3 Tn-seq Data Analysis

1. Computational analysis will require an internet connection to download DNA sequence data and analysis programs. Some programs require knowledge of the command line, and it is often recommended to use a Linux operating system or Unix-based operating system like MacOS.
2. A file (FASTA) containing genome sequence of interest for mapping Illumina reads. This can be obtained from NCBI or from annotation performed in KBase [27]. For *R. sphaeroides*, we used genome GCF_000012905.2 (Assembly: ASM1290v2), and for *Z. mobilis*, we used genome GCF_003054575.1 (Assembly: ASM305457v1).
3. There are many software packages for nucleic acid sequence read alignment. We have had success using Bowtie1 [28], Bowtie2 [29], and BWA-MEM [30]. Galaxy [31] and KBase [27] offer access to these programs in a user-friendly manner.
4. There are several programs that will assist in trimming the DNA sequence reads and the FASTQ splitting that is required for this analysis. We have had success using the following programs: FASTX-Toolkit (specifically FASTQ/A Trimmer and FASTQ/A Barcode Splitter [32]), Cutadapt [33], and Trimmomatic [34]. Galaxy [31] and KBase [27] offer access to these programs in a user-friendly manner.
5. To visualize the reads across the genome, the results generally need to be converted to a WIG format using a program such as bam2wig.pl (https://github.com/tjparnell/biotoolbox) or using the tools within Galaxy [31] or KBase [27]. The results can be viewed using any number of genome browsers can be used following the individual instructions for each including: MochiView [35], IGV [36], and UCSC Genome Browser [37].
6. There are different methods or programs to predict gene essentiality and/or genes that are important for growth under different conditions. We have experience with eight programs that are available and were able to be used for Tn-seq analyses: ARTIST [38], Bio-TraDIS [39], MAGenTA [40], PIMMS [41], Tn-seq Explorer [42], TnSeq2 [43], TRANSIT [44], and TSAS [18].

3 Methods

The following protocol for Tn-seq libraries (Fig. 1b) was developed using *R. sphaeroides* as the bacterial species of interest. A slightly different protocol was published earlier [18], and here, we improved and optimized the method with fewer steps. This protocol has successfully generated data for the *R. sphaeroides* and

Z. mobilis (manuscript in preparation). A detailed protocol for generating a bacterial transposon library using Mariner-based transposons has recently been published [45], and we recommend it for those interested in using this technology. The following protocol uses conjugation as a method for transfer of the transposon into the target organism. However, this protocol can easily be adapted for using transformation [46, 47] or electroporation [48] to introduce the transposon into the target strain. It is also easily adjusted to accommodate other media, the incubation times, growth temperature, and the conditions used to prepare or propagate the mutant library.

3.1 Mutant Library Construction

1. Mix 1 mL of *E. coli* WM6439 cells containing the pRL27 plasmid (donor strain) and 1 mL of cells of the recipient strain (i.e., *R. sphaeroides* or *Z. mobilis*) cells in late-exponential phase at a 1:1 ratio. It is recommended to use 2–5 tubes to obtain the maximum number of exconjugants.
2. Pellet the cells by centrifugation (17,300 × *g* for 30 s at room temperature), decant the supernatant, and resuspend in rich media (i.e., LB or ZRMG) supplemented with 0.1 mM DAP (*see* **Note 1**).
3. With a pipette, spot the entire resuspended cells in the center of a rich media plate (i.e., LB or ZRMG) supplemented with 0.1 mM DAP and incubate at 30 °C (one plate per conjugation mixture). Duration of incubation is dependent on bacterial species used; *R. sphaeroides* requires 3 h incubation, [18] while *Z. mobilis* requires 12–15 h of incubation [23] (*see* **Note 5**).
4. Add 1 mL of liquid rich media (i.e., LB or ZRMG) to the rich media plates and recover the cells from the plate by scraping the colonies into a separate tube. Vortex the recovered cells to break the conjugation pilus and separate donor cells from recipient, pellet the cells, decant the supernatant, and resuspend in rich media (i.e., LB or ZRMG). Combine all the conjugation mixtures into one tube and add glycerol to 25% vol/vol final concentration. Remove 100–500 μL of the conjugation mixture and immediately freeze the remaining mixture in liquid nitrogen or a dry ice-ethanol bath and store at −80 °C.
5. Prepare serial dilution of the suspension from **step 4** in LB and plate 100 μL of the 10^{-4} to 10^{-7} dilution suspensions on rich media agar plate (i.e., LB or ZRMG) with kanamycin to measure the total number of CFUs/mL of the conjugation mixture suspension. Record the CFU/mL from the plating after incubation.

6. Thaw conjugation mixture suspension and prepare dilution in rich media (i.e., LB or ZRMG) using the CFU/mL information from **step 5**. Plate on a 100 mm diameter rich media (i.e., LB or ZRMG) agar plate supplemented with kanamycin to obtain 200–2000 CFUs per plate. The surviving cells will be those exconjugants that have undergone transposition and are resistant to kanamycin (e.g., the recipient cells with successful conjugation). Incubate for 7 days at 30 °C. Duration of incubation is dependent on bacterial species. It is recommended to plate different volumes of conjugation mixture to obtain a range of CFU/plate, with the goal of more than 200,000 CFUs total (*see* **Note 6**).
7. Recover the cells from the plate by scraping the colonies that grow (and are resistant to kanamycin) from the plates and into a separate tube. Spin at 6000 × *g* for 10 min at 4 °C. Decant supernatant. Wash the cells once with 25 mL of minimal media (i.e., SIS or glucose-minimal media) and combine all resuspended cell pellets into a single tube. Spin combined cells at 6000 × *g* for 10 min at 4 °C, decant supernatant, and resuspend pellet in approximately equal volume minimal media (i.e., SIS or glucose-minimal media) as the cell pellet to obtain a very dense cell suspension. Add glycerol to 25% vol/vol final concentration, aliquot the cells (100 μL) into labeled 1.5 mL microfuge tubes, freeze in liquid nitrogen or a dry ice-ethanol bath, and store at −80 °C. This is termed the "transposon insertion library" (*see* **Note 7**).
8. Thaw three random aliquots from **step 7**, prepare dilutions in rich (i.e., LB or ZRMG) media, and plate 100 μL on rich media (i.e., LB or ZRMG) with kanamycin to determine the total number of CFUs/mL in the transposon insertion library.

3.2 Transposon Library Propagation

1. Inoculate 10^6 CFUs (at least 100 times the total number of genes in the organism to ensure an average of ~100 insertions per gene is possible) of the transposon insertion library from Subheading 3.1, **step 8** into 2 mL to 50 mL of desired media and grow in the desired conditions. The amount of media required will depend on the experiment and kind of growth apparatus used; for multiwell plates, less media are required, while for shaking flasks, more media can be used if desired. For example, to identify essential genes in the transposon insertion library (that was generated and propagated in rich media (i.e., LB or ZRMG), grow the library in the same media. To compare essential genes during aerobic or anaerobic growth for *R. sphaeroides*, simply change the media and growth conditions. For aerobic growth, sparge with 69% N_2, 30% O_2, and 1% CO_2. For photosynthetic growth, sparge with 99% N_2 and 1% CO_2 while being exposed to light [18]. Similarly, to identify

genes encoding biosynthetic functions, grow the transposon insertion library in defined minimal media instead of rich media.

2. Grow for about seven doublings, as monitored by changes in optical density at 600 nm (or some suitable wavelength for the test organism).
3. Collect 2–5 mL of cells by centrifugation for 10 min at 5000 × *g* and decant supernatant. Wash the cells with 2× mL of PBS, recollect cells by centrifugation for 5 min at 5000 × *g*, and decant supernatant.
4. Resuspend the cells with 2 mL of PBS mixed with 25% glycerol (vol/vol final concentration) and freeze in liquid nitrogen or a dry ice-ethanol bath. Store at −80 °C.

3.3 Tn-seq Sample Preparation

1. The following description assumes that the Tn5-based mutant library will be prepared for sequencing on an Illumina sequencing platform, and it can be modified as needed to accommodate additional DNA sequencing units. Furthermore, other detailed protocols have been published describing how to prepare samples from Mariner-based transposon libraries [45].
2. Thaw samples and isolate genomic DNA using a DNeasy Blood and Tissue kit (Qiagen) including the RNase A digestion step, following the manufacturer's protocol.
3. Shear 2–5 μg of DNA using a Misonic S4000 sonicator with cup horn 431C using 50% amplitude and 6× 30 s on/30 s off settings. The DNA should be sheared to fragments with an average length of 300–500 bp. The number of sonication cycles required will depend on the sonicator used. It is recommended to standardize the sonication protocol, and sonication can be evaluated by analyzing sheared DNA after every protocol change (i.e., number of sonication cycles) using gel electrophoresis (Fig. 1c) or a Bioanalyzer.
4. Analyze the size of the sheared DNA using gel electrophoresis (Fig. 1c) or a Bioanalyzer. Measure the concentration of sheared DNA. At least 2 μg of sheared DNA is required for the next step. Low yields (<2 μg) may necessitate the use of multiple sonications from different starting materials. If this is needed, the sheared DNA can be combined, purified, and concentrated. For this, use a QIAquick PCR purification kit (Qiagen) following the manufacturer's protocol.
5. After verifying that the sheared DNA is of the appropriate length, end repair the sheared DNA for each sample. Keep all reagents on ice until ready and then mix and keep reaction on ice as much as possible: 85 μL of sheared DNA (2–5 μg total), 10 μL of 10× NEBNext End Repair Reaction Buffer (included

with NEBNext End Repair Enzyme Mix purchase), and 5 μL of NEBNext End Repair Enzyme Mix (NEB). Incubate in a thermal cycler for 30 min at 20 °C.

6. Purify sheared end-repaired DNA using AxyPrep Mag PCR Clean-up Kit (1.6×, 160 μL) (*see* **Note 8**). Add 0.2 volumes of AxyPrep beads (20 μL) to the reaction and mix by pipetting up and down multiple times. Incubate the tubes at room temperature for 5–15 min (*see* **Note 9**) to bind the DNA fragments to the beads. Place the tubes on a magnet to capture the beads and incubate until the liquid is clear. Carefully remove and discard the supernatant without disturbing the beads. While the tubes are still on the magnet, add 200 μL of freshly prepared 80% ethanol. Incubate tubes on the magnet at room temperature for at least 30 s. Carefully remove and discard supernatant without disturbing the beads. While the tubes are still on the magnet, add 200 μL of freshly prepared 80% ethanol. Incubate tubes on the magnet at room temperature for at least 30 s. Carefully remove and discard supernatant without disturbing the beads. Dry the beads at room temperature for 3–5 min or until all ethanol as evaporated (*see* **Note 10**). Remove the tubes from the magnet. Resuspend the beads thoroughly in 35 μL of elution buffer or Qiagen EB buffer. Incubate the tubes at room temperature for 2 min to elute the DNA off the beads. Place the tubes on a magnet to capture the beads and incubate until the liquid is clear. Transfer the clear supernatant to a new tube.

7. C-tailing is performed on the sheared DNA for each sample in order to facilitate the library amplification in **steps** 7 and **8**. Keep all reagents on ice until ready and then mix and keep reaction on ice as much as possible: 18.6 μL of purified end-repaired sheared DNA, 2.8 μL of fresh solution of dCTP/ddCTP solution, 5.6 μL of 5× TdT reaction buffer (Included with Terminal Deoxynucleotidyl Transferase purchase), and 1.0 μL of Terminal Deoxynucleotidyl Transferase (Promega). Incubate for 60 min at 37 °C and then incubate for 20 min at 75 °C. Purify DNA using Performa DTR gel filtration column (Edge Biosystems). Centrifuge cartridge 120 s at 750 × *g* and move to a new 1.5 mL microfuge tube, add C-tailing reactions to column matrix, and centrifuge for 120 s at 750 × *g*.

8. Ligate DNA sequencing adapters using the KAPA DNA library preparation kit. Two qPCR reactions are performed in a real-time thermocycler (**steps** 7 and **8** below). The first uses a primer (Tn_Binding_Amp) that binds to the transposon sequence and contains one of the Illumina adapters and a primer that binds to the C-tail added in **step 6** (olj376), thus only amplifying those fragments with the transposon and

C-tails present. The second qPCR is a diagnostic assay to determine the optimal conditions for the actual library sequencing that will be performed in **step 9**. This second qPCR uses a primer that binds to the transposon (Tn_Binding_Illumina_Adapter) with a sequencing adapter added to the 5′ end and a primer that contains the unique index sequence and the second Illumina adapter and recognizes the new bases added in the first PCR (Index_Illumina_Adapter). The Index_-Illumina_Adapter primers can be used to multiplex sequence of many different experimental samples in a single lane of the sequencing based on unique barcode sequence within each primer. A different Index_Illumina_Adapter primer should be used for each sample. We recommend using unique indexes in the Tn_Binding_Illumina_Adapter and the Index_Illumina_Adapter primers for each sample to ensure proper indexing no matter which read is used for downstream analysis (*see* **Note 4**). This qPCR is performed in a real-time thermocycler in order to estimate the proper number of cycles required for optimal PCR amplification for sequencing to avoid PCR artifacts. An initial PCR (PCR1) is performed to amplify regions of the genome containing transposon sequences and C-tails. This should be run in a real-time thermocycler to monitor progress. For PCR1, keep all reagents on ice until ready and then mix and keep reaction on ice as much as possible: 7.4 μL of purified C-tailed DNA, 25 μL of 2× KAPA SYBR FAST qPCR Master Mix (*see* **Note 3**), 3.0 μL of 10 μM primer olj376 (binds to C-tail), 1.0 μL of 10 μM primer, Tn_Binding_Amp (binds within Transposon sequence), and 13.6 μL of PCR-grade water. Incubate in a real-time thermocycler with the following protocol: 1 cycle at 95 °C for 2 min; 24 cycles at 98 °C for 30 s, 64 °C for 30 s, and 72 °C for 1.5 s (read); 1 cycle at 72 °C for 2 min; and 10 °C hold for up to 72 h.

9. Using the PCR1 product, perform a qPCR to estimate the optimal number of cycles for the actual reaction in **step 9**. Perform this PCR (PCR2) in a real-time thermocycler to monitor number of cycles required to read the amplification inflection point. Note that you will NOT use the products from PCR2 for sequencing—this is a diagnostic PCR only. For PCR2, keep all reactions and reactions on ice until ready: 1.2 μL of PCR1 product, 25 μL of 2× KAPA SYBR FAST qPCR Master Mix (*see* **Note 3**), 3.0 μL of 10 μM primer, Index_Illumina_Adapter primer with unique multiplexing bode (use a different Index_Illumina_Adapter primer with unique barcode for each sample planned for multiplexing), 3.0 μL of 10 μM primer, Tn_Binding_Illumina_Adapter (binds to transposon sequence and has Illumina sequencing adapter at 5′ end), and 17.8 μL of PCR-grade water. Incubate

in a real-time thermocycler with the following protocol: 1 cycle at 95 °C for 2 min; 14 cycles at 98 °C for 30 s, 64 °C for 30 s, and 72 °C for 1.5 min (read); 1 cycle at 72 °C for 2 min; and 10 °C hold. Evaluate the results of the qPCR using standard software and determine the number of cycles to reach the amplification inflection point in the qPCR curve.

10. Perform the PCR for sequence analysis using the information obtained from PCR2 in **step 8**. Repeat PCR2 using the number of cycles estimated to reach the amplification inflection point determined from **step 8** for each sample. The results of PCR2 are used to determine how many cycles to use. This PCR can be performed in a standard thermocycler using PCR1 product from **step** 7. We recommend using KAPA HotStart HiFi Polymerase ReadyMix instead of the KAPA SYBR FAST qPCR MasterMix (*see* **Note 3**). In order to obtain enough amplified DNA, it is recommended that two separate 50 μL reactions are run for each sample. After the PCR, each of the runs per sample will be size selected and purified as in **step 10**. The two samples will be treated separately and combined into one tube in step 3.3.10.8 of the size selection protocol.

11. The PCR reaction products are size selected using AxyPrep Mag PCR Clean-up Kit. Dual size selection, 0.6×–0.8×, is recommended. Exclude fragments corresponding to library inserts larger than ~350 bp by adding 30 μL of AxyPrep beads to 50 μL of digested DNA. Mix thoroughly by pipetting up and down multiple times. Incubate the tubes at room temperature for 5–15 min (*see* **Note 9**) to bind the DNA fragments to the beads. Place the tubes on a magnet to capture the beads and incubate until the liquid is clear. Transfer the supernatant to new tubes, being sure to not disturb or transfer any beads. Discard the beads. Perform the second size selection by adding 0.2 volumes of KAPA Pure Beads (10 μL) to the transferred supernatant. Mix thoroughly by pipetting up and down multiple times. Incubate the tubes at room temperature for 5–15 min (*see* **Note 9**) to bind the DNA fragments to the beads. Place the tubes on a magnet to capture the beads and incubate until the liquid is clear. Transfer the supernatant to new tubes, being sure to not disturb or transfer any beads. Discard the beads. Carefully remove and discard the supernatant without disturbing the beads. While the tubes are still on the magnet, add 200 μL of freshly prepared 80% ethanol. Incubate tubes on the magnet at room temperature for at least 30 s. Carefully remove and discard supernatant without disturbing the beads. While the tubes are still on the magnet, add 200 μL of freshly prepared 80% ethanol. Incubate tubes on the magnet at room temperature for at least 30 s. Carefully remove and discard supernatant without disturbing the beads.

Dry the beads at room temperature for 3–5 min or until all ethanol as evaporated (*see* **Note 10**). Remove the tubes from the magnet. Resuspend the beads thoroughly in 35 μL of elution buffer or Qiagen EB buffer. Incubate the tubes at room temperature for 2 min to elute the DNA off the beads. Place the tubes on a magnet to capture the beads and incubate until the liquid is clear. Transfer the clear supernatant to a new tube.

12. Final QC can be performed using a Bioanalyzer or using agarose gel electrophoresis to confirm the expected size of the fragments. Quantity of libraries can be determined using a Qbit or KAPA Illumina library quantification protocol following manufacturer's protocol (*see* **Note 11**).

3.4 DNA (Illumina) Sequencing

1. After quantity control, the samples should be sequenced using the standard procedure for the Illumina instrument you will use.
2. Successful sequencing has been performed using an HiSeq 2000 and an HiSeq 2500 sequencer, but other Illumina instruments and other sequencing platforms can, in principle, also be used [49–51].
3. It is recommended you obtain at least 1× 150 bp reads for maximum alignment potential (*see* **Note 12**).
4. It is recommended to obtain enough sequence reads to have at least 100,000 unique insertions. We have found that obtaining at a minimum of two million reads typically results in adequate insertions. The reads will contain both genomic sequence and transposon sequence. It may be required to remove the transposon sequence from the reads to promote genome alignment. This can be done using tool such as FASTX trimmer, and remove 38 bases from the 5′ end of each read. The trimmed reads must be long enough and of high enough quality to align to the genome sequence (at least 36 bp reads with quality scores ≥ 30).
5. If multiplexing, it is critical that a record is made of which index sequence (from the Index_Illumina_Adapter primers) was used for each sample. If this is not done, it will be impossible to separate the sequencing data and the experiment will fail to yield useful results.

3.5 Tn-seq Data Analysis

There are two main approaches to identifying essential genes. One identifies essential genes using a single growth condition (using one dataset), while the other compares data from two different growth conditions (using two datasets) to identify conditionally essential genes. Currently, there appears to be no algorithm that will perform more complex analyses beyond single growth essentiality and two conditions conditional essentiality. Future work will likely result in algorithms that can perform advanced statistical tests to determine conditional essentiality when comparing many growth conditions or time points. These algorithms could be based on the linear models used by edgeR [20] or machine learning to compare multiple growth conditions for essential genes.

1. The initial steps for processing the FASTQ files obtained in a Tn-seq experiment are similar to any high-throughput nucleic acid sequencing project. A genome sequence (in FASTA format) for the organism of interest should be either downloaded from NCBI (if available) or generated from separate sequencing projects. Genome annotation should also be obtained to help in identifying essential genes. This can be downloaded from NCBI (for example GFF or GenBank files) or obtained from KBase [27] or generated from additional sequencing projects. Note the format of the file may need to be adjusted to fit the analysis tool you choose to use.
2. There have been many Tn-seq analysis programs or tools published, and it is recommended to use one of these due to their robust statistical analyses and peer-reviewed status (**step 6** below). However, the selection of the analysis program will likely depend on several criteria:
   (a) Which transposon have you used? Some programs can only analyze data from libraries generated using Tn5 as a transposon, and some can only analyze data from libraries generated using the Mariner transposon, while some can analyze both.
   (b) Do you want to examine essential genes, conditional essentiality, or both in a given dataset? Several programs are set up to do one of the two tests, and other programs can analyze both types of data.
   (c) What is your skill level using the command line? Several programs require at least basic knowledge of the command line, while many programs have a GUI (graphical user interface) to facilitate use.
   (d) What is your skill level using common sequencing programs such as Bowtie or samtools? Several programs require aligned DNA sequencing data, meaning the user has to be able to use an aligner program before analysis. If the user has little experience with this, there are programs that can perform all stages of the analysis.
   (e) Do you have replicate data? Several programs have built-in statistics to handle replicate data, while other programs do not.

Table 1
Summary of Algorithms for Tn-seq Analysis

Algorithm	Transposon library compatibility	Essential genes and/or conditional essentiality	Interface of algorithm	Format of input files[a]	Replicate data support
ARTIST [38]	Tn5-based and mariner-based	Essential genes and conditional essentiality	MatLab	WIG file [from Split, trimmed, and aligned sequencing FASTQ file]	None
Bio-TraDIS [39]	Tn5-based	Essential genes and conditional essentiality	Command line	Unsplit sequencing FASTQ file	None
MAGenTA [40]	Tn5-based and mariner-based	Conditional essentiality	Command line or galaxy	Unsplit sequencing FASTQ file	Built-In
PIMMS [41]	Tn5-based	Conditional essentiality	Command line	Unsplit sequencing FASTQ file	None
Tn-seq Explorer [42]	Tn5-based and mariner-based	Essential genes and conditional essentiality	Java graphical user Interface	Unsplit sequencing FASTQ file	None
TnSeq2 [43]	Tn5-based	Essential genes and conditional essentiality	Command line	WIG file [from Split, trimmed, and aligned sequencing FASTQ file]	Built-In
TRANSIT [44]	Tn5-based and mariner-based	Essential genes and conditional essentiality	Command line	WIG file [from Split, trimmed, and aligned sequencing FASTQ file]	Built-In
TSAS [18]	Tn5-based	Essential genes and conditional essentiality	Command line	Aligned BAM or SAM file [from trimmed sequencing FASTQ file]	Built-In

[a]Generating files other than Unsplit Sequencing FASTQ Files will likely require use of command line tools

3. We will briefly describe some of the available tools for use in analyzing Tn-seq data and specifically address each of the questions and how each program addresses them (Table 1). Note that while there are additional tools in the literature, we were unable to get several of them to function, so we will not discuss them here.
4. ARTIST [38] by Pritchard et al. has been cited by at least 50 articles. ARTIST can analyze data from mutant libraries generated using either Tn5 or Mariner transposons. ARTIST can analyze both single growth condition essentiality (EL-ARTIST) and conditional essentiality (CON-ARTIST).

ARTIST is written in and uses MatLab, and minimal MatLab experience is needed as there is an extensive User's Manual included. ARTIST requires MatLab to be installed on the user's computer. This may require the user to purchase and install this program. ARTIST requires a WIG file. This means the user must split the FASTQ DNA sequence files based on the index sequence (using fastx-splitter [52]), remove the transposon sequence (using cutAdapt [33]), trim the sequence for quality (using Trimmomatic [34]), align the filtered sequencing data to the genome using an aligner (such as Bowtie [53], Bowtie2 [29], or BWA-MEM [30]), then clean and organize the aligned file (using Picard Tools [54] and samtools [55]), and finally convert the aligned file into a WIG file (using custom scripts or bedtools [56]). Therefore, it is not recommended for users with little knowledge of DNA sequence analysis or use of the command line. ARTIST lacks any statistics for replicate samples, and thus, each replicate must be run separately and the user must decide to include the intersection of the replicates (those essential genes found in all the replicate samples), the union of the replicates (those essential genes found in any of the replicate samples), or another method of combining replicate data.

5. Bio-TraDIS [39] by Barquist et al. has been cited by at least 31 articles. Bio-TraDIS can only analyze data from mutant libraries generated using the transposon Tn5. Bio-TraDIS can analyze both single growth condition essentiality (tradis_essentiality.R) and conditional essentiality (tradis_comparison.R). Bio-TraDIS requires knowledge of the command line to download and run the different scripts. Bio-TraDIS requires BWA-MEM [31], smalt [57], samtools [55], tabix [58], R [59], and Bioconductor [60]. It can also be downloaded using Anaconda [61]. Bio-TraDIS can take the original, unsplit FASTQ file and perform all the necessary steps from splitting, trimming, alignment, and analysis without the user having to use separate programs. Bio-TraDIS lacks any internal statistical tests for replicate samples, and thus, each replicate must be run separately and the user must decide to include the intersection of the replicates (those essential genes found in all the replicate samples), the union of the replicates (those essential genes found in any of the replicate samples), or another method of combining replicate data.
6. MAGenTA [40] by McCoy et al. has been cited by at least four articles. MAGenTA can analyze data from libraries generated with either Tn5 or Mariner transposons. MAGenTA is built to analyze conditional essentiality experiments only. MAGenTA is available either as a command line tool or as a Galaxy [31] implemented tool. MAGenTA can be run by users with

knowledge of the command line or via Galaxy [62] for users with little desire to run the program on the command line. MAGenTA can take the original, unsplit FASTQ file and perform all the necessary steps from splitting, trimming, alignment, and analysis without the user having to use separate programs. MAGenTA has built-in statistical tools to combine replicate samples.

7. PIMMS [41] by Blanchard et al. has been cited by at least one article. PIMMS can only analyze data from libraries generated with transposon Tn5. PIMMS is built to only analyze data from conditional essentiality experiments. PIMMS is available as a command line tool, and thus, some experience with the command line is required. PIMMS can take the original, unsplit FASTQ file and perform all the necessary steps from splitting, trimming, alignment, and analysis without the user having to use separate programs. PIMMS appears to lack a built-in method to combine replicate samples.
8. Tn-seq Explorer [42] by Solaimanpour et al. has been cited by at least 17 articles. Tn-seq Explorer can analyze data from libraries generated with either Tn5 or Mariner transposons. Tn-seq Explorer is built to analyze both single growth essentiality and conditional essentiality experiments. Tn-seq Explorer is available as a Java application, meaning it has a GUI, and therefore, no use of the command line is required. Tn-seq Explorer can take the original, unsplit FASTQ file and perform all the necessary steps from splitting, trimming, alignment, and analysis without the user having to use separate programs. Tn-seq Explorer requires a gene annotation file format (.rnt file) and a protein annotation file (.ptt file) that are obsolete and no longer available from NCBI. Tn-seq Explorer contains access to files available on the NCBI file server, but if the organism is not present, the user will have to use a command line tool to create these files from existing annotation files (such as GenBank files). Tn-seq Explorer appears to lack a built-in method to combine replicate samples.
9. TnSeq2 [43] by Turner et al. has been cited by at least five articles. TnSeq2 can only analyze data from libraries generated using the transposon Tn5. TnSeq2 can analyze both single growth essentiality and conditional essentiality experiments. TnSeq2 requires knowledge of the command line to download and run the different scripts. TnSeq2 requires R [59], Perl, and R libraries DESeq [63] and dplyr [64]. TnSeq2 can analyze the original, unsplit FASTQ file and perform all the necessary steps from splitting, trimming, alignment, and analysis without the user having to use separate programs. TnSeq2 has built-in methods to combine replicate samples.

10. TRANSIT [44] by DeJesus et al. has been cited by at least 23 articles. TRANSIT can analyze data from libraries generated with either the Tn5 or Mariner transposons. TRANSIT uses a method called Tn5_Gaps for data from Tn5-based mutant libraries and one of two included methods called Gumbel and HMM for data from Mariner-based mutant libraries. TRANSIT can analyze both single growth essentiality and conditional essentiality experiments. TRANSIT requires knowledge of the command line to download and run the different scripts. TRANSIT requires Python and numerous Python modules: Numpy [65], Statsmodels [66], Scipy [67], matplotlib [68], Pillow [69], wxpython [70], and PyPubSub [71]. TRANSIT requires a WIG file. This means the user must split the FASTQ DNA sequence files based on the index sequence (using fastx-splitter [52]), remove the transposon sequence (using cutAdapt [33]), trim the sequence for quality (using Trimmommatic [34]), align the filtered sequencing data to the genome using an aligner (such as Bowtie [53], Bowtie2 [30], or BWA-MEM [31]), then clean and organize the aligned file (using Picard Tools [54] and samtools [55]), and finally convert the aligned file into a WIG file (using custom scripts or bedtools [56]). Therefore, it is not recommended for users with little knowledge of DNA sequence analysis or use of the command line. TRANSIT has built-in methods to combine replicate samples.
11. TSAS [18] by Burger et al. has been cited by at least four articles. TSAS can analyze data from libraries generated using the transposon Tn5. TSAS can analyze both single growth essentiality (Analysis_1) and conditional essentiality experiments (Analysis_2). TSAS requires knowledge of the command line to download and run the different scripts. TSAS requires Java, and therefore, Java must be installed on the user's computer. TSAS requires an aligned file (BAM or SAM format). This means the user must split the FASTQ DNA sequence files based on the index sequence (using fastx-splitter [52]), remove the transposon sequence (using cutAdapt [33]), trim the sequence for quality (using Trimmommatic [34]), align the filtered sequencing data to the genome using an aligner (such as Bowtie [53], Bowtie2 [30], or BWA-MEM [31]), and then clean and organize the aligned file (using Picard Tools [54] and samtools [55]). TSAS has built-in methods to combine replicate samples.

4 Notes

1. The DAP is added to the media to allow viability of the auxotrophic *E. coli* donor strain.

2. A bacterial strain will be a good candidate for Tn-seq analysis using the described methods if it is genetically tractable, has a genome sequence available, and is sensitive to kanamycin (or alternative drug resistance genes present on other Tn5 derivatives [19]).
3. Note that the DNA polymerases in this master mix were chosen because of the high GC content of *R. sphaeroides* [25, 26]. While they functioned well for *Z. mobilis*, which has ~47% GC content, other DNA polymerases may be required or better suited for libraries derived from bacteria with different GC content [25, 26].
4. The index primers can be custom-made or purchased. It is important to have unique index barcodes for each primer so you can separate the sequences. We recommend using unique index barcodes for each adapter primer to ensure maximum ability to separate the sequences.
5. It is recommended to use a wide-bore pipette tip at this stage.
6. The number of total CFUs will depend on species and genome size. Larger genomes require more CFUs to ensure adequate transposon insertion coverage in all genes.
7. The aliquots will fill ~500–800 1.5 mL tubes, and so it is recommended to label and prepare the tubes before starting the procedure to ensure rapid freezing.
8. The AxyPrep bead method described in Subheading 3.3, **step 5** generally provides more robust purification, but alternatively, you can purify the DNA using a MinElute PCR Purification Kit (Qiagen) following the manufacturer's protocol. Elute in 35 μL of elution buffer/Qiagen EB buffer.
9. We typically incubated for 15 min.
10. Over-drying the beads may result in a reduced yield.
11. Recommended controls include samples without the collector probe, without Ampligase, and without exonucleases.
12. It is unnecessary to obtain paired end reads for Tn-seq analysis. If the sequencing center used provides paired end reads, separate the FASTQ (if necessary) and use the R1 read for further analysis. The sequencing reads from the C-tail end are usually of low quality, and additional trimming may be required.

Acknowledgements

This material is based upon work supported by the Great Lakes Bioenergy Research Center, U.S. Department of Energy, Office of Science, Office of Biological and Environmental Research under Award Number DE-SC0018409.

References

1. Loman NJ, Constantinidou C, Chan JZ et al (2012) High-throughput bacterial genome sequencing: an embarrassment of choice, a world of opportunity. Nat Rev Microbiol 10:599–606
2. Galperin MY, Koonin EV (2010) From complete genome sequence to 'complete' understanding? Trends Biotechnol 28:398–406
3. Baba T, Ara T, Hasegawa M et al (2006) Construction of *Escherichia coli* K-12 in-frame, single-gene knockout mutants: the Keio collection. Mol Syst Biol 2:2006.0008
4. Yamamoto N, Nakahigashi K, Nakamichi T et al (2009) Update on the Keio collection of *Escherichia coli* single-gene deletion mutants. Mol Syst Biol 5:335
5. Koo B-M, Kritikos G, Farelli JD et al (2017) Construction and analysis of two genome-scale deletion libraries for *Bacillus subtilis*. Cell Syst 4:291–305.e7
6. van Opijnen T, Camilli A (2013) Transposon insertion sequencing: a new tool for systems-level analysis of microorganisms. Nat Rev Microbiol 11:435–442
7. Pettitt SJ, Krastev DB, Pemberton HN et al (2017) Genome-wide barcoded transposon screen for cancer drug sensitivity in haploid mouse embryonic stem cells. Sci Data 4:170020
8. Friedrich MJ, Rad L, Bronner IF et al (2017) Genome-wide transposon screening and quantitative insertion site sequencing for cancer gene discovery in mice. Nat Protoc 12:289–309
9. Yum S-Y, Lee S-J, Kim H-M et al (2016) Efficient generation of transgenic cattle using the DNA transposon and their analysis by next-generation sequencing. Sci Rep 6:27185
10. Bai J, Li K, Tang W et al (2019) A high-throughput screen for genes essential for PRRSV infection using a piggyBac-based system. Virology 531:19–30
11. Ni TK, Landrette SF, Bjornson RD et al (2013) Low-copy piggyBac transposon mutagenesis in mice identifies genes driving melanoma. Proc National Acad Sci U S A 110:E3640–E3649
12. Friedel RH, Friedel CC, Bonfert T et al (2013) Clonal expansion analysis of transposon insertions by high-throughput sequencing identifies candidate cancer genes in a PiggyBac mutagenesis screen. PLoS One 8:e72338
13. Choi J, Landrette SF, Wang T et al (2014) Identification of PLX4032-resistance mechanisms and implications for novel RAF inhibitors. Pigment Cell Melanoma Res 27:253–262
14. Goodman AL, McNulty NP, Zhao Y et al (2009) Identifying genetic determinants needed to establish a human gut symbiont in its habitat. Cell Host Microbe 6:279–289
15. van Opijnen T, Bodi KL, Camilli A (2009) Tn-seq: high-throughput parallel sequencing for fitness and genetic interaction studies in microorganisms. Nat Methods 6:767–772
16. Gawronski J, Wong S, Giannoukos G et al (2009) Tracking insertion mutants within libraries by deep sequencing and a genome-wide screen for *Haemophilus* genes required in the lung. Proc National Acad Sci U S A 106:16422–16427
17. Langridge G, Phan DM, Turner D et al (2009) Simultaneous assay of every *Salmonella Typhi* gene using one million transposon mutants. Genome Res 19:2308–2316
18. Burger BT, Imam S, Scarborough MJ et al (2017) Combining genome-scale experimental and computational methods to identify essential genes in *Rhodobacter sphaeroides*. Msystems 2:e00015–e00017
19. Chihiro S, Masanosuke Y (1987) A series of Tn5 variants with various drug-resistance markers and suicide vector for transposon mutagenesis. Gene 56:283–288
20. Sistrom W (1960) A requirement for sodium in the growth of *Rhodopseudomonas spheroides*. J Gen Microbiol 22:778–785
21. Bertani G (1951) Studies on lysogenesis. I. The mode of phage liberation by lysogenic Escherichia coli. J Bacteriol 62:293–300
22. Martien JI, Hebert AS, Evenson D et al (2019) Systems-level analysis of oxygen exposure in *Zymomonas mobilis*: implications for isoprenoid production. Msystems 4:e00284–e00218
23. Lal P, Wells FM, Lyu Y et al (2019) A Markerless method for genome engineering in *Zymomonas mobilis* ZM4. Front Microbiol 10:2216
24. Larsen R, Wilson M, Guss A et al (2002) Genetic analysis of pigment biosynthesis in *Xanthobacter autotrophicus* Py2 using a new, highly efficient transposon mutagenesis system that is functional in a wide variety of bacteria. Arch Microbiol 178:193–201
25. Dabney J, Meyer M (2012) Length and GC-biases during sequencing library amplification: a comparison of various polymerase-buffer systems with ancient and modern DNA sequencing libraries. BioTechniques 52:87–94
26. Aird D, Ross MG, Chen W-S et al (2011) Analyzing and minimizing PCR amplification bias in Illumina sequencing libraries. Genome Biol 12:R18

27. Arkin AP, Cottingham RW, Henry CS et al (2018) KBase: the United States Department of Energy Systems Biology Knowledgebase. Nat Biotechnol 36:566–569
28. Langmead B, Trapnell C, Pop M et al (2009) Ultrafast and memory-efficient alignment of short DNA sequences to the human genome. Genome Biol 10:R25
29. Langmead B, Salzberg SL (2012) Fast gapped-read alignment with bowtie 2. Nat Methods 9:357–359
30. Li H, Durbin R (2009) Fast and accurate short read alignment with burrows–wheeler transform. Bioinformatics 25:1754–1760
31. Goecks J, Nekruenko A, Talor J (2010) Galaxy: a comprehensive approach for supporting accessible, reproducible, and transparent computational research in the life sciences. Genome Biol 11:1–38
32. Hannon GJ (2010) FASTX-Toolkit
33. Martin M (2011) Cutadapt removes adapter sequences from high-throughput sequencing reads. EMBnet J 17:10–12
34. Bolger AM, Lohse M, Usadel B (2014) Trimmomatic: a flexible trimmer for Illumina sequence data. Bioinformatics 30:2114–2120
35. Homann OR, Johnson AD (2010) MochiView: versatile software for genome browsing and DNA motif analysis. BMC Biol 8:49
36. Robinson JT, Thorvaldsdóttir H, Winckler W et al (2011) Integrative genomics viewer. Nat Biotechnol 29:24–26
37. Kent JW, Sugnet CW, Furey TS et al (2002) The human genome browser at UCSC. Genome Res 12:996–1006
38. Pritchard JR, Chao MC, Abel S et al (2014) ARTIST: high-resolution genome-wide assessment of fitness using transposon-insertion sequencing. PLoS Genet 10:e1004782
39. Barquist L, Mayho M, Cummins C et al (2016) The TraDIS toolkit: sequencing and analysis for dense transposon mutant libraries. Bioinformatics 32:1109–1111
40. McCoy K, Antonio ML, van Opijnen T (2017) MAGenTA: a galaxy implemented tool for complete Tn-Seq analysis and data visualization. Bioinformatics 33:2781–2783
41. Blanchard AM, Leigh JA, Egan SA et al (2015) Transposon insertion mapping with PIMMS - pragmatic insertional mutation mapping system. Front Genet 6:139
42. Solaimanpour S, Sarmiento F, Mrázek J (2015) Tn-seq explorer: a tool for analysis of high-throughput sequencing data of transposon mutant libraries. PLoS One 10:e0126070
43. Turner KH, Wessel AK, Palmer GC et al (2015) Essential genome of *Pseudomonas aeruginosa* in cystic fibrosis sputum. Proc Natl Acad Sci U S A 112:4110–4115
44. DeJesus MA, Ambadipudi C, Baker R et al (2015) TRANSIT--A software tool for Himar1 TnSeq analysis. PLoS Comput Biol 11:e1004401
45. Goodman AL, Wu M, Gordon JI (2011) Identifying microbial fitness determinants by insertion sequencing using genome-wide transposon mutant libraries. Nat Protoc 6:1969–1980
46. Mandal RK, Jiang T, Kwon Y (2017) Essential genome of *Campylobacter jejuni*. BMC Genomics 18:616
47. Nilsson M, Givskov M, Tolker-Nielsen T (2019) Transposon mutagenesis in *Streptococcus* species. Methods Mol Biol 2016:39–49
48. Lin T, Gao L (2018) Genome-wide mutagenesis in *Borrelia burgdorferi*. Methods Mol Biol 1690:201–223
49. Caruccio N (2011) Preparation of next-generation sequencing libraries using Nextera™ technology: simultaneous DNA fragmentation and adaptor tagging by in vitro transposition. Methods Mol Biol 733:241–255
50. Gorbacheva T, Quispe-Tintaya W, Popov VN et al (2015) Improved transposon-based library preparation for the ion torrent platform. BioTechniques 58:200–202
51. Solonenko SA, Ignacio-Espinoza CJ, Alberti A et al (2013) Sequencing platform and library preparation choices impact viral metagenomes. BMC Genomics 14:320
52. Gordon A (2008) FASTX-Toolkit
53. Langmead B (2010) Aligning short sequencing reads with Bowtie. Curr Protoc Bioinformatics. Chapter 11:Unit 11.7
54. Institute B (2019) Picard toolkit
55. Li H, Handsaker B, Wysoker A et al (2009) The sequence alignment/map format and SAMtools. Bioinformatics 25:2078–2079
56. Quinlan AR and Kindlon N (2019) Bedtools
57. Ponstingl H (2015) SMALT
58. Li H (2011) Tabix: fast retrieval of sequence features from generic TAB-delimited files. Bioinformatics 27:718–719
59. Team CR (2013) R: a language and environment for statistical computing
60. Bates D, Carey V, Dettling M, et al (2002) Bioconductor
61. Anaconda (2016) Anaconda Software Distribution

62. Afgan E, Baker D, Batut B et al (2018) The galaxy platform for accessible, reproducible and collaborative biomedical analyses: 2018 update. Nucleic Acids Res 46:gky379
63. Anders S, Huber W (2010) Differential expression analysis for sequence count data. Genome Biol 11:R106
64. Wickham H, François R, Henry L, et al (2018) Dplyr: a grammar of data manipulation
65. Walt S, Colbert CS, Varoquaux G (2011) The NumPy array: a structure for efficient numerical computation. Comput Sci Eng 13:22–30
66. Seabold S and Perktold J (2010) Statsmodels: econometric and statistical modeling with python
67. Virtanen P, Gommers R, Oliphant TE, et al (2019) SciPy 1.0--fundamental algorithms for scientific computing in python
68. Hunter JD (2007) Matplotlib: a 2D graphics environment. Comput Sci Eng 9:90–95
69. wiredfool, Clark A, Hugo, et al (2016) Pillow: 3.1.0
70. Team T (2019) wxPython
71. Schoenborn O (2016) PyPubSub
72. Goryshin I, Reznikoff WS (1998) Tn 5 in vitro transposition. J Biol Chem 273:7367–7374

Chapter 13

Identification and Analysis of Essential Genes in *Streptococcus mutans* with Transposon Sequencing

Alejandro R. Walker and Robert C. Shields

Abstract

Transposon sequencing (Tn-seq) has greatly accelerated the rate at which gene function can be profiled in microbial organisms. This technique has been applied to the study of the dental caries pathogen *Streptococcus mutans* where it has been used to generate large transposon mutant libraries. Coupled with high-throughput sequencing and bioinformatics tools, culture of these transposon mutant libraries has facilitated the identification of essential and conditional essential genes. In this chapter, we describe a procedure for performing Tn-seq studies in *S. mutans* that covers pooled transposon mutant construction, in vitro culture, and DNA library sequencing and data analysis.

Key words Transposon mutagenesis, Essential genes, *Streptococcus mutans*, Tn-seq, *Himar1*, In vitro transposition

1 Introduction

In recent years, screening of gene function in bacteria has been enhanced by a technique known as transposon sequencing (Tn-seq [1]; related Tn-seq methodologies are known as TraDIS [2], INSeq [3], RB-TnSeq [4], HITS [5], SMarT TnSeq [6], and Tn-seq circle [7]). This technique combines saturating transposon mutagenesis with high-throughput DNA sequencing. It permits the simultaneous measurement of transposon insertion frequency across large mutant libraries as they are grown in one or more environments. Changes in the relative abundance of insertions indicate genes and pathways that are essential, or dispensable, in different environments. This technique was originally developed for the study of *Streptococcus pneumoniae* [1]. However, it has more broad applicability to bacterial species that are readily transformable, including other streptococci. Taking advantage of the broader applicability of the technique, we have recently adapted and applied the Tn-seq methodology to the study of the dental caries pathogen *Streptococcus mutans*. Application of Tn-seq in *S. mutans* has led to the

Ren Zhang (ed.), *Essential Genes and Genomes: Methods and Protocols*, Methods in Molecular Biology, vol. 2377, https://doi.org/10.1007/978-1-0716-1720-5_13,

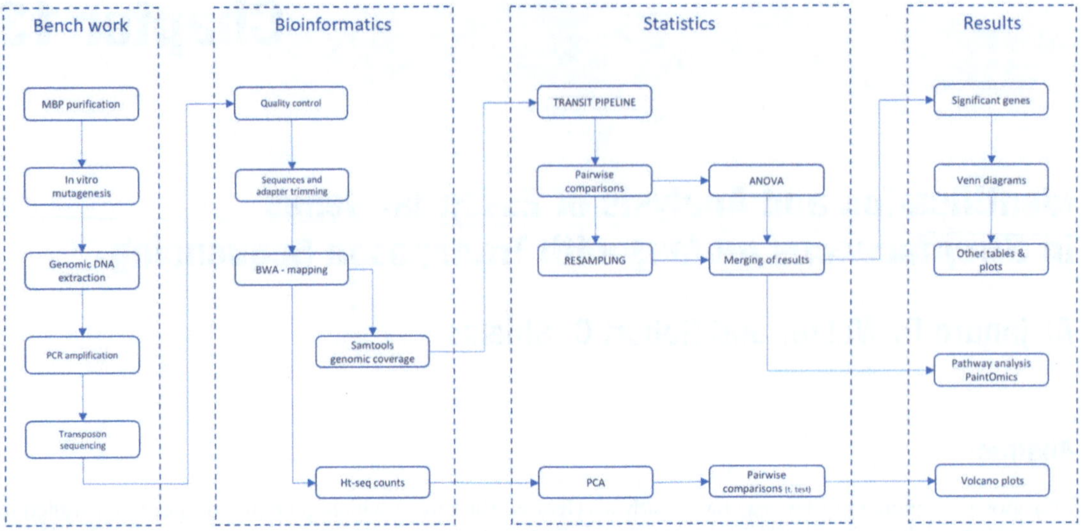

Fig. 1 Flowchart of Tn-seq methodology in *S. mutans*, covering bench work, bioinformatics, statistics, and results generation

identification of genes involved in competence regulation, essential genes in vitro, and possible virulence factors in vivo [8, 9].

Although the general workflow is similar to Opijnen et al., [10] our methodology includes optimized methods for *S. mutans* DNA transformation, DNA extraction, increased sample sequencing (additional oligonucleotide adapter barcodes), and a bioinformatics pipeline. The following protocol describes the steps required to perform Tn-seq in *S. mutans*. These steps include creation of the library by in vitro transposition (Subheadings 3.1 and 3.2), in vitro culture of the transposon library (Subheading 3.3), isolation of genomic DNA (Subheading 3.4), construction of barcoded libraries (Subheading 3.5), Illumina sequencing (Subheading 3.6), and data analysis (Subheading 3.7). A general workflow describing each step is presented in Fig. 1.

2 Materials

2.1 In Vitro Transposition

1. *E. coli* 10-beta (for routine cloning) and BL21 (for recombinant transposase expression).
2. Transposon containing plasmid pMagellan6 [1] (*see* **Note 1**).
3. Transposase-containing plasmid pMalC9 [11].
4. *S. mutans* genomic DNA; a method for isolation of genomic DNA can be found under Subheading 3.4.
5. 2× buffer A: 5 μL of 80% glycerol (v/v), 4 μL of 20 mM DTT, 4 μL of 10 mg/mL acetylated BSA, 8 μL of 5× salt solution (prepare fresh before use, which can be scaled to a larger volume).

6. 5× salt solution: 100 μL of 125 mM HEPES (pH 7.9), 11.5 μL of 5 M NaCl, and 5.75 μL of 1 M $MgCl_2$; prepare fresh before use, which can be scaled to a larger volume.
7. 10× buffer B: 500 mM Tris·Cl (pH 7.8), 100 mM $MgCl_2$, and 10 mM DTT—aliquot and store up to 1 year at −20 °C.
8. 10 U/μL *E. coli* DNA ligase (New England Biolabs).
9. 3 U/μL T4 DNA polymerase (New England Biolabs).
10. 0.52 mM nicotine adenine dinucleotide (New England Biolabs); purchased as a 50 mM solution.
11. 2 mM dNTPs (New England Biolabs).
12. 10 mg/mL acetylated BSA (Promega).
13. Heating block or water bath capable of 12, 16, 30, and 75 °C.

2.2 Transformation of S. mutans Transposon Library

1. Transformable *S. mutans*; transformation efficiency, and the ability to stimulate transformation with signaling peptides is strain dependent.
2. Nutrient-rich growth media, e.g., Brain heart infusion (BHI).
3. Spectinomycin (Sigma-Aldrich).
4. Competence stimulating peptide (18-CSP; SGSLSTFFRLFN RSFTQA); alternatively, transformation can be induced with *comX*-inducing peptide (XIP; GLDWWSL).
5. Nutrient-rich agar plates (e.g., blood or BHI agar) containing 1 mg/mL spectinomycin.
6. BHI containing 25% (v/v) glycerol.

2.3 In Vitro Library Selection

1. Tn-seq library.
2. Growth medium for selection, e.g., BHI, defined media, media plus selective compound, etc.

2.4 Genomic DNA Isolation

1. Epicentre MasterPure Gram-Positive DNA purification kit: this kit provides good yields of DNA from Gram-Positive bacteria. The kit contains a lysis solution, a protein precipitation solution, Proteinase K, RNase A, and lysozyme. Some of these components can be made or bought separately without the use of a kit.
2. Mutanolysin reconstituted to 10,000 U/mL (Sigma-Aldrich).
3. Spheroplasting buffer (20 mM Tris·Cl pH 6.8, 10 mM $MgCl_2$, and 26% (w/v) Raffinose.$5H_20$). Make the Tris-$MgCl_2$ solution 10×, autoclave, and store at room temperature. Take an aliquot and add raffinose accordingly. Store in aliquots at −20 °C.
4. Acid-washed glass beads (0.1 mm) (Sigma-Aldrich).

5. Bead beater (e.g., BioSpec Mini-Beadbeater-16).
6. Screw-cap Eppendorf tubes.
7. 100% isopropanol.
8. 75% ethanol.
9. Nuclease-free water (Thermo Fisher Scientific).
10. Nanodrop spectrophotometer (or alternative, e.g., Qubit).

2.5 Transposon Sequencing Library Preparation

1. Adapter oligonucleotides; these are identical except for an 8-base pair 'barcode' that gives each library a unique ID.
2. Library amplification primers: LibEnrichF and LibEnrichR.
3. 1 mM Tris·Cl, pH 8.3.
4. 2 U/μL *Mme*I restriction enzyme (New England Biolabs).
5. 32 mM *S*-adenosyl methionine (SAM; New England Biolabs).
6. CutSmart® buffer (New England Biolabs).
7. Shrimp Alkaline Phosphatase (rSAP; New England Biolabs).
8. 70 and 100% ethanol.
9. 3 M sodium acetate pH 5.2.
10. 24:23:1 (v/v/v) phenol/chloroform/isoamyl alcohol (Sigma-Aldrich).
11. 400 U/μL T4 DNA ligase and buffer (New England Biolabs).
12. Q5 High-Fidelity 2× Master Mix (New England Biolabs).
13. SYBR Safe DNA Gel Stain (Thermo Fisher Scientific).
14. 50-bp DNA ladder (size range 50–1350 bp; New England Biolabs).
15. 1.8% agarose gel.
16. QIAquick gel extraction kit (Qiagen).
17. Thermocycler machine.

2.6 Transposon DNA Library Sequencing

1. Amplified barcoded library DNA.
2. Illumina genomic DNA sequencing primer.
3. Library size quantification machine (e.g., Agilent 2200 Tapestation System).
4. DNA quantification tool (e.g., Qubit 4 Fluorometer).
5. Illumina NextSeq500.

2.7 Data Analysis

The bioinformatics, data preparation, and analyses were all conducted in the High Performance Cluster (HiPerGator) at the University of Florida, running under Red Hat Linux server release 7.6 (Maipo). A comprehensive list of the programs/commands used during the development of this work is listed in Table 1.

Table 1
Programs and commands used during Tn-seq data analysis

Software Tools	Action	Command
BARCOSEL	Barcode selection	
FASTX_TOOLKIT	Demultiplexing	fastx_barcode_splitter.Pl
FastQC	Quality control	
BWA	Reads mapping	Bwa-aln
HT-seq	Gene mapped reads count	Htseq-count
R statistical software	Plotting coverage	Plot
SAMTOOLS	Alignment file manipulation	View, sorting, and depth
IGV	Visualization of counts	
TRANSIT	Analysis of Tn-seq data	ANOVA and resampling

3 Methods

3.1 In Vitro Transposition of S. mutans Genomic DNA

This is a slightly modified version of a protocol originally developed in *S. pneumoniae* [10]. The protocol should work in any transformable *Streptococcus* species.

1. Generate recombinant MarC9 transposase. We follow a previously described protocol [10, 12] with a typical yield of 100–500 μg/mL. MarC9 must be kept in small aliquots (e.g., 10 μL) as thawing more than once lowers transposition efficiency. We have found that it is relatively stable when stored at −80 °C for long periods of time (>1 year); it is therefore optimal to make a large amount of enzyme and to use it over the course of several experiments.
2. For each transposition reaction, combine *S. mutans* genomic DNA, pMagellan6 plasmid DNA (1:1 ratio of genomic DNA to plasmid DNA), 2× buffer A, distilled water, and MarC9 transposase. The MarC9 protein should be added last. Example reactions are displayed in Table 2 (*see* **Note 2**).
3. The in vitro transposition reaction should be incubated for 1 h at 30 °C in a water bath. Afterward, the MarC9 protein should be heat inactivated for 10 min at 75 °C in a heating block and then transferred to ice for 5 min.
4. Precipitate DNA by adding 1/10 volume of 3 M sodium acetate (pH 5.2) and 2 volumes of 100% ice-cold ethanol and incubating for 20 min at 4 °C.
5. Microcentrifuge for 15 min at maximum speed, 4 °C, to form a pellet and then wash pellet with room temperature 70% ethanol

Table 2
Reagents used during in vitro transposition

Reagents	Small scale	Large scale
S. mutans genomic DNA	1 μg	15 μg
pMagellan6 plasmid	1 μg	15 μg
2x buffer A	10 μL	150 μL
Distilled water	Up to 19.5 μL	Up to 292.5 μL
MarC9	0.5 μL	7.5 μL

Table 3
Reagents used for transposon junction repair

Reagents	Small scale (μL)	Large scale (μL)
10× buffer B	2	30
10 mg/mL BSA	2	30
2 mM dNTPs	1	15
Distilled water	3	45
T4 DNA polymerase	1	15
DNA	11	165

and air dry. Dissolve pellet in 11 μL of water (per 20 μL of original reaction).

6. Next, the transposon junctions must be repaired. On ice, combine the following for each transposition reaction: 10× buffer B, BSA (10 mg/mL), dNTPs (2 mM), T4 DNA polymerase, distilled water, and DNA from **step 5**. Example reactions are shown in Table 3.
7. Incubate for 20 min at 12 °C with a water bath and then inactivate the reaction for 10 min at 75 °C.
8. To each 20 μL of heat killed T4 DNA polymerase reaction mixture add the following mixture: 1 μL (or 15 μL) of 0.52 mM NAD and 1 μL of (or 15 μL) *E. coli* DNA ligase.
9. Incubate overnight at 16 °C with a water bath. The ligation reactions can be stored at −80 °C, but we typically transform *S. mutans* the following day.

3.2 Transformation of S. mutans *Transposon Library*

1. Streak an *S. mutans* strain on a BHI agar plate and incubate for 2 days at 37 °C with 5% CO_2.
2. Pick a single colony, grow in 1 mL of BHI, and incubate overnight at 37 °C with 5% CO_2.

Table 4
Example transposon colonies obtained after transformation

Library	CFU/Plate	Plates/50,000 CFU
1	1279	40
2	879	57
3	1384	36
4	713	70

3. The following day inoculate 5 mL of BHI with 100 μL of the overnight culture (1:50 dilution). Incubate until the $OD_{600} = 0.1$, which usually takes 1–2 h.
4. To each 200 μL of culture volume add 2 μM CSP (final concentration) and 20 μL of transposon library DNA (*see* Subheading 3.1). Please *see* **Note 3** for an alternative protocol using *comX* inducing peptide (XIP).
5. Incubate for 2.5 h at 37 °C with 5% CO_2.
6. Spread 50–100 μL of transformation culture per blood agar plate (containing 1 mg/mL spectinomycin) and incubate for 2 days at 37 °C with 5% CO_2. Table 4 presents results from actual transposon library construction experiments, with estimates of CFUs per plates, and number of plates needed for 50,000 CFU.
7. Estimate library size from colony counting on plates. A gel dock imaging station is acceptable for imaging the plates, and colonies can be counted manually or using appropriate software.
8. Suspend colonies from one plate in 2 mL of BHI containing 1 mg/mL spectinomycin and combine colonies from multiple plates in a 50 mL falcon tube.
9. Outgrow for 30–60 min at 37 °C with 5% CO_2. This is to select for spectinomycin resistance.
10. Centrifuge at 3500 × *g* for 15 min at room temperature and discard the supernatant. Resuspend the cell pellet in 5 mL of BHI containing 25% glycerol.
11. Aliquot into twenty (250 μL each) 2 mL screw-cap microcentrifuge tubes and store at −80 °C.

3.3 Library Selection

1. Prior to library selection determine the CFU/mL of the library aliquots (they will likely contain high CFU/mL). This is required so that during outgrowth of the library in the chosen condition, cultures can be inoculated with a predetermined CFU/mL (e.g., 10^6 CFU/mL).

2. Thaw Tn-seq library on ice and target an inoculum concentration that facilitates a ca. 100-fold dilution to the desired CFU/mL. When making the inoculum, be careful to avoid bottlenecks (*see* **Note 4**).
3. Example conditions and inoculum are as follows: 50 mL of BHI inoculated with 100 μL of 1×10^9 CFU/mL will yield a final CFU/mL of 2×10^6 cells. If this culture grows back to a CFU/mL of 1×10^9 cells (e.g., late exponential), this would amount to the bacteria having divided nine times (or generations). The amount of selective pressure depends on the conditions of growth and must be carefully considered.
4. After growth of the culture, pellet by centrifugation at $3500 \times g$ for 10 min at 4 °C. Aspirate and discard the supernatant and freeze the pellet at −80 °C for subsequent DNA isolation (*see* Subheading 3.4).

3.4 Genomic DNA Isolation

1. Thaw cell pellet on ice and suspend cells in 150 μL of spheroplasting buffer.
2. Add lysozyme (final concentration 250 μg/mL) and mutanolysin (final concentration 500 U/mL). Incubate at 37 °C for 30 min.
3. Add 150 μL of MasterPure Gram Positive Cell Lysis Solution (2×) to each sample and transfer to a screw cap Eppendorf that contains 25–50 mg of acid washed glass beads (0.1 mm).
4. Homogenize sample with a cell lysis machine. Keep samples on ice.
5. Add 1 μL of Proteinase K (50 μg/μL) and incubate at 65 °C for 30 min, vortexing every 5 min.
6. Cool the samples to 37 °C and add 1 μL of RNase A (5 μg/μL), incubated at 37 °C for 30 min. Place the samples on ice for 3–5 min before proceeding with next step.
7. Add 175 μL of MPC Protein Precipitation Reagent to lysed sample and vortex mix for 10 s.
8. Pellet the precipitated debris by centrifugation at $>10{,}000 \times g$ for 10 min at 4 °C. Keep samples on ice.
9. Transfer the supernatant (ca., 350 μL) to clean microcentrifuge tube and discard the pellet. Add 500 μL of isopropanol to the supernatant to precipitate the DNA (invert tube 20–40 times). Precipitated DNA should be visible. Place on ice for at least 10 min.
10. Pellet the DNA by centrifugation at $>10{,}000 \times g$ for 10 min at 4 °C.

11. Remove the isopropanol. Wash twice with 75% EtOH (ice cold), with centrifuge steps in-between (10 min at 4 °C). Remove residual EtOH and air dry for 10 min.
12. Resuspend DNA pellet in 100 μL of nuclease-free water.
13. Measure the concentration and purity of the DNA using a Nanodrop spectrophotometer or similar technique. Store samples at −20 °C.

3.5 Transposon Sequencing Library Preparation

1. Order Tn-seq barcode adapter oligonucleotides. These must be PAGE purified, and B oligonucleotides are phosphorylated on both ends. Oligonucleotide sequences are shown in Table 5. Using these adapters, it is possible to sequence 18 samples per Illumina flow cell (NextSeq500). We used the tool, BARCOSEL, to design optimal barcodes that should minimize sequencing errors [13].
2. Resuspend library barcode adapter oligonucleotides in 0.2 mM 1 mM Tris·Cl pH 8.3. Mix equal volumes of A and B barcodes in a PCR tube. Heat for 2 min at 96 °C using a heating block, then move heating block to the laboratory bench, and allow to cool to room temperature, which will cause the oligonucleotides to anneal. Store annealed barcodes at −20 °C.
3. Next, digest the genomic DNA prepared previously (*see* Subheading 3.4). Digest for at least 4 h in the following reaction: 6 μg of genomic DNA, 3 μL of (6 U) MmeI, 0.44 μL of 32 mM SAM, 20 μL of CutSmart buffer, and 2 μL of rSAP, made to a total volume of 200 μL with distilled water. Afterward, heat inactivate the reaction at 65 °C for 15 min.
4. Add one volume of 24:23:1 (v/v/v) phenol/chloroform/isoamyl alcohol to the digested sample and vortex thoroughly for 2 min (to denature protein). Centrifuge at room temperature for 5 min at >10,000 × *g*. Remove the upper aqueous phase and transfer to a fresh microcentrifuge tube. Precipitate the DNA by adding the following: 1/10 volume of 3 M sodium acetate pH 5.2 and 2 volumes of 100% EtOH. Mix and freeze (−20 °C overnight or −80 °C for 1 h) before centrifuging at 4 °C for 5 min at >10,000 × *g*. After removing the supernatant, wash once with 70% EtOH and centrifuge at 4 °C for 5 min at >10,000 × *g*. Air dry the DNA pellet and resuspend in 25 μL of distilled water.
5. Ligate the adapter barcodes by mixing the following: 25 μL of extracted DNA, 1 μL of annealed adapter (from **step 2**), 1 μL of 400 U/μL T4 DNA ligase, and 3 μL of T4 DNA ligase buffer. Incubate in water bath overnight at 16 °C.
6. Amplify the Tn-seq libraries by combining the following: 25 μL of Q5 High-Fidelity 2× Master Mix, 1.25 μL of 10 μM LibEnrichF, 1.25 μL of 10 μM LibEnrichR, 1.5 μL of ligation mix,

Table 5
Tn-seq barcode adapter oligonucleotides

Name	Sequence (5′–3′)
TnLibBarcode1A	TTCCCTACACGACGCTCTTCCGATCT **TATAGCCT** NN
TnLibBarcode1B	p-**AGGCTATA** AGATCGGAAGAGCGTCGTGTAGGGAAAGAG-p
TnLibBarcode2A	TTCCCTACACGACGCTCTTCCGATCT **ATAGAGGC** NN
TnLibBarcode2B	p-**GCCTCTAT** AGATCGGAAGAGCGTCGTGTAGGGAAAGAG-p
TnLibBarcode3A	TTCCCTACACGACGCTCTTCCGATCT **CCTATCCT** NN
TnLibBarcode3B	p-**AGGATAGG** AGATCGGAAGAGCGTCGTGTAGGGAAAGAG-p
TnLibBarcode4A	TTCCCTACACGACGCTCTTCCGATCT **GGCTCTGA** NN
TnLibBarcode4B	p-**TCAGAGCC** AGATCGGAAGAGCGTCGTGTAGGGAAAGAG-p
TnLibBarcode5A	TTCCCTACACGACGCTCTTCCGATCT **AGGCGAAG** NN
TnLibBarcode5B	p-**CTTCGCCT** AGATCGGAAGAGCGTCGTGTAGGGAAAGAG-p
TnLibBarcode6A	TTCCCTACACGACGCTCTTCCGATCT **TAATCTTA** NN
TnLibBarcode6B	p-**TAAGATTA** AGATCGGAAGAGCGTCGTGTAGGGAAAGAG-p
TnLibBarcode7A	TTCCCTACACGACGCTCTTCCGATCT **CAGGACGT** NN
TnLibBarcode7B	p-**ACGTCCTG** AGATCGGAAGAGCGTCGTGTAGGGAAAGAG-p
TnLibBarcode8A	TTCCCTACACGACGCTCTTCCGATCT **GTACTGAC** NN
TnLibBarcode8B	p-**GTCAGTAC** AGATCGGAAGAGCGTCGTGTAGGGAAAGAG-p
TnLibBarcode9A	TTCCCTACACGACGCTCTTCCGATCT **AACTGCGC** NN
TnLibBarcode9B	p-**GCGCAGTT** AGATCGGAAGAGCGTCGTGTAGGGAAAGAG-p
TnLibBarcode10A	TTCCCTACACGACGCTCTTCCGATCT **CGCTAAGC** NN
TnLibBarcode10B	p-**GCTTAGCG** AGATCGGAAGAGCGTCGTGTAGGGAAAGAG-p
TnLibBarcode11A	TTCCCTACACGACGCTCTTCCGATCT **GGTTGACC** NN
TnLibBarcode11B	p-**GGTCAACC** AGATCGGAAGAGCGTCGTGTAGGGAAAGAG-p
TnLibBarcode12A	TTCCCTACACGACGCTCTTCCGATCT **GTTGAATA** NN
TnLibBarcode12B	p-**TATTCAAC** AGATCGGAAGAGCGTCGTGTAGGGAAAGAG-p
TnLibBarcode13A	TTCCCTACACGACGCTCTTCCGATCT **TCGGCTAA** NN
TnLibBarcode13B	p-**TTAGCCGA** AGATCGGAAGAGCGTCGTGTAGGGAAAGAG-p
TnLibBarcode14A	TTCCCTACACGACGCTCTTCCGATCT **TGGATTCA** NN
TnLibBarcode14B	p-**TGAATCCA** AGATCGGAAGAGCGTCGTGTAGGGAAAGAG-p
TnLibBarcode15A	TTCCCTACACGACGCTCTTCCGATCT **TGGTTGGC** NN
TnLibBarcode15B	p-**GCCAACCA** AGATCGGAAGAGCGTCGTGTAGGGAAAGAG-p
TnLibBarcode16A	TTCCCTACACGACGCTCTTCCGATCT **TTGCCGAT** NN

(continued)

Table 5
(continued)

Name	Sequence (5′–3′)
TnLibBarcode16B	p-**ATCGGCAA** AGATCGGAAGAGCGTCGTGTAGGGAAAGAG-p
TnLibBarcode17A	TTCCCTACACGACGCTCTTCCGATCT **AACCAACG** NN
TnLibBarcode17B	p-**CGTTGGTT** AGATCGGAAGAGCGTCGTGTAGGGAAAGAG-p
TnLibBarcode18A	TTCCCTACACGACGCTCTTCCGATCT **CAAGGCTG** NN
TnLibBarcode18B	p-**CAGCCTTG** AGATCGGAAGAGCGTCGTGTAGGGAAAGAG-p

Table 6
Tn-seq library enrichment oligonucleotides

Name	Sequence (5′–3′)
LibEnrichF	CAA GCA GAA GAC GGC ATA CGA GAT AGA CCG GGG ACT TAT CAT CCA ACC TGT
LibEnrichR	AAT GAT ACG GCG ACC ACC GAG ATC TAC ACT CTT TCC CTA CAC GAC GCT CTT CCG ATCT

and 21 μL of distilled water. Oligonucleotide sequences are shown in Table 6. PCR settings should be set at 98 °C (30 s), 15–20× [98 °C (10 s), 56 °C (30 s), 72 °C (10 s)], and 72 °C (5 min). The cycle number is variable because amplification should be tested across several cycles to ensure it remains linear to minimize amplification biases (*see* **Note 5**).

7. Run PCR products on a 2% agarose gel using SYBR safe gel DNA stain and a 50-bp ladder. Excise the 120-bp band under UV light. Purify the DNA from the agarose gel using the QIAquick gel extraction kit.
8. Measure the DNA concentration using a Nanodrop spectrophotometer (expected concentration 5–15 ng/μL).

3.6 Transposon DNA Library Sequencing

1. Measure library DNA concentration using a fluorometric method (e.g., Qubit 4 Fluorometer) and library size (e.g., Agilent 2200 TapeStation System). The library size should be a consistent 120-bp and not contain any additional bands.
2. Pool the barcoded libraries by equal molar mass. The final library pool should be >5 nM.
3. Sequence on an Illumina platform (NextSeq500 is recommended because it allows for enough reads for downstream analysis). A standard Illumina genomic DNA sequencing primer is used (5′ ACACTCTTTCCCTACACGACGCTCTTCCGATCT 3′). The barcoded pool should be

Table 7
Sequencing metrics (reads, mapped reads, and reads within genes) obtained after Tn-seq library processing

Sample ID	Total read count	Mapped reads	Percentage of mapped reads	Mapped reads within gene space
S1C10001	21,044,026	21,015,976	99.87%	15,670,839
S1C10002	15,820,071	15,797,840	99.86%	11,704,728
S1C20001	15,512,337	15,472,655	99.74%	11,190,256
S1C20002	11,434,317	11,418,098	99.86%	8,527,921
S1C30001	16,167,445	16,139,956	99.83%	12,024,478
S1C30002	20,435,536	20,401,168	99.83%	15,238,288
S2C10001	19,075,630	18,899,021	99.07%	13,738,211
S2C10002	18,773,364	18,643,673	99.31%	13,657,966
S2C20001	15,071,577	15,044,528	99.82%	11,398,159
S2C20002	18,869,633	18,846,889	99.88%	14,185,465
S3C10002	13,973,336	13,950,459	99.84%	10,460,705
S3C20001	19,038,069	19,014,367	99.88%	14,741,811
S3C20002	15,579,653	15,552,203	99.82%	11,663,545
S4C10001	16,910,420	16,892,884	99.90%	13,021,518
S4C10002	15,526,451	15,515,910	99.93%	11,714,038
S4C20001	18,646,062	18,628,201	99.90%	14,400,825
S4C20002	31,951,128	31,913,679	99.88%	24,819,317
S5C10001	11,142,502	11,118,131	99.78%	7,830,173
S5C10002	16,073,204	16,061,330	99.93%	12,187,700
S5C20001	17,103,888	17,091,235	99.93%	13,169,285
S5C20002	20,253,516	20,237,417	99.92%	15,308,043
S6C10001	14,338,717	13,806,073	96.29%	11,272,922
S6C10002	10,963,377	10,949,700	99.88%	8,582,504
S6C20001	12,332,293	12,325,219	99.94%	10,270,635
S6C20002	12,963,537	12,947,446	99.88%	10,077,760

run with a 15–25% PhiX spike-in control to monitor sequencing quality. Sequencing parameters: single-read, 51 cycles.

4. The total amount of reads is usually >300 M, with approximately 20–30% of these reads aligned to PhiX. Expect 15–30 M reads for each barcoded sample. Table 7 shows a summary of the sequencing results after QC.

3.7 Data Analysis

1. Read files generated from Illumina sequencing should be concatenated into a single file. This is achieved by creating a new file and concatenating the data from four lanes. The Linux command to do this would be: for lane in {001.0.004}; do cat data_run_1_${lane}.fastq.gz >> data_run_1.fastq.gz; done. Note: even though the files are compressed (*.gz), this action would not compromise the binary structure of neither the individual nor resulting files.
2. Next, demultiplex the reads according to barcode identifiers. The barcode splitter tool from the FASTX_TOOLKIT [14] suit can be used (allowing for no mismatches).
3. The quality of the sequencing reads should be assessed using FastQC [15]. Typically, it shows a quality baseline between Phred scores 34 and 36 for almost the full length of the reads. Based on this, reads do not require quality trimming, which is normally the case; nevertheless, if quality drops considerably, trimming should be considered (quality scores <20 should be removed).
4. After FastQC, reads are processed so that nongenomic DNA sequences (primers, transposon DNA, barcodes, etc.) are removed. The genomic DNA sequences were isolated from each read in the data again using bash commands (e.g., awk -v len=${len} '{ if(NR%2==1) print $1; if(NR%2==0) print substr($1,9,len) }' < sequence_read.fasta, where "len" is the number of nucleotides to keep starting from position 9). This isolates the genomic DNA sequence, which is expected to be ~20 bp in length. Our internal testing has shown that genomic DNA reads have a variable length from 10 to 20 bases. 16-bp reads are chosen for further analyses based on the total number of successfully mapped reads.
5. The processed reads are now ready to be mapped to a reference genome. The reference genome for *Streptococcus mutans* UA159 has the RefSeq assembly accession number GCF_000007465.2. The reference genome should be downloaded from NCBI with the features file (GFF). Several programs can map reads to a reference genome; we use a BWA-ALN [16] fast short reads alignments pipeline. The output should be a Sequence Alignment Map format file (SAM).
6. After mapping reads to a reference genome, reads within genes should be counted with HT-Seq [17] with the following settings: --stranded no, --mode intersection-strict, --type gene, and --idattr locus_tag. *See* **Note 6** for an alternative approach to read mapping.
7. Next, we parse the read counts dataset into R [18] statistical software for principal component analysis and plotting. This can give valuable insights on how the samples and strains are

clustering for the different conditions. It will also show if any samples have aberrant deviations compared with others in similar conditions, which would result from poor sequencing quality or biological variability.

There are several ways to analyze Tn-seq data after sequencing alignments and/or read counting. We use three different methods that we present below:

8. Visual assessment of read count distribution across the genome is a useful way to interpret Tn-seq findings. Genome browsers allow the user to scan visually for regions of the genome with certain attributes; for Tn-seq, this could be over- or under-represented areas that may indicate essentiality or transposon-mediated polarity effects. We use the Integrative Genomics Viewer (IGV) [19]. To visualize data within IGV, users must upload a genome file (created from FASTA and annotation files that can be downloaded from NCBI) and then align sequenced reads in BAM format (created using SAMTOOLS [20]).
9. Another way to assess Tn-seq data is with the TRANSIT command line pipeline [21]. Preparing data for TRANSIT requires additional steps that include sorting the alignment files with the command sort from the SAMTOOLS [20] suit (output should be in BAM format) and later generating a depth of coverage map (command depth also from SAMTOOLS), which should be a text file containing three columns; first is the reference genome ID, second is the genomic coordinate, and third is the count of reads mapping at each coordinate. These coverage files are the input data for TRANSIT. The files names and experimental design for pairwise comparisons of two conditions listed in a metadata file would be parsed to the subsequent subroutine analyses (e.g., ANOVA and resampling). The generation of metadata, datasets, and analyses can all be automated in batch scripts (*see* **Note 7**). In order to achieve better results and to avoid over-representation of genes due to sequences mapping in the intergenic region upstream and downstream that also mapped to genes, the GFF gene coordinates were modified by increasing the starting coordinate by a 5% of the gene length and reducing the ending coordinate by a 10% of the gene length. Significant genes can be called using two methods: ANOVA and Resampling. We process the data using both statistical methods; the lists of significant genes from both methods are combined, and only those genes that were significant in both methods are considered further. The resulting dataset can then be processed for graphical or tabular display.
10. Alternatively, genes required for fitness in certain environments can be determined using the much simpler method 'dval' [22]. Using the read count data generated from HT-Seq,

dval can be calculated for each gene. First, calculate the expected number of reads per gene based on gene size and transposon library size, assuming total randomness of Tn insertion (normalized to one million reads). The observed reads per gene (normalized to one million reads) are then divided by the expected reads obtaining a ratio. For a gene, a ratio of 0.01 to 0.1 indicates compromised fitness, with a ratio less than 0.01 indicating essentiality of that gene. A dval of 1 indicates that mutations within that gene have no fitness cost. This method has proven to be effective for determining gene essentiality in *S. mutans* with subsequent validation with CRISPR interference gene silencing [9, 23]. However, it does not include assessments of significance across conditions.

4 Notes

1. The Magellan6 mariner transposon contains the spectinomycin resistance gene *aad9*. In addition, it lacks transcriptional terminators, and this should allow the transcription of downstream genes. We have noticed that the Magellan6 transposon can have polar effects on the expression of downstream genes under certain conditions. This is likely related to expression of the spectinomycin resistance gene. It is prudent when using a transposon library for screening/enrichment purposes to keep the expansion of the library low because overexpression of certain genes could result in mutants that quickly overtake the population.
2. It is best to perform a 'small scale' experiment prior to making a saturated transposon library. This will be a good indicator for how active the transposase is and the efficiency of transformation. In our studies, the 'large scale' experiment provided enough transposed DNA to make a library of 50,000–100,000 mutants (plated on 50 blood agar plates). However, information gained from the 'small scale' experiment will facilitate scaling accordingly.
3. Transformation with XIP is carried out exactly as with CSP except that the cells must be grown in a defined media (e.g., FMC or CDM) as peptides present in rich media block XIP signaling. Transformation efficiency with XIP in CDM may exceed that of CSP in rich media under certain conditions [24].
4. Bottleneck effects are less likely to be an issue during in vitro culture than while running in vivo experiments. However, care should be taken to minimize bottleneck effects (insertion mutants lost by chance rather than fitness defects). Chao et al. [25] provide a good discussion on the design of Tn-seq experiments.

5. The enrichment PCR step can introduce bias during library preparation [26]. Due to differences in amplifications reagents, thermocycler machines, etc., it is difficult to follow a standardized protocol for library amplification. As a result, PCR conditions will need to be optimized on a case-by-case basis.
6. Galaxy, an online open source bioinformatics platform, is an alternative for researchers who are not knowledgeable in command line [27]. Galaxy hosts all the tools required to complete Subheading 3.7, **step 1–6**:

   Concatenate datasets: join Illumina sequencing files together.

   Barcode Splitter: split files according to the barcode sequence.

   FastQC: assessment of read quality.

   Trimmomatic: removal of the first 8 bases of the sequencing read (barcode).

   Clip: transposon sequences clipped from the 3′ end of reads (e.g., ACAGGTTGGATGATAA).

   Bowtie2/Map with BWA: mapping genomic reads to a reference genome.

   htseq-count: calculate transposon insertion reads per gene.
7. Transit pipeline bash script:
   (a) ANOVA analysis:

```
--
--
--
--
--
--
--
--
--
--
--
--
--
--
--
--
--
--
--
--
--
--
--
```

```
- -
- -
- -
- -
- -
- -
- -
```

```
----------------------------------------------------------------
#!/usr/bin/bash
### Creating time stamp variable
start='date +%s'
### Loading transit module
module load transit
###FormattingsystemvariablesfromSLURMarrayvariableandexterna-
l files
pos1=$(( ${SLURM_ARRAY_TASK_ID} + (${SLURM_ARRAY_TASK_ID}-1) *
3 ))
pos2=$(( $pos1 + 3 ))
strain=$( cat ./transit_anova_list.txt | sed -n "${pos1},
${pos1}p" | gawk '{ print $1 }' )
files=$( cat ./transit_anova_list.txt | sed -n "${pos1},
${pos2}p" | cut -f 4 | tr '\n' ',' | sed 's/,*$//g')
### Verifying if metadata file exist and deleting it in case it
does
if [ -e ../metadata/${strain}_metadata_5-10.txt ]
then
echo "deleting metadata"
rm ../metadata/${strain}_metadata_5-10.txt
fi
### Reformatting new metadata file
echo "ID Condition Filename" > ../metadata/${strain}_metada-
ta_5-10.txt
cat ./transit_anova_list.txt | sed -n "${pos1},${pos2}p" |
gawk '{ print $2"\t"$3"\t"$4 }' >> ../metadata/${strain}
_metadata_5-10.txt
### Combining all wig files
transit export combined_wig ${files} ../transit_UA159_5-
10_perc_trimmed.gff.prot_table ./${strain}_combine_data.wig
### Running ANOVA method on the data
transit anova ./${strain}_combine_data.wig ../metadata/
${strain}_metadata_5-10.txt ../transit_UA159_5-10_perc_-
trimmed.gff.prot_table ./${strain}_results_TTR.txt
### Print run time
end='date +%s'
runtime=$((end-start))
out=$(echo "scale=2; (${runtime}/1)" | bc); echo "${out}
seconds"
```

(b) Resampling analysis.

```
- -
- -
- -
- -
- -
- -
- -
- -
- -
- -
- -
- -
- -
- -
- -
- -
- -
- -
```

```
------------------------------------------------------------
#!/usr/bin/bash
### Creating time stamp variable
start='date +%s'
### Loading transit module
ml transit
###FormattingsystemvariablesfromSLURMarrayvariableandexternal
files
pos1=$(( ${SLURM_ARRAY_TASK_ID} + (${SLURM_ARRAY_TASK_ID}-1) *
3 ))
pos2=$(( $pos1 + 1  ))
pos3=$(( $pos1 + 2  ))
pos4=$(( $pos1 + 3  ))
strain=$( cat transit_resampling_list.txt | sed -n "${pos1},
${pos1}p" | gawk '{ print $1 }' )
files1=$( cat transit_resampling_list.txt | sed -n "${pos3},
${pos4}p" | cut -f 4 | tr '\n' ',' | sed 's/,*$//g')
files2=$( cat transit_resampling_list.txt | sed -n "${pos1},
${pos2}p" | cut -f 4 | tr '\n' ',' | sed 's/,*$//g')

### Running Resampling method (default) on the data
transit resampling ${files1} ${files2} ../transit_UA159_5-
10_perc_trimmed.gff.prot_table ${strain}_resampling_default.
txt -h

### Print run time
end='date +%s'
runtime=$((end-start))
out=$(echo "scale=2; (${runtime}/1)" | bc); echo "${out}
```

```
seconds"
- -
- -
- -
- -
- -
- -
- -
- -
- -
- -
- -
- -
- -
- -
- -
- -
- -
- -
- -
- -
- -
- -
- -
- -
- -
- -
- -
- -
- -
- -
____________________________________________________________-
------------------------------------------------------------
```

Acknowledgements

We thank Nick Jakubovics at Newcastle University for help with the genomic DNA isolation protocol; Andrew Camilli at Tufts University for providing us the Tn-seq plasmids (pMalC9 and pMagellan6); Lin Zeng at the University of Florida for assistance with adapting the original Tn-seq protocol to work with *S. mutans* and help with editing this chapter; and David Moraga at the Interdisciplinary Center for Biotechnology Research at the University of Florida for help with troubleshooting and optimization of Tn-seq DNA sequencing.

References

1. van Opijnen T, Bodi KL, Camilli A (2009) Tn-seq: high-throughput parallel sequencing for fitness and genetic interaction studies in microorganisms. Nat Methods 6(10):767–772
2. Langridge GC, Phan M-D, Turner DJ, Perkins TT, Parts L, Haase J et al (2009) Simultaneous assay of every *Salmonella Typhi* gene using one million transposon mutants. Genome Res 19:2308–2316. https://doi.org/10.1101/gr.097097.109
3. Goodman AL, McNulty NP, Zhao Y, Leip D, Mitra RD, Lozupone CA et al (2009) Identifying genetic determinants needed to establish a human gut symbiont in its habitat. Cell Host Microbe 6(3):279–289
4. Wetmore KM, Price MN, Waters RJ, Lamson JS, He J, Hoover CA et al (2015) Rapid quantification of mutant fitness in diverse bacteria by sequencing randomly bar-coded transposons. mBio 6(3):e00306–e00315
5. Gawronski JD, Wong SMS, Giannoukos G, Ward DV, Akerley BJ (2009) Tracking insertion mutants within libraries by deep sequencing and a genome-wide screen for *Haemophilus* genes required in the lung. Proc Natl Acad Sci U S A 106(38):16422–16427
6. Dale JL, Beckman KB, Willett JLE, Nilson JL, Palani NP, Baller JA et al (2018) Comprehensive Functional Analysis of the *Enterococcus faecalis* Core Genome Using an Ordered, Sequence-Defined Collection of Insertional Mutations in Strain OG1RF. mSystems 3(5): e00062–e00018
7. Gallagher LA, Shendure J, Manoil C (2011) Genome-Scale Identification of Resistance Functions in *Pseudomonas aeruginosa* Using Tn-seq. mBio 2(1):e00315–e00310
8. Shields RC, O'Brien G, Maricic N, Kesterson A, Grace M, Hagen SJ et al (2018) Genome-wide screens reveal new gene products that influence genetic competence in *Streptococcus mutans*. J Bacteriol 200(2):16 e00508-17
9. Shields RC, Zeng L, Culp DJ, Burne RA (2018) Genomewide identification of essential genes and fitness determinants of *Streptococcus mutans* UA159. mSphere 3(1):e00031-18
10. van Opijnen T, Lazinski DW, Camilli A (2014) Genome-wide fitness and genetic interactions determined by Tn-seq, a high-throughput massively parallel sequencing method for microorganisms. Curr Protoc Mol Biol 106:7.16.1–7.16.24
11. Lampe DJ, Akerley BJ, Rubin EJ, Mekalanos JJ, Robertson HM (1999) Hyperactive transposase mutants of the Himar1 mariner transposon. Proc Natl Acad Sci U S A 96 (20):11428–11433
12. Lampe DJ, Churchill ME, Robertson HM (1996) A purified mariner transposase is sufficient to mediate transposition *in vitro*. EMBO J 15:5470–5479
13. Somervuo P, Koskinen P, Mei P, Holm L, Auvinen P, Paulin L (2018) BARCOSEL: a tool for selecting an optimal barcode set for high-throughput sequencing. BMC Bioinformatics 19(1):257
14. Gordon A, Hannon GJ (2014) Gordon. FASTX-Toolkit. http://hannonlab.cshl.edu/fastx_toolkit
15. Andrews S. (2015) FASTQC: a quality control tool for high throughput sequence data. Babraham Institute
16. Li H, Durbin R (2009) Fast and accurate short read alignment with burrows-wheeler transform. Bioinformatics 25(14):1754–1760
17. Anders S, Pyl PT, Huber W (2015) HTSeq-A Python framework to work with high-throughput sequencing data. Bioinformatics 31(2):166–169
18. Team RDC. (2010) R: a language and environment for statistical computing. Vienna, Austria: R Foundation for Statistical Computing
19. Robinson JT, Thorvaldsdóttir H, Winckler W, Guttman M, Lander ES, Getz G et al (2011) Integrative genomics viewer. Nat Biotechnol 29(1):24–26
20. Li H, Handsaker B, Wysoker A, Fennell T, Ruan J, Homer N et al (2009) The sequence alignment/map format and SAMtools. Bioinformatics 25(16):2078–2079
21. DeJesus MA, Ambadipudi C, Baker R, Sassetti C, Ioerger TR (2015) TRANSIT - a software tool for Himar1 TnSeq analysis. PLoS Comput Biol 11(10):e1004401
22. Valentino MD, Foulston L, Sadaka A, Kos VN, Villet RA, Santa Maria J et al (2014) Genes contributing to Staphylococcus aureus fitness in abscess- and infection-related ecologies. mBio 5(5):e01729–e01714
23. Shields RC, Walker AR, Maricic N, Chakraborty B, Underhill SAM, Burne RA (2020) Repurposing the *Streptococcus mutans* CRISPR-Cas9 system to understand essential gene function. PLoS Pathog 16:e1008344. https://doi.org/10.1371/journal.ppat.1008344
24. Morrison DA, Khan R, Junges R, Åmdal HA, Petersen FC (2015) Genome editing by natural genetic transformation in *Streptococcus mutans*.

J Microbiol Methods 119:134–141. https://doi.org/10.1016/j.mimet.2015.09.023

25. Chao MC, Abel S, Davis BM, Waldor MK (2016) The design and analysis of transposon insertion sequencing experiments. Nat Rev Microbiol 14(2):119–128
26. Aird D, Ross MG, Chen W-S, Danielsson M, Fennell T, Russ C et al (2011) Analyzing and minimizing PCR amplification bias in Illumina sequencing libraries. Genome Biol 12(2):R18
27. Afgan E, Baker D, Batut B, van den Beek M, Bouvier D, Čech M et al (2018) The galaxy platform for accessible, reproducible and collaborative biomedical analyses: 2018 update. Nucleic Acids Res 46(W1):W537–W544

Chapter 14

The Use of TnSeq to Identify Essential Alphaproteobacterial Genes Reveals Operational Variability in Conserved Developmental and Cell Cycle Systems

Gayatri Sharma and Patrick D. Curtis

Abstract

A powerful method for examining genetic fitness and function on a large scale is to couple saturating transposon mutagenesis with high-throughput sequencing (TnSeq). By mapping where transposon insertions can be tolerated in a genome, it is possible to analyze the fitness of every gene in a genome simultaneously under a given growth condition. While this technique can describe genes as essential or nonessential under those growth conditions, sufficient mutagenesis and sequencing depth can provide more subtle differences in fitness. In this paper, TnSeq was used to analyze gene fitness of two Alphaproteobacteria from different environments: the freshwater oligotroph *Brevundimonas subvibrioides* (*Caulobacterales*) and the soil plant pathogen *Agrobacterium tumefaciens* (*Rhizobiales*) for the purpose of comparing conservation of gene function.

Key words Essential genes, TnSeq, Transposome, Transposon mutagenesis, High-throughput sequencing, Alphaproteobacteria

1 Introduction

The advent of high-throughput sequencing methods has led to a boom in bacterial genome sequence information with accompanying bioinformatics analyses. Tremendous scientific strides have been made analyzing the conservation of gene content. However, physiological characterization of gene content does not proceed at the same rapid pace, likely due to the scarcity of high-throughput physiological characterization techniques. One technique that provides physiological data on a global scale is TnSeq [1–3]. Here, an organism is mutagenized with a transposon and hundreds of thousands of mutants are pooled into a single library. High-throughput sequencing is used to map as many transposon insertion sites as possible, and transposon density in a given gene is inversely correlated with the fitness that gene provides under the mutagenesis conditions (i.e., the lower the transposon density, the

Ren Zhang (ed.), *Essential Genes and Genomes: Methods and Protocols*, Methods in Molecular Biology, vol. 2377,
https://doi.org/10.1007/978-1-0716-1720-5_14,

more important that gene is for growth). With sufficient mutagenesis and sequencing depth, genes can be characterized with greater finesse than simply essential vs. nonessential and instead can be characterized on a fitness gradient. Moreover, TnSeq can be performed on the same organisms under different conditions to identify genes with conditional fitness or, as described here, TnSeq can be performed in different organisms to query the conservation of gene function across evolutionary distance. In this study, TnSeq was performed on two Alphaproteobacteria, *Brevundimonas subvibrioides* (*Caulobacterales*) and *Agrobacterium tumefaciens* (*Rhizobiales*), for the purpose of comparing results with published TnSeq data on *Caulobacter crescentus* (*Caulobacterales*). TnSeq is a robust and powerful tool that generates a large amount of physiological data, which can be useful to researchers in many different areas. While this study used a freshwater oligotroph and a plant pathogen, TnSeq has been performed on a number of other organisms, from human pathogens [4–6] to methanogenic archaea [7, 8]. Other uses are being devised, such as the examination of organisms grown under in situ conditions [9]. Clearly, this technique is broadly applicable to multiple areas of interest.

2 Materials

Refer to (*see* **Note 1**) for the Materials Section.

1. *Brevundimonas subvibrioides* (ATCC 15264).
2. *Agrobacterium tumefaciens* (ERM77).
3. PYE broth: peptone (2 g/l), yeast extract (1 g/l), $MgSO_4 \cdot 7H_2O$ (0.3 g/l), and $CaCl_2 \cdot 2H_2O$ (0.0735 g/l). 2XPYE is made by doubling all components per liter.
4. LB broth: tryptone (10 g/l), yeast extract (5 g/l), and NaCl (10 g/l).
5. PYE agar plates: peptone (2 g/l), yeast extract (1 g/l), $MgSO_4 \cdot 7H_2O$ (0.3 g/l), agar (15 g/l), and 150 mm petri dishes.
6. LB agar plates: tryptone (10 g/l), yeast extract (5 g/l), NaCl (10 g/l), agar (15 g/l), and 150 mm petri dishes.
7. Kanamycin (10 mg/ml in water). Kanamycin is used at 5 μg/ml where necessary in PYE media and 50 μg/ml where necessary in LB media.
8. Sterile distilled water.
9. Sterile 10% glycerol in water.
10. EZ-Tn5 <Kan-2> transposome: purified transposon with transposase associated (Epicentre).

11. Electroporator.
12. Electroporation cuvettes (0.1 cm gap width).
13. Round-bottom culture tubes.
14. Centrifuge with appropriate rotors for bottles and tubes.
15. Centrifuge tubes (35 ml, sterile).
16. Centrifuge bottles (250 ml, sterile).
17. Spreader.
18. Qiagen Maxi prep kit.
19. TE buffer (1.21 g/l Tris, 0.372 g/l EDTA).
20. DNA7500 chip (Agilent Technologies).
21. Nebulizer.
22. Thermocycler and reagents for conducting PCR.
23. QIAquick PCR Purification Kit.
24. DNA Ligase Buffer 2×.
25. DNA Ligase.
26. Index PE Adapter Oligo Mix.
27. Accustart PCR mix (Quantabio).
28. Primers: Transleft (AATGATACGGCGACCACCGAGATCTACACATCCTCTAGAGTCGACCTGCAGGCATGCAAGC), Seq-out-transposon (CCTGCAGGCATGCAAGCTTCAGGGTTGAGATGTGTA), Index 2.0 primer, and Illumina multiplex primers.
29. Illumina GAP2-PEM2X instrument (Illumina).
30. Illumina RTA 1.9.35.
31. BWA mapping tool.

3 Methods

3.1 Preparation of Transposon Libraries

1. Grow 25 ml overnight cultures of *B. subvibrioides* and *A. tumefaciens*. *B. subvibrioides* is grown in 2XPYE at 30 °C, while *A. tumefaciens* is grown in LB at 26 °C. For both cultures, the target OD_{600} should be approximately 1.0 as measured on a standard spectrophotometer.
2. Harvest cells by centrifugation (12,000 × *g*, 4 °C, 5 min). From now on, cells should be maintained at chilled temperatures.
3. Prepare cells for electroporation. Remove supernatant from both cultures. Resuspend *B. subvibrioides* in 25 ml ice-cold distilled water. Resuspend *A. tumefaciens* in 25 ml of ice-cold 10% glycerol. Centrifuge again, remove supernatant, and resuspend cells in 25 ml of ice-cold water (*B. subvibrioides*) or

ice-cold 10% glycerol (*A. tumefaciens*). Centrifuge again and resuspend cells in 0.25 ml of ice-cold water (*B. subvibrioides*) or ice-cold 10% glycerol (*A. tumefaciens*).

4. To each culture add 1 μl of EZ-Tn5 <KAN-2> transposome (Epicentre). This contains the EZ-Tn5 transposon, with the R6Kgori deleted and preloaded with transposase (*see* **Note 2**). Mix transposome gently throughout the cell culture.
5. Equally distribute *B. subvibrioides* cell culture into 5 chilled 0.1 cm gap electroporation cuvettes. Equally distribute *A. tumefaciens* cell culture into 5 additional chilled 0.1 cm electroporation cuvettes.
6. Electroporate each cuvette (*see* **Note 3**). For *B. subvibrioides*, use 1.5 kV, 25 μF, and 400 Ω for the electroporation conditions. For *A. tumefaciens*, use 2.4 kV, 25 μF, and 200 Ω for the electroporation conditions.
7. Resuspend cells in each cuvette in 1 ml of growth media. For *B. subvibrioides*, use 2XPYE. For *A. tumefaciens*, use LB. Transfer cells from each cuvette into individual round-bottom tubes.
8. Incubate tubes with transformed cells on a shaker. For *B. subvibrioides*, incubate at 30 °C for 3 h. For *A. tumefaciens*, incubate at 26 °C for 3 h.
9. Plate transformations. Distribute culture from each tube across 3 agar plates (~333 μl). For *B. subvibrioides*, use PYE + Kan 150 mm agar plates. For *A. tumefaciens*, use LB + Kan 150 mm agar plates.
10. Sterilely spread culture on plates.
11. Incubate plates until colonies arise. For *B. subvibrioides*, incubate at 30 °C for ~5–6 days. For *A. tumefaciens*, incubate at 26 °C for ~2–3 days.
12. Store plates at 4 °C.
13. Repeat **steps 1–12** twenty times, for a total of 100 transformations and 300 plates per organism.
14. Count the colonies on every fifth plate to estimate the total number of transformants in the library.
15. Harvest cells. Pipet 1.5 ml of sterile growth media (PYE for *B. subvibrioides* and LB for *A. tumefaciens*) on to each plate. Use a sterile spreader to resuspend colonies into the growth media. If the plate is dry and absorbs a significant portion of the liquid, more growth media may need to be added. Tilt the plate so the cell suspension collects at the bottom. Use the spreader to scrape residual cell suspension into the pool at the bottom of the plate. Collect the cell suspension with a pipettor and place into a 250 ml centrifuge bottle on ice. More growth

media may be added followed by scraping to collect residual cells. Repeat for all 300 plates for one organism. Do the same for the other organism, collecting in a separate centrifuge bottle.

16. Once cells are collected, seal and shake each bottle to evenly mix the transposon pool.
17. Remove 900 μl aliquots (as many as desired) and mix each aliquot with 100 μl of DMSO. Freeze these viable cell aliquots at −80 °C for future experiments.
18. Centrifuge the remaining cell suspension (12,000 × *g*, 4 °C, 15 min) and discard the supernatant. Freeze cell pellets at −80 °C.

3.2 DNA Extraction, Preparation, and Sequencing

1. Extract genomic DNA from frozen cell pellets. The same protocol is used for either *B. subvibrioides* or *A. tumefaciens*. Genomic DNA extraction is performed using the Qiagen Maxi prep kit (Qiagen). To remove cells from the frozen cell pellet for genomic DNA extraction, add 11 ml of Buffer B1 (with RNase A) and gently shake the solution on top of the frozen pellet, allowing frozen cells to thaw and enter the buffer solution. Do this until the buffer solution has reached an $OD_{600} = 2.0$. Continue with the genomic DNA extraction as described in the kit manual (*see* **Note 4**).
2. Genomic DNA can be quantified using a spectrophotometer at this time, if desired.
3. Resuspend 10 μg of extracted genomic DNA in a final volume of 100 μl of TE buffer and then nebulize at 45 psi for 1 min 15 s to shear the DNA into different sized fragments.
4. Purify the nebulized DNA using QIAquick PCR Purification Kit and assay on a DNA 7500 chip (Agent Technologies). Divide aliquots of nebulized DNA so that each library has approximately 5 μg of starting material.
5. Perform end repair of fragmented DNA to get blunt-ended fragments with 5′-phosphorylated ends as described in the Multiplexing Sample Preparation Guide (Illumina).
6. Clean up the samples using 1.8× Ampure XP (Beckman Coulter) followed by a 1× Ampure XP cleanup to size select the end repaired DNA (fragments of ~200 bp in length).
7. Assay the sample on D1K High Sensitivity TapeStation tape (Agilent Technologies) to verify the end repair.
8. Perform A-tailing to adenylate the 3′ Ends and get 3′-dA overhang. Perform Ampure XP purification (1.8×) according to Multiplexing Sample Preparation Guide.

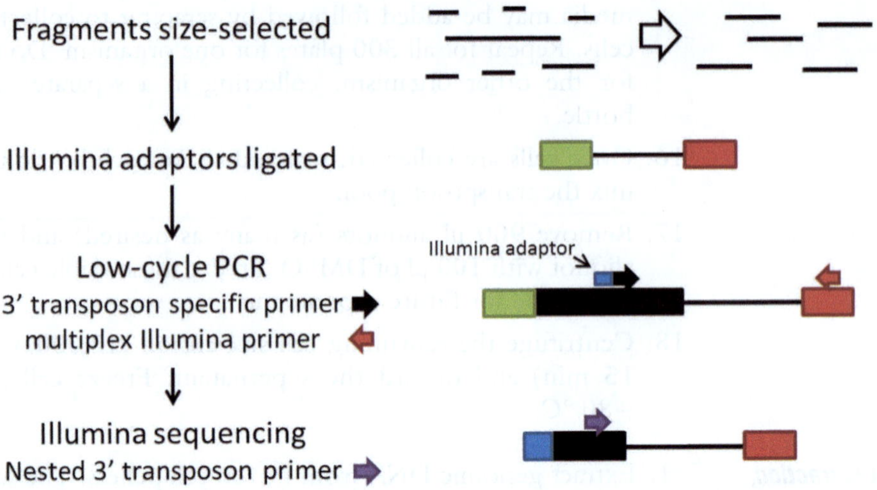

Fig. 1 Schematic diagram of the Low cycle PCR strategy (source: [10] supplementary data)

9. Ligate Illumina paired-end adapters to create adapter modified ends of DNA fragments. For ligation, prepare the ligation mixture in order of (DNA fragment (10 μl), DNA ligase 2× buffer (25 μl), Index PE Adapter Oligo Mix (10 μl), and DNA ligase (5 μl)) on ice. Incubate on the thermal cycler for 15 min at 20 °C. After incubation, purify the samples with a 1× ratio of Ampure XP followed by assay on TapeStation using D1K Hifg Sensitivity tapes to verify adaptor ligation.
10. Enrich for transposon containing sequences within the library. This is done with a low-cycle PCR and specialized primers. As outlined in Fig. 1, one primer (Transleft) was designed to anneal to a unique position near the 3′ end of the transposon and contained an Illumina adaptor sequence. The other primer is a multiplex Illumina primer containing a barcode. Amplify adapter ligated fragments using 1 μl of multiplex primer, 1 μl of Transleft primer (25 μM), 1 μl of IN2.0 primer (0.5 μM), 28 μl of DNA template, 10.3 μl of water, 5 μl of 10× Accustart Buffer, 2 μl of 50 mM $MgSO_4$, 1 μl of 10 mM dNTP mix, and 0.7 μl of Accustart Hifi Taq. Perform PCR with the following thermocycler conditions: 94 °C 1:30; 5 cycles of 94 °C for 15 s, 58 °C for 45 s and 68 °C for 30 s, 15 cycles of 94 °C for 15 s, 65 °C for 45 s and 68 °C for 30 s, and final extension of 68 °C for 10 min.
11. Purify the reaction using a 1× ratio of Ampure XP beads and elute in 28 μl of EB buffer. Then, assay the samples on the TapeStation using D1K High Sensitivity tapes.
12. Sequence samples using an Illumina GAP2-PEMA2X instrument. Perform sequencing run using v8.3px and v4 Illumina cluster kit (120 tiles per lane) along with a v5 Illumina SBS run

reagent kit. Sequence using the Seq-out-transposon primer (0.05 mM); this primer anneals within the transposon over the junction of the end of the inner transposon and the mosaic sequence.

3.3 Data Analysis

1. Remove sequence reads not containing transposon sequence. Any sequencing read originating in a transposon using the Seq-out-transposon primer will have the same final 10 bp of the transposon mosaic end. Cull reads that do not start with these 10 bp. Trim the transposon sequence from reads of interest prior to mapping and filter out low-quality sequencing reads.
2. Align the short reads generated to a reference genome with BWA mapping tool, producing sequence alignment/map (SAM) files. From the alignment file, list the insertion site of each read.
3. Calculate the number of reads at any given unique insertion site and annotate based on the ORF co-ordinates to determine the number and precise position of sequence reads for each individual gene (*see* **Note 5**).
4. Calculate the transposon density for each ORF. From each ORF, remove 20% of sequence from the 5′ end and discard any transposons in that region; autoannotation of genomes can incorrectly predict ORF start sites, and this region may appear to tolerate insertions when the rest of the sequence does not (*see* **Note 6**). Also, remove 20% of sequence from the 3′ end and discard any transposons in that region; many ORFs can tolerate insertions in the 3′ end without significantly disrupting gene function, which can artificially elevate transposon density. Divide the number of unique transposon insertion sites detected in the internal 60% of each ORF by the number of base pairs in the internal 60% of that ORF. Note that this metric does not include the frequency of insertion site detection, only the number of different insertion sites detected in the internal 60% of the ORF.
5. Tabulate information of interest in a program of choice (e.g., Excel). Include locus tag, locus coordinates, ORF length, number of unique insertion sites, and transposon density, as well as any other pertinent information desired, such as common gene name, annotation information, predicted COG, etc.
6. Copy information to a separate file and order ORFs by increasing transposon density.
7. Visualize the data by histogram. Adjust the bin size to best visualize the data. Commonly, transposon densities will form two peaks in a histogram, with a smaller peak close to the *Y*-axis that contains essential genes and a much larger peak distal to

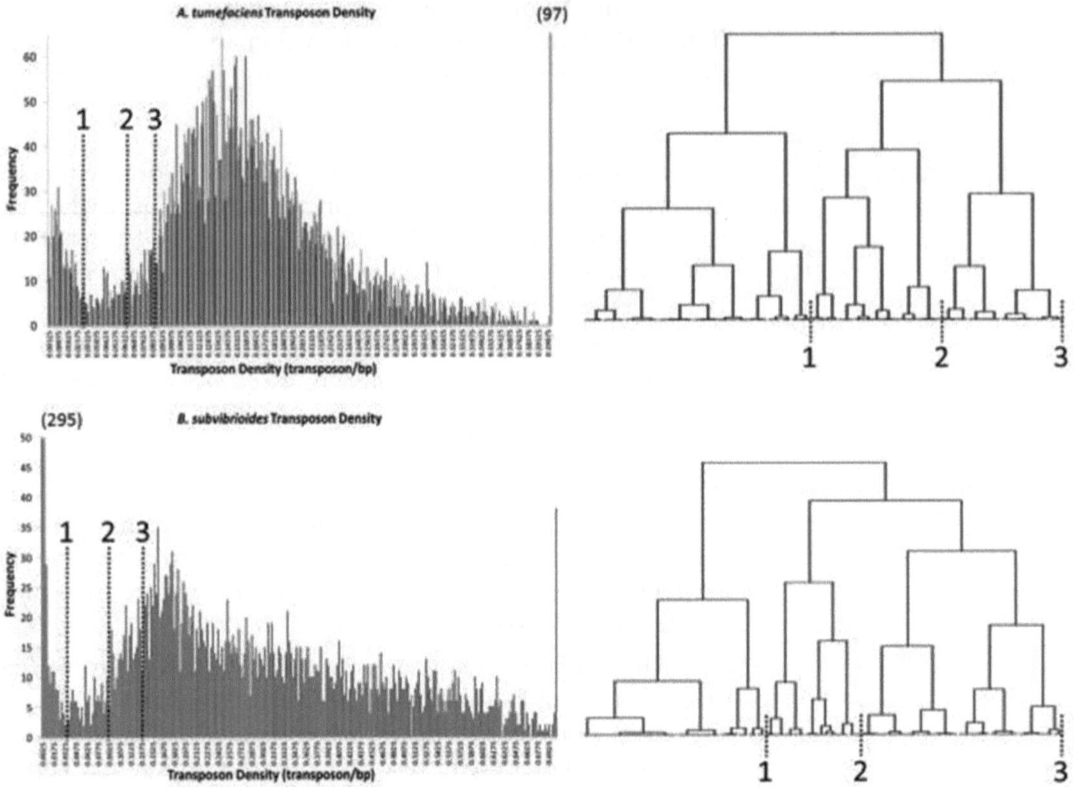

Fig. 2 Transposon densities distributions. In the right of each distribution is a portion of the ward's cluster analysis performed using the internal transposon densities (source [10])

the *Y*-axis that contains nonessential genes. The nonessential gene peak will often have a long tail of increasing transposon density. The trough region between the peaks may or may not have a large number of ORFs. Trough genes are likely nonessential genes that have a growth defect when disrupted (*see* **Note** 7).

8. To tentatively categorize ORFs, subject transposon densities to Ward's cluster analysis (*see* **Note 8**). Visualize transposon density clusters and correlate them with areas of the transposon density histogram (Fig. 2). Label ORFs in the left peak cluster as essential, in the trough cluster as unresolved, and all other ORFs as nonessential.
9. Perform other analysis of interest (*see* **Note 9**). In this study, fitness of genes of interest (including developmental regulators and cell division genes) from *B. subvibrioides*, *A. tumefaciens*, and *C. crescentus* was compared to identify genes that had changes in fitness despite their presence being conserved in all three organisms.

4 Notes

1. Prepare all culture media using deionized water and analytical grade chemical constituents. Follow all waste disposal regulations when disposing waste materials.
2. There are different methods or kits available for random transposon mutagenesis. We utilized the EZ: Tn5 transposon kit (Epicentre) for this study, but any transposon system can be used, as long as the transposon sequence is known. When using a different transposon system, adjust the primers for sequencing.
3. Different transformation approaches can be used to mutagenize a target organism. Here, electroporation was used to deliver a prepared transposome. Alternatively, conjugal mating can be used. Conjugal mating may provide a higher transformation efficiency than electroporation, which is desired in TnSeq (the more the transposon mutants, the better). However, conjugal mating may result in donor organism or donor plasmid residing on transformation plates, which can be harvested and carried along in sequencing and may compose a noticeable portion of sequencing results. If so, these donor plasmid sequences will need to be culled from the sequence data. In this study, electroporation of the transposome resulted in sufficient mutagenesis, but other TnSeq studies have successfully used conjugal mating, [11–13] demonstrating that either mutagenesis method is viable.
4. For genomic DNA extraction of a mutant library, any of the common techniques or DNA extraction kits can be used as long as it provides DNA with enough quantity and quality for high-throughput sequencing.
5. Depending on how thorough the colony resuspension is performed and how deep the sequencing of the library is achieved, it is possible to recover close to 100% of the estimated transposon pool. In our previous studies, we have recovered 67%, 90%, and 98% of our transposon libraries [10, 14]. However, these numbers should be viewed with some skepticism. First, the total transposon library is only an estimate based on counting a subset of transformation plates. Second, transposon insertions in cells that do not produce colonies are still present on the plate and can be harvested and sequenced in downstream steps. Examples can be seen as single-read insertions in genomic areas otherwise lacking insertions. These types of insertions have little bearing on fitness categorization because they do not significantly increase the transposon density, but in total, they can inflate total unique insertion sites detected. In theory, it

would be possible to recover greater than 100% of the total estimated transposon pool because of these types of insertions. One could cull single-read insertion sites from the dataset, but single-read insertions can compose up to 50% of the unique insertion sites, so single-read insertions clearly are not just false positives and instead represent a significant portion of legitimate colony-producing insertions. For these reasons, total library recovery percentages are simply used as a rough metric of sequencing depth.

6. When viewing TnSeq data, sometimes, large numbers of insertions can be seen at the 5′ end of ORFs that otherwise lack transposons. This is usually the result of a misannotated start site. The TnSeq method can be a relatively useful mechanism for mapping incorrect start sites on a global scale, although it obviously works best for ORFs that have fitness costs for disruption (i.e., essential genes).

7. More ORFs in the trough region appear to correlate with the size of the transposon library, library recovery, and sequencing depth. We believe ORFs in the trough region to be ORFs that are nonessential, but disruption causes reduced growth, which has been experimentally demonstrated for one ORF [15]. We believe this method does not define ORFs in a strict essential vs. nonessential status, but orders them on a fitness gradient.

8. In our studies, we have used relatively simple metrics for analysis of ORF fitness. As the TnSeq method has risen in popularity, several software suites have been developed to analyze TnSeq data and also perform comparative TnSeq. Such suites include ESSENTIALS [16], TnSeq Explorer [17], and TRANSIT [18].

9. We typically perform two additional analyses of the total dataset. First is to analyze transposon insertion sites for potential insertion bias (Fig. 3a). Some transposons are reported to have biases for the genomic region characteristics they insert into (e.g., EZ-Tn5 is thought to prefer low GC sites) [1]. Insertion site bias could impact TnSeq results. We analyzed the GC content of 100 bp windows around each unique insertion site and compared the distribution against the distribution of random 100 bp windows. We have found the EZ-Tn5 transposon to have no-to-minimal bias for insertion site by GC content [10, 14]. It is possible that the transposon does have a bias, but so many mutants were made that the bias was overcome. The second analysis we perform is to correlate insertion orientation with ORF orientation (Fig. 3b). If the vast majority of transposons insert such that the antibiotic resistance gene is in the

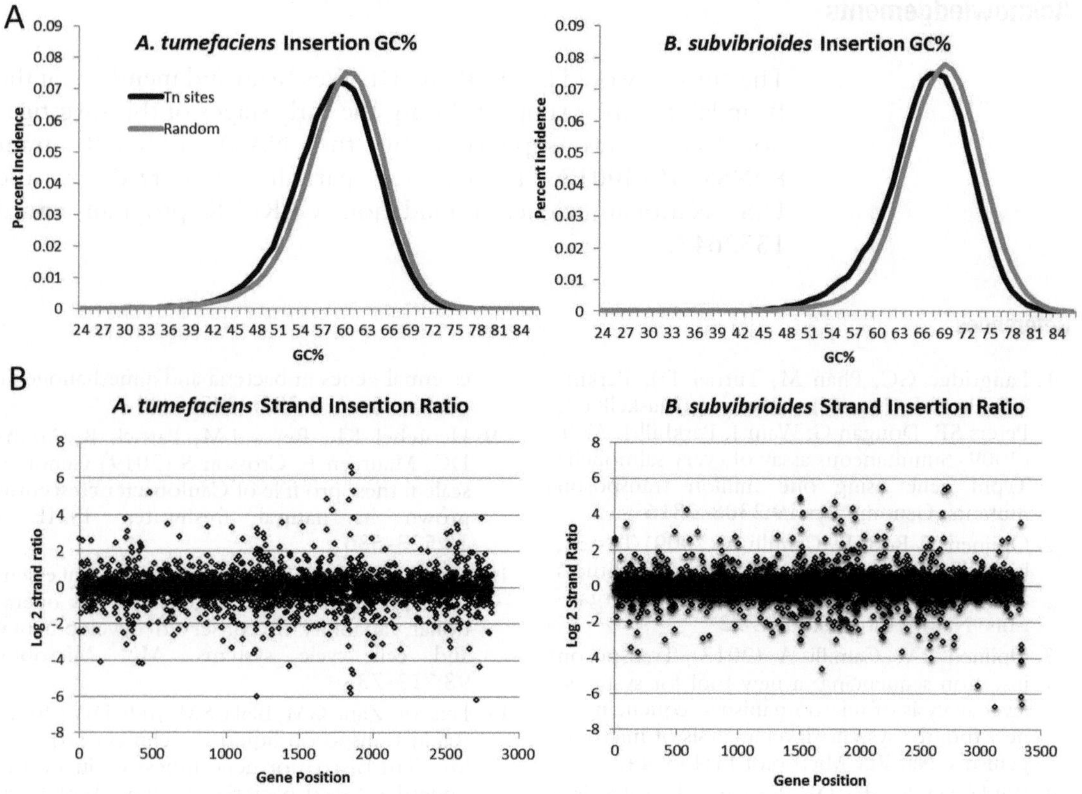

Fig. 3 Analyses of transposon insertion bias (**a**) and correlation of insertion orientation with ORF orientation (**b**) (source [10], supplementary data)

same direction as the ORF inserted into, it is likely a result of the antibiotic resistance gene promoter being nonfunctional in the target organism [19]. In that case, antibiotic resistance (and therefore colony formation and transposon recovery) would be strongly dependent on the expression levels of the ORF inserted into and would significantly bias the TnSeq results. Optimally, a small pilot study would be performed prior to large-scale mutagenesis to determine the suitability of a transposon for a target organism. In our studies, individual ORFs may have a bias for insertion orientation, but no systemic bias was detected. Individual ORFs may have a bias for insertion orientation because expression of the antibiotic resistance gene in the opposite direction of the inserted gene may disrupt expression of the inserted gene, and if expression of that gene is necessary for cell function, insertions of opposite orientation would have fitness defects during colony growth. While we have not specifically analyzed such ORFs, it is potentially another way to utilize TnSeq data.

Acknowledgements

The authors would like to thank Dr. Yves Brun and members of the Brun lab for their support during the early stages of this investigation. G.S. was supported by the NASA EPSCoR grant 80NSSC19M0013. P.D.C. was partially supported by the U.S. National Science Foundation CAREER program award 1552647.

References

1. Langridge GC, Phan M, Turner DJ, Perkins TT, Parts L, Haase J, Charles I, Maskell DJ, Peters SE, Dougan G, Wain J, Parkhill J, Ak T (2009) Simultaneous assay of every salmonella Typhi gene using one million transposon mutants. Genome Res 19:2308–2316
2. Opijnen T, Bodi K, Camilli A (2009) Tn-seq; high-throughput parallel sequencing for fitness and genetic interaction studies in microorganisms. Nat Methods 6:767–772
3. Opijnen TV, Camilli A (2013) Transposon insertion sequencing: a new tool for systems-level analysis of microorganisms sequencing: a new tool for systems-level analysis of microorganisms. Nat Rev Microbiol 11:435–442
4. Wilde AD, Snyder DJ, Putnam NE, Valentino MD, Hammer ND, Lonergan ZR, Hinger SA, Aysanoa EE, Blanchard C, Dunman PM, Wasserman GA, Chen J, Shopsin B, Gilmore MS, Skaar EP, Cassat JE (2015) Bacterial hypoxic responses revealed as critical determinants of the host-pathogen outcome by TnSeq analysis of *Staphylococcus aureus* invasive infection. PLoS Pathog 11:1–24
5. de Vries SPW, Eleveld MJ, Hermans PWM, Bootsma HJ (2013) Characterization of the molecular interplay between Moraxella catarrhalis and human respiratory tract epithelial cells. PLoS One 8(8):e72193
6. Klein BA, Duncan MJ, Hu LT (2015) Defining essential genes and identifying virulence factors of Porphyromonas gingivalis by massively parallel sequencing of transposon libraries (Tn-seq). Methods Mol Biol 1279:25–43
7. Zhang C, Phillips APR, Wipfler RL, Olsen GJ, Whitaker RJ (2018) The essential genome of the crenarchaeal model Sulfolobus islandicus. Nat Commun 9:1–11
8. Grazziotin AL, Vidal NM, Venancio TM (2015) Uncovering major genomic features of essential genes in bacteria and a methanogenic archaea. FEBS J 282:3395–3411
9. Hentchel KL, Reyes LM, Patrick R, Aretha DC, Maureen F, Crosson S (2019) Genome-scale fi tness pro fi le of Caulobacter crescentus grown in natural freshwater. ISME J 13:523–536
10. Curtis P, Brun Y (2014) Identification of essential alphaproteobacterial genes reveals operational variability in conserved developmental and cell cycle systems. Mol Microbiol 93:713–735
11. Fels SR, Zane GM, Blake SM, Judy DW (2013) Rapid transposon liquid enrichment sequencing (TnLE-seq) for gene fitness evaluation in underdeveloped bacterial systems. Appl Environ Microbiol 79:7510–7517
12. Opijnen TV, Lazinski DW, Camilli A (2015) Genome-Wide Fitness and Genetic Interactions Determined by Tn-seq, a High-Throughput Massively Parallel Sequencing Method for Microorganisms. Curr Protoc Microbiol 106:1–32
13. Brutinel ED, Gralnick JA (2012) Anomalies of the anaerobic tricarboxylic acid cycle in Shewanella oneidensis revealed by Tn-seq. Mol Microbiol 86:273–283
14. Curtis PD (2016) Essential genes predicted in the genome of Rubrivivax gelatinosus. J Bacteriol 198:2244–2250
15. Curtis PD (2017) Stalk formation of Brevundimonas and how it compares to Caulobacter crescentus. PLoS One 12:1–24
16. Zomer A, Burghout P, Bootsma HJ, Hermans PWM, Hijum SAFTV (2012) Essentials: software for rapid analysis of high throughput transposon insertion sequencing data. PLoS One 7:1–9
17. Solaimanpour S, Sarmiento F, Mrázek J (2015) Tn-seq explorer: a tool for analysis of high-

throughput sequencing data of transposon mutant libraries. PLoS One 10:1–15

18. DeJesus MA, Ambadipudi C, Baker R, Sassetti C, Ioerger TR (2015) TRANSIT - a software tool for Himar1 TnSeq analysis. PLoS Comput Biol 11:1–17

19. Gallagher LA, Ramage E, Jacobs MA, Kaul R, Brittnacher M, Manoil C (2007) A comprehensive transposon mutant library of Francisella novicida, a bioweapon surrogate. Proc Natl Acad Sci U S A 104:1009–1014

Chapter 15

A Proposed Framework to Identify Dispensable and Essential Functions in Bifidobacteria: Case Study *of Bifidobacterium breve* UCC2003 as a Prototype of Its Genus

Lorena Ruiz, Francesca Bottacini, Lucie Semenec, Amy Cain, and Douwe van Sinderen

Abstract

Functional genomics of bacteria commonly aims at establishing genotype-phenotype links in microorganisms of industrial, technological and biomedical relevance. In this regard, random transposon mutagenesis coupled to high-throughput next-generation sequencing approaches, termed transposon-insertion sequencing (TIS), has emerged as a robust, genome-wide alternative to perform functional genome analysis. Though these approaches have been used in a large number of studies involving pathogenic and clinically relevant bacteria, they have received little attention in the fields of commensal and potentially beneficial bacteria, including probiotic microorganisms. In this chapter, we describe the implementation of the TIS method *Tra*nsposon-*D*irected *I*nsertion *S*equencing to describe the set of essential genes in a representative strain of a genus encompassing several commensal and potentially probiotic bacteria and discuss considerations when applying similar methodological approaches to other *Bifidobacterium* species/strains of interest.

Key words *Bifidobacterium*, Commensal, Probiotic, Transposon, TraDIS, Essential genes

1 Introduction

The genus *Bifidobacterium* encompasses a substantial number of commensal microbial species and strains whose presence in the human gastrointestinal tract has been associated with a range of health-promoting effects [1, 2]. Given the importance of bifidobacteria as a core member of a healthy microbiome, significant scientific efforts have been devoted to discern the mechanisms and underlying genes being responsible for these purported beneficial traits and for allowing cross-talk with the human intestinal ecosystem [3].

Gene inactivation coupled to phenotypic profiling of mutated strain derivatives represents the gold standard methodology to perform functional genomic analyses and unequivocally prove the function of targeted genes. However, molecular biology techniques in bifidobacteria are still challenging due to their low transformation and conjugation efficiencies which hamper the generation of random or targeted mutants [4]. Therefore molecular information on members of this important genus is sorely lacking. Most *Bifidobacterium* strains remain recalcitrant to genetic manipulation with only a few species and strains having been successfully mutated [5–8]. The increasing number of genome-based studies in bifidobacteria have generated knowledge necessary to facilitate genetic engineering in this group of microorganisms and to design tailor-made tools to achieve gene inactivation in various *Bifidobacterium* species/strains, including some random mutagenesis strategies [6, 7, 9–12].

Transposon-based mutagenesis represents a random mutagenesis strategy which allows the generation of large collections of untargeted gene inactivation derivatives for a given model microorganism [13]. Classical transposon-based mutagenesis strategies have relied on the in vivo delivery of a transposon-encompassing piece of DNA, including a selective marker, in a conditional replicative vector, which also provides a transposase-encoding gene. Once introduced in the target microorganism, expression of the transposase causes the transposon to jump into near-random positions in the target genome, and further loss of the delivery vector stabilizes the insertion event [14]. Initial transposon mutagenesis attempts in bifidobacteria were met with major limitations as there were no robust conditional replicative vectors available for bifidobacteria. Thus, the development of in vitro transposon delivery strategies represented an important advancement as they offered attractive opportunities to tackle transposon-based random mutagenesis in microorganisms recalcitrant to genetic manipulation, such as bifidobacteria. Such approaches are based on the in vitro assembly of stable and functional transposon-transposase complexes, which can be directly electroporated into the targeted host, where they successfully transpose into the host DNA [15].

Large transposon-mutant collections created by any of the means described above enable functional genetic screenings at a genome-wide scale [16], including the functional characterization of genes with previously unknown function [17]. In this context, the advent of high-throughput next-generation sequencing (NGS) approaches coupled to dense, random transposon insertion libraries, termed transposon-insertion sequencing (TIS), has offered a powerful alternative to perform relatively rapid high-throughput identification of an organism's essential genes, as well as information on gene functionality and gene interaction networks under various growth conditions. Genes unable to tolerate (and therefore lacking) transposon insertions are categorized as being essential and those with a significantly lower insertion frequency depending on the condition tested are considered to be required to grow under that condition. There are four variations of the TIS method that differ in their sequencing protocol and library preparation procedure, namely: Transposon-Directed Insertion Sequencing (TraDIS) [18], Transposon Sequencing (TnSeq) [19], High-throughput Insertion Tracking by deep Sequencing (HITS) [20] and INsertion Sequencing (INSeq) [21]. Notably, one has to keep in mind that due to the rapid advancements in TIS and the multitude of different methods and tools already available for transposon mutagenesis and sequencing analysis, the specific approach chosen can differ depending on specific characteristics of the genome being studied, the specific scientific aim of the study or the sequencing platform. Besides, it is worth remarking that prediction of essential genes can be tricky due to stochastic effects during Tn mutant library preparation and sequencing. If a library is not saturated (i.e. not all TA sites contain an insert), further challenges arise in accurate essential gene prediction. Thus a cautious design of the experimental approach and analytical methods to be applied may help to overcome some of these limitations. Further details on these general considerations prior to designing a TIS approach are described in **Note 1**.

TIS has been applied to just a few representative species of human commensal microorganisms [21–23]. Besides, identification of the full complement of essential genes in a given strain coupled to comparative genomic analysis in phylogenetically close and distant strains provides novel insights into the evolutionary history and niche adaptation of the corresponding microorganism [24]. Characterization of essential genes in representative strains of the genus *Bifidobacterium* has recently been achieved through TIS studies, but is currently limited to two species *B. breve* and *B. longum* [12, 22, 25] using the TraDIS and Tn-Seq methodology, respectively. Besides, essential gene prediction from transposon mutants pools, can be performed either under optimal conditions, where the genes with a low number of insertions are deemed 'essential' for growth under any condition, or under specific conditions where those genes would be classified as 'conditionally essential'.

Herein, we will focus on the description of the experimental approach and analytical tools utilized to analyze and interpret sequencing results from TIS experiments so as to achieve essential gene prediction under optimal conditions, as already performed in bifidobacterial representative model strains, with the aim to provide a practical guidance reference to assist when implementing such analyses to other strains. A schematic overview of the main steps required for such analysis is represented in Fig. 1.

For *B. breve*, essential genes were characterized using a TraDIS Tn5 mutant library constructed in the *B. breve* strain UCC2003 by employing the methodologies described in the following sections.

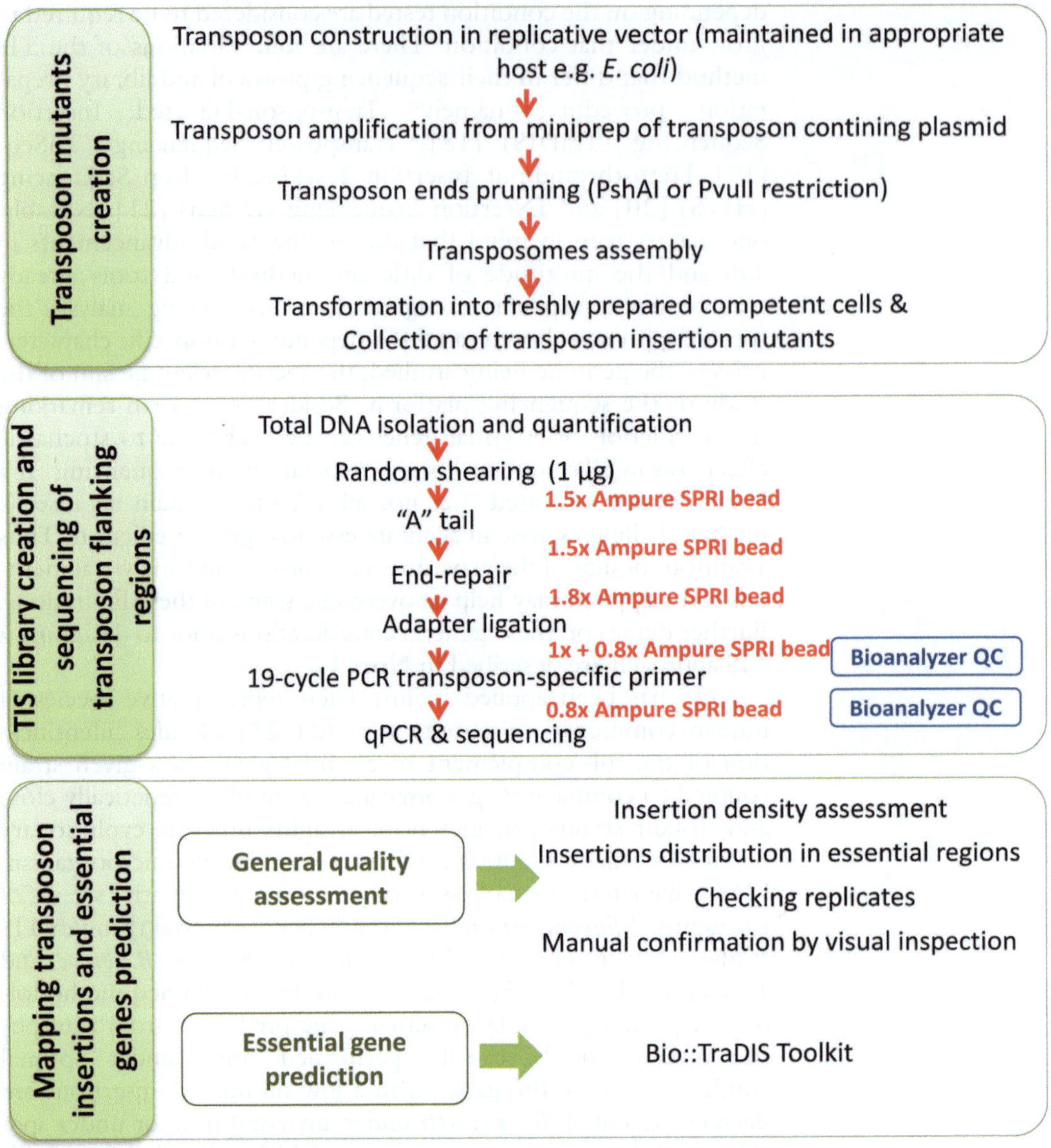

Fig. 1 Schmatic diagram displaying the main steps followed to perform essential genes prediction experiment in *B. breve* UCC2003, from generation of a collection of transposon random mutants to generation of a TIS dataset and essential gene calls prediction

2 Materials

2.1 Basic Microbiology Reagents

1. Luria Bertani (LB) media: 10 g/L tryptone (Merck, Darmstadt, Germany), 5 g/L yeast extract (Merck), 10 g/L NaCl (Merck), sterilized by autoclaving to grow *E. coli* cells or, supplemented with 10 μg/ml tetracycline in order to select for *E. coli* strains harbouring plasmid pMOD2-TetW.

2. de-Man-Rogosa-Sharpe (MRS): Proteose Peptone (10 g/L) (Difco, Franklin Lakes, NJ, US), beef extract (10 g/L) (Difco), yeast extract (5 g/L) (Difco), polysorbate (tween 80) (1 ml/L) (Sigma Aldrich; St Louis, MO, US), $C_6H_{17}N_3O_7$ (2 g/L) (Sigma Aldrich), $MgO_4S{\cdot}7H_2O$ (0.525 g/L) (Merck), $MnO_4S{\cdot}4H_2O$ (0.12 g/L) (Merck), K_2HPO_4 (3 g/L) (Merck), KH_2PO_4 (3 g/L) (Merck), pyruvic acid (0.2 g/L) (Sigma Aldrich), cysteine-HCl (0.3 g/L) (Sigma Aldrich), $FeSO_4{\cdot}7H_2O$ (0.034 g/L) (Sigma Aldrich), adjusted to pH 6.8 and sterilized by autoclaving.
3. 10% D-Lactose solution (sterilized by filtration through 0.2 μm diameter pore) (*see* **Note 2**).
4. 10% L-cysteine-HCl solution (sterilized by filtration through 0.2 μm diameter pore).
5. Electrocompetent cells washing buffer: 1 mM citrate, 0.5 M sucrose buffer, pH 5.8, sterilized by autoclaving.
6. Reinforced Clostridial Medium (RCM) (Oxoid) and Reinforced Clostridial Agar (RCA) (Oxoid) plates supplemented with appropriate antibiotic concentrations for transposon insertion selection. If using the Tn5-TetW transposon described for *B. breve* strains, use a final concentration of 10 μg/ml of tetracycline (*see* **Note 3**).
7. 10 mg/ml tetracycline stock solution dissolved in 50% ethanol and sterilized by filtration (0.2 μm diameter pore).
8. Plasmid capable of replication in the selected *Bifidobacterium* host, preferably harbouring the same antibiotic resistance cassette as the transposon, to be used as a positive control to test transformation efficiency of each competent cell batch. If using pMOD2 TetW, the *E.coli Bifidobacterium* shuttle vector pAM5 can be used as a control plasmid.

2.2 Basic Molecular Biology Reagents

1. Standard molecular biology reagents required for transposon preparation; chromosomal DNA isolation and Tn sequencing library preparation: high-fidelity PCR master mix; SYBR green qPCR (Kapa Biosystems); appropriate restriction enzymes for transposon end pruning (PshAI or PvuII if using pMOD2-TetW); plasmid mini-preparation kits; PCR purification kits; agarose; 1× TAE buffer (40 mM Tris–HCl pH 7.6, 20 mM acetic acid, 1 mM EDTA); ethidium bromide or alternative nucleic acid dye; TE buffer (10 mM Tris, 1 mM EDTA, pH 8.0); TSE buffer (TE buffer supplemented with 25% sucrose); sodium dodecyl sulphate; lysozyme (Sigma-Aldrich); proteinase K (Sigma-Aldrich); DNAse-free RNAse A (Sigma-Aldrich), phenol-chloroform (1:1, pH 8.0), 3 M $C_2H_3NaO_2$ pH 5.2; 70% ethanol; isopropanol.

2. Replicative vector containing a Tn*5* transposon including a selection marker, e.g. pMOD2-TetW from [12] including a tetracycline resistance gene (tetW) which has previously been shown to be a good selection marker in tetracycline-sensitive *Bifidobacterium* strains or, alternatively, another customized Tn*5* transposon (*see* **Note 4**). This vector can be stably maintained into appropriate *E. coli* cells (*endA*$^-$ strains are recommended).
3. Oligonucleotides to amplify the Tn*5* transposon from the replicative vector which contains it (e.g. pMOD2-TetW). If using pMOD2-TetW, or any other pMOD2-based construct, pMOD<MCS>Fw (5′-*ATTCAGGCTGCGCAACTGT*-3′) and pMOD<MCS>Rev (5′-*GTCAGTGAGCGAGGAAGCGGAA G*-3′) from the transposase provider.
4. Purified EZ-Tn*5* transposase (commercially available by Lucigen Corporation, Middleton, WI, US—www.lucigen.com).
5. Molecular grade glycerol.
6. Reagents for the sequencing library preparation: Ampure SPRI beads (Beckman Coulter Genomics); NEBNext End Repair Module (NEB); NEBNext dA-Tailing Module (NEB); NEBNext Quick Ligation Module (NEB).

2.3 Equipment

Thermocycler, Real-Time PCR, Microcentrifuge, Fume hood, Electrophoresis unit and transilluminator for DNA visualization (BioRad), Electroporator (e.g. Gene Pulser II Porator Electroporation System from Bio Rad) and electroporation cuvettes (0.2 cm gap) (BioRad), Device for spectrophotometric DNA quantification, e.g. Qubit systems (Lifetechnologies), Rotatory shaker incubator, Anaerobic work station (gas composition: 10% H_2, 10% CO_2, 80% N_2) (Don Whitley Scientific), Refrigerated centrifuge with rotor for bottles with a volume of at least 50 ml, Covaris' Adaptive Focused Acoustics (AFA) technology for DNA fragmentation (Covaris, cat. no. 520045), Agilent Bioanalyzer (Agilent Technologies) and High Sensitivity DNA kit, and Sequencing platform (for instance MiSeq or Hiseq2500 Illumina platforms).

3 Methods

3.1 Transposon Mutagenesis Strategy for B. breve UCC2003

Random transposon mutagenesis in selected *B. breve* strains can be achieved through the design of a custom Tn*5* transposon, coupled to in vitro transposome assembly and optimization of transformation procedures in the corresponding host strain [12]. Description of the methods required to accomplish these steps is presented in the following sections.

3.1.1 Transposon Construction and Preparation

1. Perform plasmid mini-preparations from *E. coli* cells harbouring the transposon-containing plasmid (e.g. pMOD2-TetW) according to guidelines of the plasmid mini-preparation kit provider, and verify extracted plasmid by restriction profiling followed by agarose gel electrophoresis visualization. For instance, pMOD2-TetW can be verified through digestion with the restriction enzymes SphI and XbaI according to indications from the restriction enzyme provider. Following restriction, mix 10 μl of restricted plasmid DNA with 2 μl of commercial loading buffer dye and load on a 1% agarose gel. Apply a voltage of 1.5 V/cm until the dye reaches about two thirds of the gel length; and stain the gel with a nucleic acid staining dye according to manufacturer's instructions. Visualization of DNA bands in a UV transilluminator should demonstrate visible bright bands of 2.5 Kb (pMOD2 backbone) and 2.8 Kb (tetracycline resistance Tn*5* transposon).
2. The transposon piece of DNA will be PCR amplified from the previously isolated plasmid in order to generate large quantities of a TetW-Tn*5* transposon ready for transposome assemblage (*see* **Note 5**). If using a transposon constructed within a pMOD2 vector, such as the one described above for bifidobacteria, oligonucleotides pMOD<MCS>Fw and pMOD<MCS>Rev (Lucigen) can be used as these anneal immediately upstream of the transposon ends in pMOD2. A suggested PCR cycling scheme is as follows: (1) denature the template at 94 °C for 2 min; (2) perform 30 cycles denaturing at 94 °C for 30 s; annealing at 60 °C for 45 s and extending at 72 °C for 1 min for every kb of the expected product.
3. To verify the PCR reactions load 5 μl of the PCR reaction in a well of a 1% agarose gel as previously described in **step 2**. The TetW-Tn*5* transposon [12] should produce a band of 2.8 kb. Each PCR product must be purified to eliminate salts, and excess of nucleotides and primers, for example, using a silica column according to the commercial provider.
4. Restriction of the TetW-Tn*5* generated with PshAI or PvuII, whose restriction sites are located immediately upstream of the transposon-associated inverted repeats, will generate phosphorylated transposon ends and maximize the efficiency of the transposition process (*see* **Note 6**). These restriction reactions must be set up according to guidelines issued by the restriction enzyme provider. Subsequently, clean up the transposon for instance using a column PCR purification system, and concentrate by eluting each DNA preparation in a small volume in the final step of the cleaning procedure (20 μl per purification column).

5. Pool all transposon preparations and adjust the concentration of transposon ends to desired values taking into consideration the transposon length, by following instructions from the Tn*5* transposase provider.
6. Assemble the transposon-transposase complexes according to instructions from the transposase provider. As a standard guideline, mix 2 μl of transposon DNA preparation; 4 μl of EZ-Tn*5* transposase (1 U/μl) and 2 μl of 100% glycerol, vortex and incubate at room temperature for 30 min. This preparation can be stored at −20 °C until use.

3.1.2 Preparation of Electrocompetent Cells of Bifidobacterium (See Note 7) and Generation of a Collection of Transposon Insertion Mutants

1. Inoculate the corresponding bacterial strain into 10 ml of RCM broth and grow anaerobically overnight (~16 h) at 37 °C in standing tubes.
2. Pre-warm and pre-reduce 50 ml tubes containing de Man-Rogosa-Sharpe broth supplemented with 0.05% L-cysteine and 1% of D-lactose (*see* **Note 2**) by storing them overnight in the anaerobic chamber.
3. Inoculate pre-reduced and pre-warmed de Man-Rogosa-Sharpe broth with the culture grown overnight (2% v/v). Incubate the standing tubes at 37 °C anaerobically until the cultures reach an OD_{600nm} of about 0.5–0.7 (5–7 h).
4. While the culture(s) are growing, keep the citrate-sucrose buffer, 1.5 ml tubes, pipettes and electroporation cuvettes refrigerated in an ice box.
5. When cultures reach the desired optical density, spin down the cells at 4300 × *g* in a cold rotor (4 °C) for 10 min. *From this point onwards the cells need to be kept ice-cold.*
6. Wash the cells twice by decanting the supernatant and gently swirling the cell pellet into the same volume (50 ml) of ice-cold 1 mM citrate–0.5 M sucrose buffer. Suspend the washed cell pellet obtained from 50 ml of culture into 500 μl of ice-cold citrate-sucrose buffer.
7. Dispense adequate amounts of Tn5 transposase-transposon complexes assembled as described in Subheading 1 in 1.5 ml Eppendorf tubes. In parallel, add 5 μl of a replicative plasmid to one of the tubes, to be used as a positive control and verify the quality of the competent cells and *Bifidobacterium* electrotransformation procedure.
8. Add to each tube containing the transposase-transposon complexes or positive control plasmid DNA, 45μl of freshly prepared electrocompetent cells. Mix by gently pipetting up and down and transfer the whole volume to an ice cold electroporation cuvette, making sure the solution is poured in

between the metallic plates of the cuvette and no air bubbles are left at the bottom of the electroporation cuvette.

9. Apply a pulse of 2.0 KV, 200 Ω, 25 μF in an electroporator, then aseptically add 950 μl of RCM, gently suspend the cells by pipetting up and down a few times and incubate the suspensions for 30 min at 37 °C anaerobically.
10. Aseptically place 100 μl aliquots of electroporated cells onto RCA plates supplemented with 10 μg/ml of tetracycline and incubate the plates anaerobically at 37 °C for 2–3 days (*see* **Note 8**).
11. Aseptically recover the antibiotic-resistant colonies grown by adding 500 μl of RCM supplemented with 30% glycerol on top of the plate, and then gently scrub the agar surface with a disposable sterile spatula. Collect the liquid containing the colonies by pipetting and pour in a sterile container. Homogenize the cells suspension and store at −80 °C.

3.2 TraDIS Sequencing Library Preparation

3.2.1 Total DNA Isolation

In order to perform a TIS approach to predict essential genes in a collection of transposon random mutants, total genomic DNA extraction from the transposon mutant pool should be conducted. For this purpose, take a volume from the homogenized cells frozen stock (corresponding to ~10^9 cells) and proceed to perform DNA isolation by using any adequate DNA isolation procedure. As an example, a common DNA isolation procedure combining enzymatic lysis and DNA purification with organic solvents is briefly described below, although commercial DNA purification kits can also be employed:

- Collect the cells by centrifugation (15,900 × *g*, 10 min, 4 °C), wash once in 500 μl of TBS solution and suspend in 500 μl of TBS supplemented with lysozyme (30 mg/ml) and DNase-free RNase A (100 μg/ml). Incubate at 37 °C for 30 min.
- Add 20 μl of proteinase K (20 mg/ml) and 25 μl of 10% SDS and mix thoroughly. Incubate at 60 °C for at least 1 h to complete the lysis.
- Perform three consecutive extraction steps with 500 μl of phenol:chloroform (1:1 v/v) pH 8.0, thoroughly mixing the suspension by vortexing, then centrifuging the cell lysate for 10 min at maximum speed at RT and transferring the upper phase to a clean tube carefully avoiding to touch the interphase.
- Add 50 μl of 3 M sodium acetate pH 5.2 and 500 μl of isopropanol, mix gently by inversion and collect the precipitated DNA by centrifugation (maximum speed, 15 min, 4 °C). Wash the pellet once with 0.5 ml of 70% ethanol (−20 °C). Finally, air dry the pellets at RT until the ethanol has fully evaporated, and dissolve in 50–100 μl of TE buffer.

3.2.2 Manual TraDIS Library Protocol

The TraDIS library construction protocol described below is based on the automation-friendly workflow as described for the TraDIS Toolkit approach (*see* Barquist et al. [26]). All reactions are performed as per the manufacturer's protocol unless otherwise stated.

For TraDIS sequencing out of one transposon end (double this amount if sequencing out of both ends), 2 μg of high purity genomic DNA is used as starting material, which is quantified using a Qubit fluorometer. This is then sheared to approximately 300 bp fragments using Covaris' Adaptive Focused Acoustics (AFA) technology commonly using settings: Duty cycle, 20%; Intensity, 5; Cycles, 200; Time, 60s for the Covaris S2 or E210 instruments. A clean-up step is performed using 1.5× Ampure SPRI beads (Beckman Coulter Genomics) and eluted with 50 μl Elution buffer (EB). Fragments subsequently undergo end-repair using NEBNext End Repair Module (NEB) and a second purification with 1.5× Ampure SPRI beads is performed, eluting in 32 μl EB buffer. Next, 'A' tailing is performed using the NEBNext dA-Tailing Module (NEB). A final 1.8× Ampure SPRI clean-up is performed, finally eluting in 20 μl EB buffer. This elution goes directly into the following adapter ligation reaction using the NEBNext Quick Ligation Module (NEB), following the manufacturer's protocol, except 1 μl of 100 μM SplA5 adapter (*see* Barquist et al. [26]) is used per sample reaction. Next, a 1× Ampure SPRI clean-up is performed, eluting in 50 μl EB followed by a second 0.8× Ampure SPRI clean-up with elution in 21 μl. Dilution of this is performed 1 in 10 in water (using 1 μl of starting) and resolved using a High Sensitivity DNA kit on an Agilent Bioanalyzer to check for correct adaptor ligation and detect any contamination.

Next, PCR enrichment of transposon-containing fragments is performed using a custom transposon-specific primer and an indexed adapter-specific primer (SplAP5.x; *see* Barquist et al. [26]). PCR is carried out using cycling conditions (95 °C for 2 min, 19 × 98 °C for 20 s, 65 °C for 30 s, 72 °C for 30 s, 72 °C for 2 min) using the Kapa HiFi HotStart ReadyMix (Kapa Biosystems) with an example reaction as follows:

Adapter-ligated DNA	10 μl (should be at least 100 ng)
Kapa HiFi 2× ReadyMix	25 μl
5′ or 3′ transposon-specific primer (100 μM)	0.5 μl
SplAP5.x index primer (10 μM)	5 μl
Water	9.5 μl

PCR products can then be purified using a 0.8× Ampure SPRI clean-up with elution in 30 μl EB buffer and 1 μl of this is diluted

1 in 10 and resolved again on an Agilent Bioanalyzer for quantification and checked for contamination. To confirm quantification of the TraDIS library amount, a SYBR green qPCR, is performed in duplicate per sample, using P5 and P7 primers (*see* Barquist et al. [26]). About 20 μl qPCR reactions can be set up as follows:

P5 ± P7 qPCR	
DNA (diluted 1:1000 or 1:5000)	4 μl
Kapa SYBR FAST 2× mix	10 μl
10 μM qPCR2.1 primer	0.4 μl
10 μM qPCR2.2 primer	0.4 μl
Water	5.2 μl

Transposon + P7 qPCR	
DNA (diluted 1:1000 or 1:5000)	4 μl
Kapa SYBR FAST 2× mix	10 μl
10 μM transposon sequencing primer	0.4 μl
10 μM qPCR2.2 primer	0.4 μl
Water	5.2 μl

NB in the transposon-specific qPCR, the transposon-specific sequencing primer is used.

Transposon-containing fragments are typically quantified as 40–90% of the total fragments that could potentially cluster, but this increases almost always to >90% in the final sequencing reads because the sequencing primer binds a transposon region downstream of where the transposon-specific PCR binds. This qPCR data generated from the transposon sequencing primer qPCR reactions can be used to guide loading of your chosen sequencing platform, for example, MiSeq or Hiseq2500 loading concentrations are generally optimal at 8pM. The TraDIS sequencing reaction is then run via custom TraDIS Illumina recipes as given in Barquist et al. [26].

3.3 Approaches to Essential Gene Prediction in Bifidobacteria from Datasets Generated Through TIS Approaches

The main objective pursued through analysis of a TIS dataset, is usually to achieve classification of genes or gene domains depending on their essential or dispensable character for cell survival (*see* further information in **Note 9**). Although there are several analytical tools to analyze TIS datasets to achieve this common goal (*see* **Note 10**), herein we will focus on those methodologies that have been applied specifically to analyze a transposon library in *B. breve* UCC2003, through using a TraDIS approach [22].

3.3.1 Sequencing and Quality Control

Once the library has been prepared and sequenced, several quality control measures need to be taken during prediction of essential genes.

1. Insertion density assessment.

   Transposon mutant data require normalization, which is achieved by dividing the number of insertions by gene length yielding an insertion index. The insertion index follows a bimodal distribution where the left-most peak represents genes with no or almost no insertions and thus deemed essential (i.e. housekeeping genes) and a peak toward the right with many insertions representing regions not involved in cell growth (e.g. transcriptional repressors). Using the bimodal distribution, a likelihood ratio (LR) for gene essentiality can be deduced. If the log2 of the LR was less than −2, the gene is considered essential. Whereas when the log2-LR is >2, the gene is likely non-essential (Fig. 2a) [18]. When insertion density (i.e. library saturation) is low, the number of essential genes that can be predicted are limited to those that do not contain any insertions (Fig. 2b). However, in a low-density library, the probability of a non-essential gene not containing an insertion by chance increases and thus may be erroneously classified as essential. For this reason, the ability to resolve non-essential genes is decreased with very low insertion densities [27]. An acceptable insertion density is 50%, whereas 20–30% is considered a low density library [28].

2. Considerations for insertions in essential regions.

   Insertions can still occur in essential genes, for example, non-essential regions that do not necessarily disrupt function. Commonly, insertions at the 3′ end of the gene or within non-essential linker regions between domains are not considered. Different computational analysis pipelines have incorporated ways to deal with this.

   – The BioTraDIS Toolkit has parameters (-trim3 and -trim5) in its tradis_gene_insert_sites function that can be adjusted to account for insertions in non-essential regions of genes (often found at the gene extremities that correspond to the N- or C-terminus of the corresponding gene product). This is especially useful if FastQC analysis indicates poor quality base calls towards the 3′ or 5′ ends. For example, one can trim ~10% of the 3′ end off and 5% of the 5′ end with the following command:

```
>tradis_gene_insert_sites -trim3 0.1 -trim5 0.05 file.embl
plot.gz
```

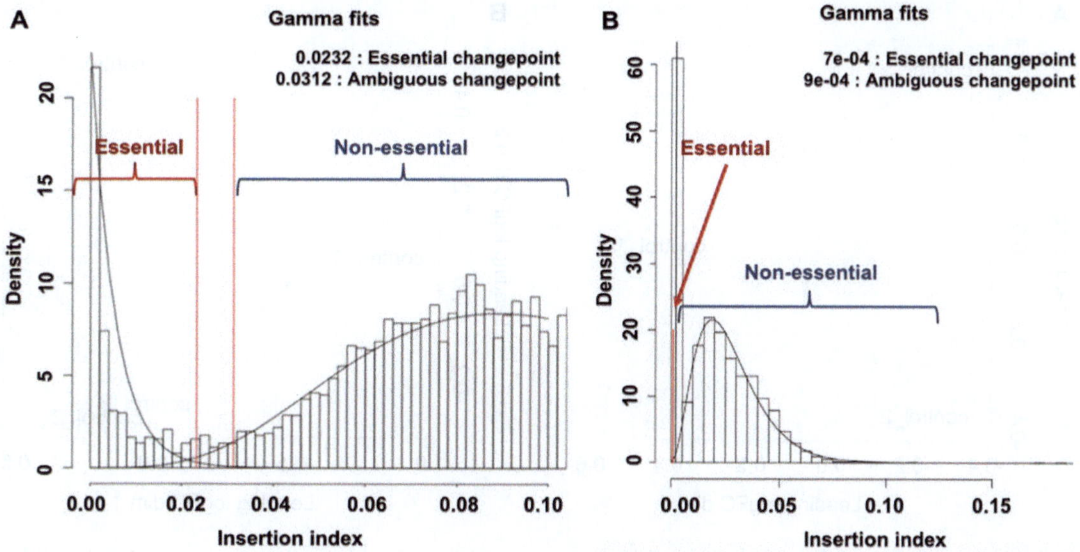

Fig. 2 Mutant library density versus insertion indices. (**a**) High saturation library improves prediction of essential genes; (**b**) low saturation library limits predicted essential genes to only those with zero insertions. The image in panel (**a**) is adapted from the BioTraDIS tutorial available on github (https:/github.com/sanger-pathogens/Bio-Tradis/blob/master/BioTraDISTutorial.pdf)

– Other tools like AlbaTraDIS, ARTIST, TRANSIT, MAGenTA and Tn-Seq Explorer have incorporated a sliding window rather than gene-centric approach to address the fact that some genes contain both non-essential and essential domains [28–31].

3. Checking replicates.

   Surveying the reproducibility of replicates via statistical plots is an important step to determine which replicates or controls should be included in the analysis. This can be achieved by performing multi-dimensional scaling (MDS) of replicate mutant libraries (Fig. 3a). Analysis can be improved either by removing outliers or by including additional controls (Fig. 3b). Several tools incorporate functions that allow for multi-dimensional scaling analysis and generation of volcano plots including BioTraDIS and TRANSIT.

4. Manual confirmation by visual inspection.

   Many software tools incorporate graphical features such as viewing insert site plots generated by BioTraDIS or the Track-View feature in TRANSIT to allow visualization of insertion patterns within the genome. It is highly recommended to inspect genes classified as essential in a genome browser such as Artemis [32] or IGV [33] to confirm predictions made for specific genes of interest or examine other potentially interesting genes or domains. For instance, you may see that some

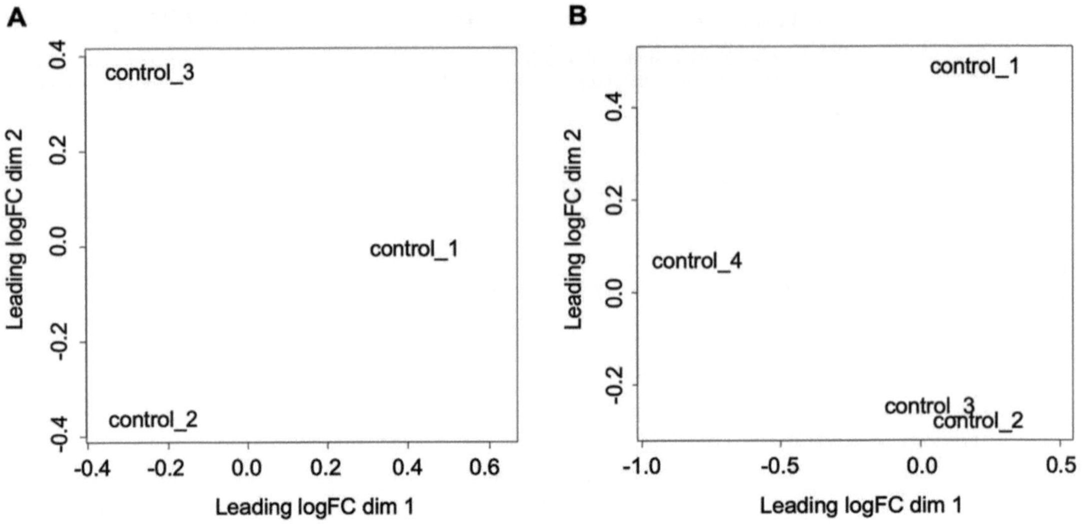

Fig. 3 MDS plot of TraDIS mutant libraries in control conditions. (**a**) Variance between controls when three controls are included. (**b**) Change in variance when an additional control is included

genes that were not called essential may contain domains that appear essential under the condition(s) tested (Fig. 4). This can be determined by performing an InterPro [34] or Pfam [35] search for conserved domains within the gene of interest. When these data are corroborated with RNA-sequencing data, this may unveil otherwise overlooked information.

3.3.2 Essential Gene Prediction Workflow Applied to TIS Experimental Approaches in Bifidobacteria

B. longum and *B. breve* are currently the only bifidobacterial species to have undergone transposon mutagenesis followed by insertion site sequencing to determine their essential gene profiles. Here we focus on the Bio::TraDIS Toolkit (https://github.com/sanger-pathogens/bio-tradis) [26] that was employed for analysis of *B. breve* UCC2003 transposon mutant libraries [22] generated by following the methodology described in previous sections. The workflow described herein is based on the BioTraDIS tutorial from [26].

Inputs required:

1. FASTQ sequencing reads with transposon tags attached.
   - In case a TraDIS dark cycle was included in the sequencing protocol, the transposon tag reads and genomic reads will be separated and can be concatenated using **add_tradis_tags** script.
2. A reference genome file in fasta and embl format.

Step 1: Check quality of FASTQ reads with FastQC tool (http://www.bioinformatics.babraham.ac.uk/projects/fastqc/).

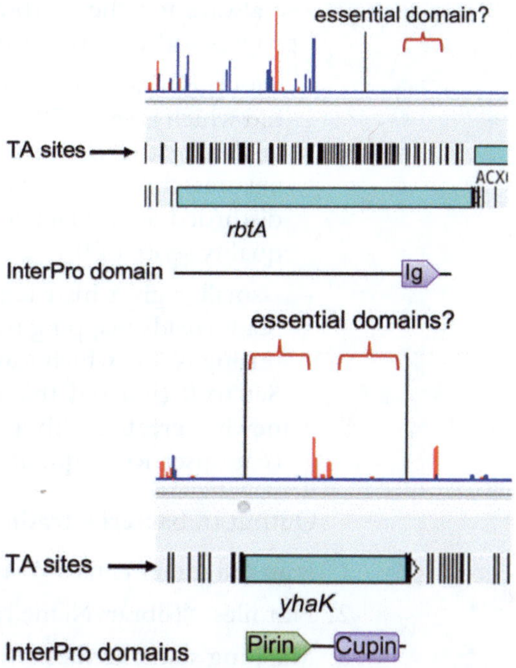

Fig. 4 Visual inspection of gene insert site plots generated by BioTraDIS and visualized in Artemis may provide insight into essential regions or genetic domains. The region with decreased insertions is labelled as a putative essential domain. TA sites are indicated by the black lines in the gene track. The bottom panel shows the domains predicted by InterPro of the *rbtA* gene. If the *rbtA* gene is overexpressed in RNA-seq data, this would provide further evidence that these domains may be essential despite the gene being called as non-essential

Given the nature of TraDIS data, many of the standard QC checks will not give expected results. The main QC checks to perform are looking at the per-base sequence quality and the adapter contamination reports.

- If most reads have low quality towards the 3′ end, PRINSEQ (http://prinseq.sourceforge.net/) can be used to trim a specified number of bases from the 3′ ends of the reads making sure to keep the transposon tags on the reads.

Step 2: Align reads to reference genome.

The bacteria_tradis script is used to align reads to a reference genome using the SMALT aligner (sanger.ac.uk/resources/software/smalt/) and to quantify the number of insertions in each region. A basic use of this script is:

- **> bacteria_tradis -v --smalt_r 0 -m 30 -f fastqs.txt -t TAAGAGACAG -r ref.fasta**
  - Create a text file with the names of the FASTQs and call it fastqs.txt.

- -v always for the 'verbose' option to be turned on so the progress of the script is displayed while it runs.
- -m option describes the minimum PHRED quality threshold which is set at 30 by default. Also 30 was the value set in analyzing the *B. breve* libraries [22]. By setting it to 0, it ensures that reads mapped to multiple locations are not discarded as unmapped, since multi-mapping reads have a quality score of 0.
- --smalt_r gives instructions on how to treat multi-mapping reads (reads mapping to more than one region). The default setting is −1, which leaves multi-mapping reads unmapped. Set to 0 to avoid this issue, which is important if using a newly created library to avoid repetitive elements (e.g. insertion sequences) from being called as essential.

Output of **bacteria_tradis**:

1. Bam (alignment) files: "(libraryName).mapped.bam".
2. Plot files: "(libraryName.repliconName).insert_site_plot.gz".
3. Mapping statistics files: "(input list name).stats".

For more information on optional parameters of bacteria_tradis refer to the Bio::TraDIS toolkit user manual.

Step 3: Generate transposon insert site statistics.

Create TraDIS gene insert site data with the -trim3 parameter set to 0.1 to trim away 10% of the 3′ end of the reads. This is done since many essential genes can still be disrupted by insertions toward the 3′ end.

```
>tradis_gene_insert_sites -trim3 0.1 reference.embl sample.
insert_site_plot.gz
```

Step 4: View insert site plots in genome browser.

These plots can be viewed in the Artemis genome browser. Open the embl file of the reference genome used for mapping the reads. Add the uncompressed insert site plots generated in step 3 by selecting the 'Add User Plot' option in the 'Graph' dropdown menu.

3.4 Comparative Analysis of Essential Gene and Core Genome Functions Across Bacteria

Comparative genome analysis allows identifying orthologous genes across bacteria and can be useful for mapping the essential gene functions identified by TraDIS with different type of datasets (e.g. comparison with other TraDIS studies, mapping of the essential genes with the core genome functions of different bacterial species, comparison with transcriptomic datasets).

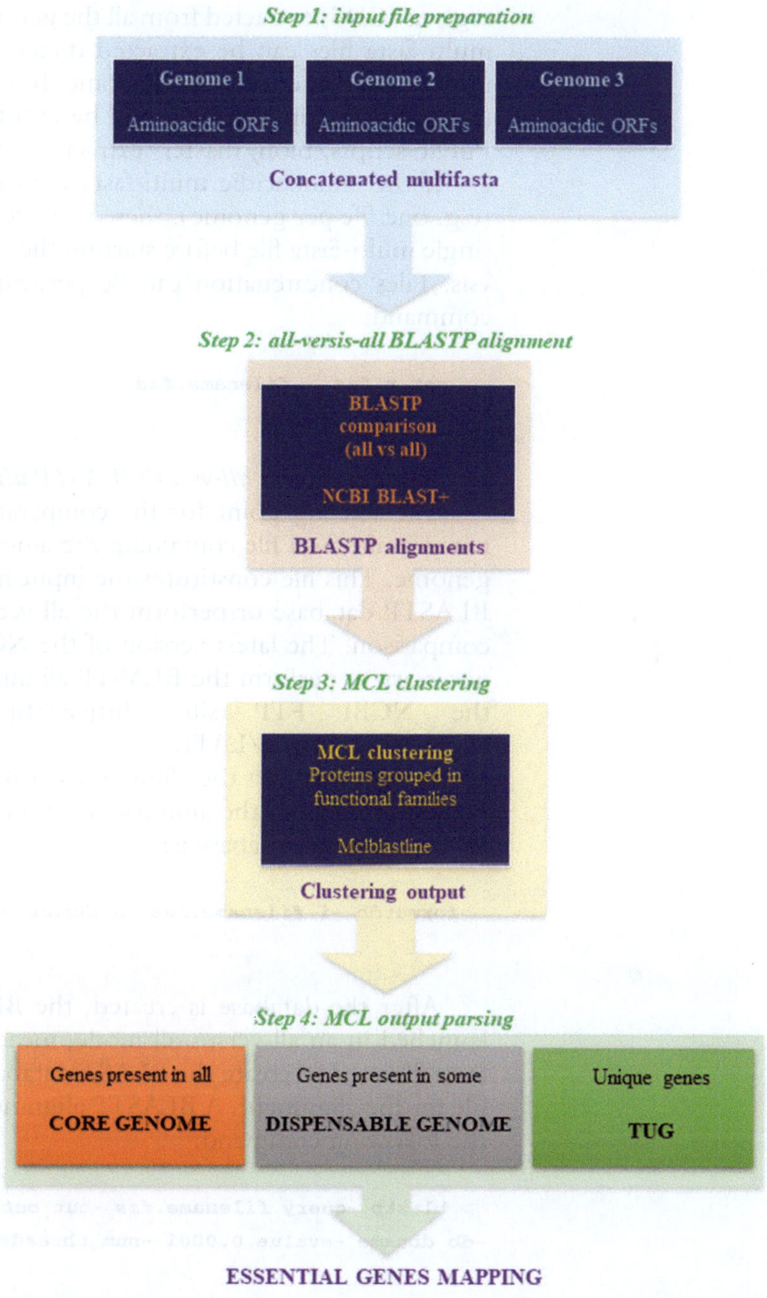

Fig. 5 Workflow representing a typical comparative genomic analysis pipeline

A typical comparative analysis pipeline is presented in Fig. 5 and is composed of four main steps, which can be performed via the Linux command line:

Step 1: Input file preparation.

The input file for the analysis is a multi-fasta (e.g. file extension *.fas or *.faa) containing the aminoacidic sequence of the coding regions (CDS) extracted from all the genomes of interest. The CDS multi-fasta files can be extracted directly from the Genbank files using either the Artemis genome browser or a custom script (an example script can be found here: https://github.com/frbot/Public-scripts/blob/master/extract_aa_nt-CDS.sh).

If the aminoacidic multi-fasta are organized in multiple files (e.g. one file per genome), these will need to be concatenated in a single multi-fasta file before starting the comparative genome analysis. Files concatenation can be performed using a simple bash command:

```
> cat *.fas > filename.fas
```

*Step 2: Perform all-*vs*-all BLASTP alignments.*

The starting point for the comparative analysis is a concatenated multi-fasta file containing the aminoacidic sequence of each genome. This file constitutes the input file for either generating a BLASTP database or perform the all-versus-all pairwise sequence comparison. The latest version of the NCBI BLAST+ executables necessary to perform the BLASTP alignments can be retrieved via the NCBI FTP site (https://ftp.ncbi.nlm.nih.gov/blast/executables/blast+/LATEST/).

In order to start the alignments, a BLASTP database needs to be generated from the aminoacidic multi-fasta file. The command for creating the database is:

```
>formatdb -i filename.fas -n dbname -p T -o T
```

After the database is created, the BLASTP alignment can be launched in an all-versus-all mode, meaning that the same multi-fasta file used to create the BLAST database is also used as an input file for the alignment. A BLASTP alignment can be launched using the following command:

```
> blastp -query filename.fas -out outfile.blastp -task blastp
-db dbname -evalue 0.0001 -num_threads numcores -outfmt 6
```

The above command generates an output file named *outfile.blastp*, containing all the BLASTP pairwise sequence alignments in tab-delimited format, which can be further filtered (based on the desired similarity cut-off) or directly used as an input file for the mcl clustering.

For more information on optional parameters of the blastp pipeline, please refer to the user manual, accessible from:

```
> blastp -h
```

Step 3: Perform MCL clustering.

The all-versus-all BLASTP output file can be used as an input to compute the clusters of functionally related families using the Markov Cluster Algorithm (MCL) mclblastline [36].

The MCL clustering allows the grouping of all protein sequences in clusters of orthology, based on their sequence similarity. The clustering allows further classification of proteins in the dataset which constitute the core genome (e.g. proteins present in all the genomes), from those that represent accessory functions (e.g. proteins present only in few representative genomes). The tab-delimited BLASTP output file obtained from BLASTP can be directly fed into the mclblastline pipeline as following:

```
>mclblastline -blast -m9 -blast-score=b -blast-sort=a -blast-but=5 -mcl-I=2.5 outfile.blastp
```

The main output file produced by the mclblastline pipeline (named dump.out.*) is a tab-delimited file, where each row represents an orthogroup and contains the gene ids (or locus_tags) of the proteins predicted to be members of the same cluster of orthology.

For more information on optional parameters of the mclblastline pipeline, please refer to the user manual accessible from:

```
> mclblastline -h
```

Further details on the MCL algorithm implemented in the mclblastline pipeline can be found in the official manual (https://micans.org/mcl/man/).

Step 4: Parsing of the MCL file and extraction of core genome functions.

The tab-delimited file obtained from the mclblastline pipeline contains information on each computed orthogroup (or families of orthology) and contains a list of orthologous locus_tags on each row. This file can be further processed by custom generated scripts in order to retrieve the number of orthologous and paralogous genes in each orthogroup. This parsing process allows the retrieval of families representing the core genome functions across the compared genomes.

An example of a custom script that counts the number of times a locus tag prefix occurs in each row (thus providing information on gene orthology and copy number for each orthogroup) can be found here: https://github.com/frbot/Comparative-genomics/blob/master/gene-frequency-count.sh.

This script accepts as input the mclblastline *dump.out* file and counts the number of times each locust_tag prefix occurs in each row. The script can be used as follows:

```
> gene-frequency-count.sh dump.out.file > outfile
```

As output, the script generates an outfile which is a tab-delimited matrix containing the orthogroup ID on the first column (indicated as FamXX), as well as the number of orthologous (genes present in one copy) and paralogous (genes present in more than one copy) per each genome in each orthogroup. Further filtering of this output matrix can be employed to extract only families or orthogroups containing genes in all the genomes (e.g. excluding rows containing the 0 value), thus constituting the core genome functions. Similarly, rows containing exclusively value 1 in each column will constitute the single-copy core genes (as they are genes present only in one copy in each genome).

Example custom scripts showing the implementation of the comparative genome analysis described above can be found here: https://github.com/frbot/Comparative-genomics.

Once the list of locus_tags assigned to the core genome, single-copy genes and essential genes are retrieved, recursive grep commands can be used to generate custom GFF files exclusively containing the genes of interest for visualization purposes. Indicated below is an example of a grep loop which can be used to extract the essential gene functions from the GFF file of the reference genome (given that the locus_tags of the essential genes to be extracted are provided as a one-column file list called *essentials.txt*):

```
> for f in $(cat essentials.txt); do grep "$f" reference.gff
>> essentials.gff ; done
```

Similarly, custom gff files can be generated containing genes representing core genomes and single-copy genes (given that the relative file lists are provided):

```
> for f in $(cat core.txt); do grep "$f" reference.gff >>
core.gff ; done
> for f in $(cat single-copy.txt); do grep "$f" reference.gff
>> single-copy.gff ; done
```

Custom GFF files containing exclusively core genome functions, single-copy genes or essential genes are a good starting point to obtain the gene coordinates of these sets of genes, necessary for visualization purposes. GFF can be used to create circular DNA maps displaying regions of interest in the reference genome. For example, GFF files can be directly imported into available visualization tools such as DNA plotter (available from http://sanger-pathogens.github.io/Artemis/DNAPlotter/) and used to highlight regions of interest as concentric circles in the reference genome.

Further details on the use and options of DNA plotter can be found in the Sanger website (https://www.sanger.ac.uk/tool/dnaplotter/).

Of course, alternative visualization programs can also be used to generate circular genome maps (e.g. Circos available at http://circos.ca), but they require further manipulation of the GFF files, in order to adapt their format for visualization. Further documentation and recipes for the use of Circos plots for microbial genomes can be found in the Circos documentation (http://circos.ca/documentation/).

4 Notes

1. General considerations when designing a TIS approach depending on various experimental variables and research goals:

   Nature of the genome being studied: In some bifidobacterial strains, their GC content is unevenly distributed with part of the genome containing a higher GC content than the rest [37]. In such a guanosine-cytosine (GC)-variable genome, it could be useful to perform a combination of Tn5-based and mariner-based transposon mutagenesis to avoid insertion biases skewing data analysis provided that appropriate mutagenesis procedures are functional in bifidobacteria.

   Sequencing approach: The TraDISort technique would be ideal to discover new efflux pump genes [38], whereas the recently developed TraDIS-Xpress should be employed if the interest is focused on characterizing genetic regulatory elements [39]. If the goal is to understand single cell phenotypes of each mutant from the transposon mutant library, droplet Tn-Seq (dTn-Seq) utilizes microfluidics to interrogate separated single cells independently using TIS [40] and to elucidate genetic interaction networks, construction of transposon mutant libraries in a mutant background can be performed [19].

   Choice of sequencing platforms:

   – For Illumina-based TIS sequencing, single end sequencing is often preferred, as it ensures efficient use of sequencing bandwidth that does not get used up generating the mate pair. The reads are simply mapped against a reference in order to identify insertion sites and not required to be paired as is necessary for complex genome assembly.

- Paired end sequencing is preferable when sheared DNA is used to generate the library for sequencing as it allows for identification of PCR jackpot events where duplicate reads from the same site of shearing are amplified [41]. These can be removed using Picard or SAMtools [42] before analysis.
- Long-read whole-genome sequencing with Oxford Nanopore's MinION platform has recently been proposed as an alternative to simplify transposon insertion site mapping [43], which normally employs engineered transposons for high-throughput transposon insertion site sequencing or utilizes Tn*5* vectors, chromosomal digestion and recapture cloning. This technique could be a preferable alternative if monetary constraints are a limiting factor; however, it is not fully optimized yet and it should be noted that sequencing depth is lost due to amplification of non-target regions.

Transposon choice: Mariner-based (TA sites) versus *Tn5-based (random sites; GC bias)*

- Mariner-based transposons generate highly complex and saturated libraries with low insertion bias since their thymine-adenine (TA) dinucleotide targets are commonly evenly distributed in the genome [41, 44].
- Tn*5*-based transposons have a slight bias for high GC content and certain insertion motifs [45]. This transposon is best used when it is too difficult to build mariner-based transposons for a particular species.

Random fragmentation compared to uniform fragment sizes:

- Several transposon insertion sequencing approaches (TnSeq and INSeq) introduce a MmeI restriction enzyme digestion step to yield uniform sizes. This can prevent any chance of sequencing bias of smaller sized fragments.
- Other approaches use genomic shearing which creates uneven fragment sizes. The benefit to these approaches is that they result in longer fragments which can provide more resolution during insertion site-mapping and better mapping due to longer read lengths [18].

Biological question under investigation: If it is known that significant experimental bottlenecks will be encountered, for instance if conditions to study select for a minute proportion of the original mutant population as seen in many host infection experiments [40, 46–48], there are several options to reduce bottlenecks, such as using less complex libraries (around

10,000 mutants) and replicates pooled or, using less complex defined libraries (around 10,000 mutants) as opposed to random mutagenesis [21]. To further validate that mutant pools are being selected for by true biological negative selection rather than stochastic bottlenecks, repeated screening of replicate pools can be performed [49]. For instance, given that the human gut is a highly selective environment where highly adapted species thrive, in vivo host colonization studies of bifidobacterial fitness should be carefully planned to mitigate potential experimental bottlenecks [50, 51].

2. D-Lactose might need to be replaced by an alternative carbon source depending on the particular needs of the strain used. In order to facilitate preparation of a 10% stock solution of D-lactose, the suspension might need to be warmed up at 37 °C, then filter sterilize (0.2 μm pore filters).

3. For tetracycline-resistant strains, the antibiotic resistance cassette in the Tn*5* transposon needs to be replaced by an appropriate one, which should be selected considering the particular antibiotic resistance profile of the corresponding strain of interest and the availability of robust selection markers capable to achieve efficient expression in the strain of interest.

4. Virtually any piece of DNA can be included into a transposon in between the inverted repeats recognized by the Tn*5* transposase. Design of a custom Tn*5* transposon would usually include an appropriate selection marker such as antibiotic resistance genes for antibiotic sensitive hosts, as well as appropriate expression signals enabling strong and constitutive expression in the selected host. Additional sequences required for further downstream applications (e.g. T7 promoters) can also be included. For further instructions and background information, please check Lucigen guidelines.

5. It is recommended to perform several PCR reactions to guarantee the production of a sufficiently large quantity of transposon DNA (at least 50–60 μg of PCR product).

6. When designing a new customized transposon construct, ensure the absence of PshAI or PvuII internal restriction sites in the constructed transposon. Alternatively, whenever these two sites have target sequences within the transposon sequence, it is possible to PCR-amplify the transposon piece of DNA by employing a single phosphorylated oligonucleotide, annealing with the Tn5 mosaic end (5′-*CTGTCTCTTA TACACATCT*-3′). If the transposon is amplified in this way from the pMOD-derived constructs, there is no need to trim the transposon ends with PshAI or PvuII restriction.

7. It is recommended to perform the electroporation using freshly prepared *Bifidobacterium* electrocompetent cells in order to maximize the transformation efficiencies. Growth media, wash buffer and antibiotic-containing plates are recommended to be prepared freshly the day before the experiment starts.
8. If no antibiotic-resistant colonies are grown following electroporation of transposome complexes, verify the presence of antibiotic-resistant colonies in the reaction performed with the positive control plasmid DNA. If transformation efficiency of this control plasmid is under 10^4 cfu/μg, repeat the experiment or consider optimizing the competent cells preparation or transformation procedure.
9. Analysis of a TIS dataset usually intends to differentiate *essential genes* (an insertion in these would putatively disrupt proper protein production and function thus causing a growth defect, and therefore those genes are confidently identified to be missing insertions not just due to chance); from *non-essential genes* (those that contain a statistically significant insertion density that does not alter bacterial fitness in the conditions tested). In contrast, sometimes it is not possible to fully differentiate those, for instance if a gene is too short or if the insertion density is too low, a call that meets statistical requirements could not be made confidently and some genes would be classified as *uncertain genes.* Further, if insertion density is sufficiently low, some tools can identify essential genetic domains and intergenic regions such as non-coding regions (i.e. non-coding RNAs) and regulatory elements by using a sliding window approach rather than a gene-centric approach, and will thus enable to identify e*ssential genomic features* (PLoS Pathog. 2012 Sep;8(9):e1002946. doi: https://doi.org/10.1371/journal.ppat.1002946). Tools that can do this include ARTIST, TRANSIT, Tn-Seq Explorer and MAGenTA (*see* further information on these and other available tools in **Note 10**) (PLoS Genet. 2014 Nov 6;10(11):e1004782. doi: https://doi.org/10.1371/journal.pgen.1004782; PLoS One. 2015 May 4;10(5):e0126070. doi: https://doi.org/10.1371/journal.pone.0126070; PLoS Comput Biol. 2015 Oct 8;11(10):e1004401. doi: https://doi.org/10.1371/journal.pcbi.1004401; Bioinformatics. 2017 Sep 1;33(17):2781–2783. doi: https://doi.org/10.1093/bioinformatics/btx320). This approach is particularly useful for identifying essential regions that have not yet been annotated.
10. A brief description of alternative available tools to predict essential genes on TIS datasets are briefly described below:

    ESSENTIALS —(PLoS One. 2012;7(8):e43012. doi: https://doi.org/10.1371/journal.pone.0043012*).*

This software tool utilizes a Negative Binomial distribution to identify and quantify statistical significance of essential genes. It has been noted to overestimate essential genes predicted while also missing certain essential genes (false negatives) due to insertions in non-essential regions like the N- or C-termini.

ARTIST—(PLoS Genet. 2014 Nov 6;10(11):e1004782. doi: https://doi.org/10.1371/journal.pgen.1004782*).*

This tool utilizes various approaches to overcome the limitations of conditional essential gene prediction. It employs a novel adaptation of hidden Markov models (HMM) to allow for annotation independent mapping of fitness-linked loci so that novel essential sequences can be discovered. Simulation-based resampling facilitates normalization and is used to model and remove noise due to experimental bottlenecks and sequence sampling errors, thus limiting the number of false positives detected. This tool comprises of two analysis tools, EL-ARTIST (*E*ssential *L*oci), which predicts all essential loci of transposon libraries grown under standard laboratory conditions and CON-ARTIST (*Con*ditionally essential loci), which predicts all conditionally essential genes where the transposon mutant library is subjected to different conditions.

TRANSIT—(PLoS Comput Biol. 2015 Oct 8;11(10): e1004401. doi: https://doi.org/10.1371/journal.pcbi.1004401*).*

TRANSIT is a software tool that contains a GUI and built-in analyses that allow for data visualization (i.e. TrackView and volcano plots), identification of essential and conditionally essential genes and regions and calculation of statistical significance. It provides two methods for essentiality analysis: (1) Bayesian and (2) HMM. The former allows for quantifying the significance of 'gaps'; long stretches of TA sites devoid of insertions while the latter accounts for local differences in read counts and enables identification of loci larger or smaller than a gene, like operons, protein domains and intergenic regions. TRANSIT is designed in a modular way where pre-processing of read counts into template counts is separate from computational analysis tools. Therefore, although pre-processing is specific to Tn-Seq libraries generated with particular barcodes, TRANSIT computational analysis tools can be applied to any Himar1 created Tn-Seq library.

Tn-Seq Explorer—(PLoS One. 2015 May 4;10(5):e0126070. doi: https://doi.org/10.1371/journal.pone.0126070*).*

This tool uses a sliding window method for identification of regions depleted of insertions with respect to the genome. This approach does not include calculation of statistical significance for the Essentiality Index (EI) and hence is not a robust method for identifying conditionally important genes (i.e. through comparative analysis of different conditions). It just utilizes simple comparison of EI values.

*TraDIS Toolkit—(*https://github.com/sanger-pathogens/Bio-Tradis*) (Bioinformatics. 2016 Apr 1;32 (7):1109–11. doi:* https://doi.org/10.1093/bioinformatics/btw022*).*

The TraDIS Toolkit comprises of a comprehensive pipeline for TraDIS analysis via a command-line interface (implemented in Perl and R languages). The TraDIS Toolkit can identify genes as essential, ambiguous or non-essential, using the observed bimodal distribution of insertion sites over genes that have been normalized for gene length. It can also perform statistical analysis of conditionally essential and significant genes from TraDIS experiments. It incorporates visualization tools for the generation of volcano, MDS and bimodal distribution plots and insertion site plots that can be loaded into the Artemis genome browser (Bioinformatics. 2012 Feb 15;28 (4):464–9. doi: https://doi.org/10.1093/bioinformatics/btr703). Its command-line approach allows for simultaneous processing of multiple insertion libraries. It incorporates the SMALT short read mapper for alignment of insertion mutant reads to reference genomes which supports alignment of contigs and plasmids. The commands incorporate flexible user defined parameters that can be adjusted for optimum analysis such as adjusting k-mer lengths, step sizes and percent identity.

MAGenTA—(Bioinformatics. 2017 Sep 1;33 (17):2781–2783. doi: https://doi.org/10.1093/bioinformatics/btx320*).*

The Microbial Assessment by Genome-Wide Tn-Seq Analysis (MAGenTA) package of software tools and pipeline available at the Galaxy ToolShed repository that allows for the prediction of growth rates from transposon mutagenesis approaches including TraDIS, Tn-Seq, IN-Seq, HITS. This enables the prediction of genes, ncRNAs and pathways involved in drug tolerance, disease, genomic organization and host factors affecting bacterial host susceptibility. The tools included allow for growth rate (fitness) calculations, sliding window analysis, bottleneck calculations and corrections, statistics to compare experiments/strains and genome-wide fitness visualization. When performing transposon-insertion sequencing experiments in vivo, sometimes the insertion is lost by chance rather than because the insertion makes the bacterium less fit. MAGenTA has incorporated a calculation for loss of insertions by chance through determining the rate of neutral insertion loss (i.e. those that do not affect fitness) in pseudogenes or degenerate genes. A sliding window approach can be used to inspect the genome for regions that have a significant impact on fitness.

AlbaTraDIS—(https://github.com/quadram-institute-bioscience/albatradis) (AlbaTraDIS: Comparative analysis of large datasets from parallel transposon mutagenesis experiments, bioRxiv, doi: https://doi.org/10.1101/593624*).*

This tool is designed to take as input TraDIS and TraDIS-Xpress experiment data (Genome Res. 2020 Feb;30 (2):239–249. doi: https://doi.org/10.1101/gr.254391.119) and aid in predicting the impact transposon insertions have on neighbouring genes in terms of expression. Normalization can be performed as a first step to account for experimental variation. A sliding window approach is utilized to perform reference and annotation independent estimation of insertion density per loci. The window sizes can be adjusted; however, the default is set to 50 bases as this was determined to be the minimum size where an insertion density of at least one in every ten bases could be found. The tool also provides the option to perform annotated reference-guided analysis if desired.

Acknowledgements

F.B. and D.v.S. are members of APC Microbiome Ireland, which is funded by Science Foundation Ireland (SFI) through the Irish Government's National Development Plan (Grant Numbers SFI/12/RC/2273-P1 and SFI/12/RC/2273-P2). L.R. has received funding from the Spanish State Research Agency (RTI2018-095021-J-I00).

References

1. Tojo R, Suárez A, Clemente MG et al (2014) Intestinal microbiota in health and disease: role of bifidobacteria in gut homeostasis. World J Gastroenterol 20:15163–15176
2. Ventura M, Turroni F, Lugli GA, van Sinderen D (2014) Bifidobacteria and humans: our special friends, from ecological to genomics perspectives. J Sci Food Agric 94:163–168
3. Bottacini F, van Sinderen D, Ventura M (2017) Omics of bifidobacteria: research and insights into their health-promoting activities. Biochem J 474:4137–4152. https://doi.org/10.1042/BCJ20160756
4. Cronin M, Ventura M, Fitzgerald GF, Van Sinderen D (2011) Progress in genomics, metabolism and biotechnology of bifidobacteria. Int J Food Microbiol 149:4–18
5. Brancaccio VF, Zhurina DS, Riedel CU (2013) Tough nuts to crack: site-directed mutagenesis of bifidobacteria remains a challenge. Bioengineered 4:197–202. https://doi.org/10.4161/bioe.23381
6. O'Callaghan A, Bottacini F, O'Connell Motherway M, van Sinderen D (2015) Pangenome analysis of *Bifidobacterium longum* and site-directed mutagenesis through by-pass of restriction-modification systems. BMC Genomics 16:832. https://doi.org/10.1186/s12864-015-1968-4
7. Hidalgo-Cantabrana C, Sánchez B, Álvarez-Martín P et al (2015) A single mutation in the gene responsible for the mucoid phenotype of *Bifidobacterium animalis* subsp. *lactis* confers surface and functional characteristics. Appl Environ Microbiol 81:7960–7968. https://doi.org/10.1128/AEM.02095-15

8. Sakaguchi K, He J, Tani S et al (2012) A targeted gene knockout method using a newly constructed temperature-sensitive plasmid-mediated homologous recombination in *Bifidobacterium longum*. Appl Microbiol Biotechnol 95:499–509. https://doi.org/10.1007/s00253-012-4090-4
9. O'Connell Motherway M, Watson D, Bottacini F et al (2014) Identification of restriction-modification systems of *Bifidobacterium animalis* subsp. *lactis* CNCM I-2494 by SMRT sequencing and associated methylome analysis. PLoS One 9:e94875. https://doi.org/10.1371/journal.pone.0094875
10. O'Connell Motherway M, O'Driscoll J, Fitzgerald GF, Van Sinderen D (2009) Overcoming the restriction barrier to plasmid transformation and targeted mutagenesis in *Bifidobacterium breve* UCC2003. Microb Biotechnol 2:321–332. https://doi.org/10.1111/j.1751-7915.2008.00071.x
11. Bottacini F, Morrissey R, Roberts RJ et al (2018) Comparative genome and methylome analysis reveals restriction/modification system diversity in the gut commensal *Bifidobacterium breve*. Nucleic Acids Res 46:1860–1877. https://doi.org/10.1093/nar/gkx1289
12. Ruiz L, Motherway MOC, Lanigan N, van Sinderen D (2013) Transposon mutagenesis in *Bifidobacterium breve*: construction and characterization of a Tn5 transposon mutant library for *Bifidobacterium breve* UCC2003. PLoS One 8. https://doi.org/10.1371/journal.pone.0064699
13. Choi KH, Kim KJ (2009) Applications of transposon-based gene delivery system in bacteria. J Microbiol Biotechnol 19:217–228
14. Picardeau M (2010) Transposition of fly mariner elements into bacteria as a genetic tool for mutagenesis. Genetica 138:551–558. https://doi.org/10.1007/s10709-009-9408-5
15. Goryshin IY, Reznikoff WS (1998) Tn5 in vitro transposition. J Biol Chem 273:7367–7374. https://doi.org/10.1074/jbc.273.13.7367
16. Scherens B, Goffeau A (2004) The uses of genome-wide yeast mutant collections. Genome Biol 5:229. https://doi.org/10.1186/gb-2004-5-7-229
17. Price MN, Wetmore KM, Waters RJ et al (2018) Mutant phenotypes for thousands of bacterial genes of unknown function. Nature 557:503–509. https://doi.org/10.1038/s41586-018-0124-0
18. Langridge GC, Phan MD, Turner DJ et al (2009) Simultaneous assay of every *Salmonella* Typhi gene using one million transposon mutants. Genome Res 19:2308–2316. https://doi.org/10.1101/gr.097097.109
19. van Opijnen T, Bodi KL, Camilli A (2009) Tn-seq: high-throughput parallel sequencing for fitness and genetic interaction studies in microorganisms. Nat Methods 6:767–772. https://doi.org/10.1038/nmeth.1377
20. Gawronski JD, Wong SMS, Giannoukos G et al (2009) Tracking insertion mutants within libraries by deep sequencing and a genome-wide screen for *Haemophilus* genes required in the lung. Proc Natl Acad Sci U S A 106:16422–16427. https://doi.org/10.1073/pnas.0906627106
21. Goodman AL, McNulty NP, Zhao Y et al (2009) Identifying genetic determinants needed to establish a human gut symbiont in its habitat. Cell Host Microbe 6:279–289. https://doi.org/10.1016/j.chom.2009.08.003
22. Ruiz L, Bottacini F, Boinett CJ et al (2017) The essential genomic landscape of the commensal *Bifidobacterium breve* UCC2003. Sci Rep 7. https://doi.org/10.1038/s41598-017-05795-y
23. Licandro-Seraut H, Scornec H, Pédron T et al (2014) Functional genomics of *Lactobacillus casei* establishment in the gut. Proc Natl Acad Sci U S A 111:E3101–E3109. https://doi.org/10.1073/pnas.1411883111
24. Hutchison CA, Chuang RY, Noskov VN et al (2016) Design and synthesis of a minimal bacterial genome. Science 351(6280):aad6253. https://doi.org/10.1126/science.aad6253
25. Sakanaka M, Nakakawaji S, Nakajima S et al (2018) A transposon mutagenesis system for *Bifidobacterium longum* subsp. *longum* based on an IS3 family insertion sequence, ISBlo11. Appl Environ Microbiol 84(17):e00824-18. https://doi.org/10.1128/AEM.00824-18
26. Barquist L, Mayho M, Cummins C et al (2016) The TraDIS toolkit: sequencing and analysis for dense transposon mutant libraries. Bioinformatics 32(7):1109–1111. https://doi.org/10.1093/bioinformatics/btw022
27. Dembek M, Barquist L, Boinett CJ et al (2015) High-throughput analysis of gene essentiality and sporulation in *Clostridium difficile*. MBio 6:2e02383–2e02314. https://doi.org/10.1128/mBio.02383-14

28. DeJesus MA, Ambadipudi C, Baker R et al (2015) TRANSIT—a software tool for Himar1 TnSeq analysis. PLoS Comput Biol 11:e1004401. https://doi.org/10.1371/journal.pcbi.1004401
29. Page AJ, Bastkowski S, Yasir M, et al (2019) AlbaTraDIS: Comparative analysis of large datasets from parallel transposon mutagenesis experiments. bioRxiv 593624. https://doi.org/10.1101/593624
30. Solaimanpour S, Sarmiento F, Mrázek J (2015) Tn-seq explorer: a tool for analysis of high-throughput sequencing data of transposon mutant libraries. PLoS One 10:e0126070. https://doi.org/10.1371/journal.pone.0126070
31. Mccoy KM, Antonio ML, Van Opijnen T (2017) MAGenTA: a galaxy implemented tool for complete Tn-Seq analysis and data visualization. Bioinformatics 33(17):2781–2783. https://doi.org/10.1093/bioinformatics/btx320
32. Carver T, Harris SR, Berriman M et al (2012) Artemis: an integrated platform for visualization and analysis of high-throughput sequence-based experimental data. Bioinformatics 28:464–469. https://doi.org/10.1093/bioinformatics/btr703
33. Robinson JT, Thorvaldsdóttir H, Winckler W et al (2011) Integrative genomics viewer. Nat Biotechnol 29:24–26
34. Mitchell AL, Attwood TK, Babbitt PC et al (2019) InterPro in 2019: improving coverage, classification and access to protein sequence annotations. Nucleic Acids Res 47: D351–D360. https://doi.org/10.1093/nar/gky1100
35. El-Gebali S, Mistry J, Bateman A et al (2019) The Pfam protein families database in 2019. Nucleic Acids Res 47:D427–D432. https://doi.org/10.1093/nar/gky995
36. Enright AJ, Van Dongen S, Ouzounis CA (2002) An efficient algorithm for large-scale detection of protein families. Nucleic Acids Res 30(7):1575–1584
37. O'Connell Motherway M, Zomer A, Leahy SC et al (2011) Functional genome analysis of *Bifidobacterium breve* UCC2003 reveals type IVb tight adherence (Tad) pili as an essential and conserved host-colonization factor. Proc Natl Acad Sci U S A 108:11217–11222. https://doi.org/10.1073/pnas.1105380108
38. Paulsen IT, Cain AK, Hassan KA (2017) Physical enrichment of transposon mutants from saturation mutant libraries using the TraDISort approach. Mob Genet Elements 7:1–7. https://doi.org/10.1080/2159256x.2017.1313805
39. Yasir M, Keith Turner A, Bastkowski S et al (2020) TRADIS-XPress: a high-resolution whole-genome assay identifies novel mechanisms of triclosan action and resistance. Genome Res 30:239–249. https://doi.org/10.1101/gr.254391.119
40. Troy EB, Lin T, Gao L et al (2013) Understanding barriers to *Borrelia burgdorferi* dissemination during infection using massively parallel sequencing. Infect Immun 81:2347–2357. https://doi.org/10.1128/IAI.00266-13
41. Chao MC, Abel S, Davis BM, Waldor MK (2016) The design and analysis of transposon-insertion sequencing experiments. Nat Rev Microbiol 14:119–128
42. Li H, Handsaker B, Wysoker A et al (2009) The sequence alignment/map format and SAMtools. Bioinformatics 25:2078–2079. https://doi.org/10.1093/bioinformatics/btp352
43. Sydenham TV, Sóki J, Hasman H et al (2015) Identification of antimicrobial resistance genes in multidrug-resistant clinical *Bacteroides fragilis* isolates by whole-genome shotgun sequencing. Anaerobe 31:59–64. https://doi.org/10.1016/j.anaerobe.2014.10.009
44. Rubin EJ, Akerley BJ, Novik VN et al (1999) In vivo transposition of mariner-based elements in enteric bacteria and mycobacteria. Proc Natl Acad Sci U S A 96:1645–1650. https://doi.org/10.1073/pnas.96.4.1645
45. Green B, Bouchier C, Fairhead C et al (2012) Insertion site preference of Mu, Tn5, and Tn7 transposons. Mob DNA 3:3. https://doi.org/10.1186/1759-8753-3-3
46. Skurnik D, Roux D, Aschard H et al (2013) A comprehensive analysis of in vitro and in vivo genetic fitness of *Pseudomonas aeruginosa* using high-throughput sequencing of transposon libraries. PLoS Pathog 9:e1003582. https://doi.org/10.1371/journal.ppat.1003582
47. Pritchard JR, Chao MC, Abel S et al (2014) ARTIST: high-resolution genome-wide assessment of fitness using transposon-insertion sequencing. PLoS Genet 10(11):e1004782. https://doi.org/10.1371/journal.pgen.1004782
48. Fu Y, Waldor MK, Mekalanos JJ (2013) Tn-seq analysis of *Vibrio cholerae* intestinal colonization reveals a role for T6SS-mediated antibacterial activity in the host. Cell Host Microbe 14:652–663. https://doi.org/10.1016/j.chom.2013.11.001

49. Chaudhuri RR, Morgan E, Peters SE et al (2013) Comprehensive assignment of roles for *Salmonella typhimurium* genes in intestinal colonization of food-producing animals. PLoS Genet 9:e1003456. https://doi.org/10.1371/journal.pgen.1003456
50. Xu J, Mahowald MA, Ley RE et al (2007) Evolution of symbiotic bacteria in the distal human intestine. PLoS Biol 5:1574–1586. https://doi.org/10.1371/journal.pbio.0050156
51. Abel S, Abel zur Wiesch P, Davis BM, Waldor MK (2015) Analysis of bottlenecks in experimental models of infection. PLoS Pathog 11:e1004823. https://doi.org/10.1371/journal.ppat.1004823

Chapter 16

The Application of Transposon Insertion Sequencing in Identifying Essential Genes in *B. fragilis*

Yaligara Veeranagouda and Hannah M. Wexler

Abstract

Essential genes are those that are indispensable for the survival of organism under specific growth conditions. Investigating *essential genes* in pathogenic bacteria not only helps to understand vital biological networks but also provides novel targets for drug development. Availability of genetic engineering tools and high-throughput sequencing methods has enabled *essential genes* identification in many pathogenic gram-positive and gram-negative bacteria. *Bacteroides fragilis* is one of the major bacteria specific of human gastrointestinal microbiota. When *B. fragilis* moves out of its niche, it turns into deadly pathogen. Here, we describe detailed method for the essential gene identification in *B. fragilis*. Generated transposon mutant pool can be used for other applications such as identification of genes responsible for drug resistance in *B. fragilis*.

Key words Essential gene, Bacteroides fragilis, Transposon insertion sequencing

1 Introduction

Bacteroides fragilis is an obligatory anaerobic gram-negative bacterium. It is one of the major bacteria found in human colon and exhibit commensal behavior. As a commensal, *B. fragilis* is essential for the metabolism of complex polysaccharides, volatile fatty acid synthesis, and host immune development [1]. Recent studies also indicated that treatment of rodents with *B. fragilis* reduced gut permeability and reduced autism spectrum disorder-like behaviors [2]. Thus, as commensal, *B. fragilis* plays vital role in human gut microbiome development.

Although *B. fragilis* is an integral part of human gut microbiota, it is considered as most deadly pathogen [1]. Pathogenic symptoms appear only when *B. fragilis* leaves it niche (gut) due to disruption of mucosal barrier, ulcers, or through surgical interventions. It is one of the most common species found in human infections. *B. fragilis* infection causes abscess formation and bacteremia [1]. *B. fragilis* infections are usually treated with antibiotics

Ren Zhang (ed.), *Essential Genes and Genomes: Methods and Protocols*, Methods in Molecular Biology, vol. 2377, https://doi.org/10.1007/978-1-0716-1720-5_16, © Springer Science+Business Media, LLC, part of Springer Nature 2022

such as aminoglycosides, b-lactams, and metronidazole. Over the past decade, use of antibiotics leads to the development of multidrug resistance [3–5]. Hence, multidrug resistant *B. fragilis* were reported from all over the world. As a result, treatment of *B. fragilis* infections becomes more challenging and requires development of new treatment options [6]. Understanding of *B. fragilis* gene function facilitates the discovery of new targets and development of novel drugs. We previously reported the identification of *essential genes* in *B. fragilis* [7]. Using transposon vector, we generated *B. fragilis* 638R mutant library and identified mutated genes by next-generation sequencing. Our data indicated that only 12.7% (550) of total genes were essential for the survival of *B. fragilis* on nutrient-rich medium. Among the 550 essential genes, 41.3% (227) of genes belong to hypothetical proteins. Interestingly, 31% of essential genes were also present in Database of Essential genes (V.10.02). Here, we report detailed procedure for the *essential genes* identification in *B. fragilis*.

2 Materials

2.1 Transposon Mutant Library (TML) Generation

1. Strains: *B. fragilis* 638R, *E. coli* S-17 λpir-pSAM_Bt (Prof. Andrew L. Goodman lab), or pYV07(Prof. Hannah M Wexler lab).
2. Culture Media: Brain heart infusion (BHI) broth or BHI-agar (Sigma) supplemented with 15 mg/L hemin (Sigma or Anaerobe Systems, Morgan Hill, CA, USA), Luria–Bertani (LB) broth or LB agar (Sigma), and LB with glycerol (20% v/v).
3. Antibiotics, Ampicillin (50 mg/L), Kanamycin (40 mg/L), Gentamicin (25 mg/L), Erythromycin (10 mg/L), and Rifampicin (10 mg/L).
4. Shaking incubator.
5. Anaerobic jars (Advanced instruments) and Anaerobic sachets (AnaeroPack.Co) or Anoxomat Mark II or upper versions (Advanced instruments) and gas cylinder containing 5% CO_2, 5% H_2, and 90% N_2.
6. Autoclave for sterilization of culture medium.
7. Pipette and nuclease-free tips.
8. Vortexer/multivial vortex shaker.
9. Glass test tubes (15 ml) and flasks (250 ml). Sterile Falcon tubes (15 ml). Petri dishes.
10. Sterile inoculating Loops, Needles, & Spreaders (plastic or glass).

11. Glycerol.
12. Spectrophotometer.

2.2 Sequencing Library Preparation and Sequencing

2.2.1 Genomic DNA (gDNA) Isolation, gDNA Shearing, and Adding C-Tail

1. Genomic DNA isolation kit (Quick-DNA™ Fungal/Bacterial Midiprep Kit, Catalog No. D6105 Zymo research).
2. Covaris E220 focused-ultrasonicator.
3. Terminal transferase kit (New England Biolabs, MA) and the chain terminator ddCTP (GE Healthcare Biosciences, NJ).
4. PCR mix cleaning kit (DNA Clean & Concentrator 25 kit, D4033, Zymo Research or Agencourt AMPure XP beads (A63880, Beckman Coulter, Inc.).
5. Thermocycler.

2.2.2 Transposon Insertion Site (TIS) Amplification

1. PCR primers (*see* Table 1).
2. PCR tubes (0.2 ml) and 1.5 ml tubes (nuclease free).
3. Phusion® High-Fidelity PCR Master Mix (New England Biolabs, MA).
4. Thermocycler.
5. PCR product cleaning kit (DNA Clean & Concentrator 25 kit, D4033, Zymo Research) or.

2.2.3 Sample Barcoding PCR, Sequencing, and Analysis

1. Barcoding primers (*see* Table 1).
2. PCR tubes (0.2 ml) and 1.5 ml tubes (nuclease free).
3. Phusion High-Fidelity PCR Master Mix (New England Biolabs, MA).
4. Qubit fluorometer and Qubit dsDNA HS Assay Kit (Thermo Fisher Scientific, cat. no. Q32851).
5. Illumina sequencer and appropriate Sequencing cartridge and flow cell.

3 Methods

3.1 Transposon Mutant Library (TML) Generation

1. Remove stock culture tubes (*B. fragilis* 638R ([8, 9]) and *E. coli* S-17 λpir-pSAM-Bt ([10])/or pYV07([11]) from −80 °C and allow them to thaw on ice for 10–15 min. Streak *B. fragilis* 638R and *E. coli* S-17 λpir-pSAM-Bt or pYV07-strains on BHI-Agar and LB-Agar/Ampicillin plate, respectively (*see* **Note 1**).
2. Incubate the *E. coli* S-17 λpir-pSAM-Bt or pYV07at 37 °C for 12–18 h. Keep BHI-Agar plate containing *B. fragilis* 638R in anaerobic jars and create anaerobic environment using anaerobic sachets or Anoxomat and incubate at 37 °C for 16–24 h.

Table 1
Primers used in sequencing library construction

Primer name	Sequence (5′–3′)
pSAMF1	CCTGACGGATGGCCTTTTTGCGTTTCTACC
IRA-PolyG	GTGACTGGAGTTCAGACGTGTGCTCTTCCGATCTGGGGGGGGGGGGGGGG
pSAMF2	AATGATACGGCGACCACCGAGATCTACACTCTTTGACCGGGGACTTATCA TCCAACCTGTTA
Barcode1	CAAGCAGAAGACGGCATACGAGATAAAAAAGTGACTGGAGTTCA GACGTGTGCTCTTCCGATCT
Barcode2	CAAGCAGAAGACGGCATACGAGATACACACGTGACTGGAGTTCA GACGTGTGCTCTTCCGATCT
Barcode3	CAAGCAGAAGACGGCATACGAGATAGAGAGGTGACTGGAGTTCA GACGTGTGCTCTTCCGATCT
Barcode4	CAAGCAGAAGACGGCATACGAGATATATATGTGACTGGAGTTCA GACGTGTGCTCTTCCGATCT
Barcode5	CAAGCAGAAGACGGCATACGAGATCACACAGTGACTGGAGTTCA GACGTGTGCTCTTCCGATCT
Barcode6	CAAGCAGAAGACGGCATACGAGATCCCCCCGTGACTGGAGTTCA GACGTGTGCTCTTCCGATCT
Barcode7	CAAGCAGAAGACGGCATACGAGATCGCGCGGTGACTGGAGTTCA GACGTGTGCTCTTCCGATCT
Barcode8	CAAGCAGAAGACGGCATACGAGATCTCTCTGTGACTGGAGTTCA GACGTGTGCTCTTCCGATCT
Barcode9	CAAGCAGAAGACGGCATACGAGATGAGAGAGTGACTGGAGTTCA GACGTGTGCTCTTCCGATCT
Barcode10	CAAGCAGAAGACGGCATACGAGATGCGCGCGTGACTGGAGTTCA GACGTGTGCTCTTCCGATCT
Barcode11	CAAGCAGAAGACGGCATACGAGATGGGGGGGTGACTGGAGTTCA GACGTGTGCTCTTCCGATCT
Barcode12	CAAGCAGAAGACGGCATACGAGATGTGTGTGTGACTGGAGTTCA GACGTGTGCTCTTCCGATCT
Barcode13	CAAGCAGAAGACGGCATACGAGATTATATAGTGACTGGAGTTCA GACGTGTGCTCTTCCGATCT
Barcode14	CAAGCAGAAGACGGCATACGAGATTCTCTCGTGACTGGAGTTCA GACGTGTGCTCTTCCGATCT
Barcode15	CAAGCAGAAGACGGCATACGAGATTGTGTGGTGACTGGAGTTCA GACGTGTGCTCTTCCGATCT
Barcode16	CAAGCAGAAGACGGCATACGAGATTTTTTTGTGACTGGAGTTCA GACGTGTGCTCTTCCGATCT
pSAM3-SeqP	ACACTCTTTGACCGGGGACTTATCATCCAACCTGTTA

3. Inoculate *E. coli* S-17 λpir-pSAM-Bt or pYV07into tube containing 2.5 ml of sterilized LB+ Ampicillin broth (in 15 ml tube, cotton plugged) and incubate the culture tubes at 37 °C shaking incubator (200–250 rpm/min) for overnight. Also, inoculate loop full of *B. fragilis* 638R into 2.5 ml of BHI broth (in 10–15 ml tube, cotton plugged) and transfer culture tubes to anaerobic jar. Create anaerobic atmosphere and anaerobic jar on 37 °C shaking incubator (200–250 rpm/min) overnight.
4. Next day morning, prepare 8–10 sterilized LB medium (5 ml) in 15 ml tubes (cotton plugged) and inoculate with 50 μl of *E. coli* S-17 λpir-pSAM-Bt or pYV07culture. For *B. fragilis* 638R, prepare four 50 ml of BHI broth in 250 ml conical flasks (flask mouth should be plugged with cotton) and inoculate with 0.5 ml of *B. fragilis* 638 culture/flask. Transfer flasks to anaerobic jar and create anaerobic atmosphere as described above (*see* **Note 2**).
5. Incubate *E. coli* S-17 λpir-pSAM-Bt or pYV07culture tubes and *B. fragilis* 638R culture flasks in anaerobic jar on shaking incubator set at 37 °C for 3–5 h. At time interval, remove 1 ml of broth from respective culture and measure the optical density (OD) at 600 nm using a spectrophotometer.
6. When culture OD reaches 0.1–0.15, remove tubes and flasks from incubator and immediately keep tubes/flaks on ice (to prevent further growth) (*see* **Note 3**).
7. Take 15 Falcon Conical Tubes (50 ml capacity). To each tube, add 2.5 ml of *E. coli* S-17 λpir-pSAM-Bt or pYV07culture broth and 10 ml of *B. fragilis* culture broth.
8. Centrifuge culture broth containing tubes at 5000 × *g* for 5 min. Discard the supernatant, add 0.1 ml of BHI broth to each tube, and resuspend cell pallet.
9. Transfer 0.1 ml of resuspend culture mix to one BHI-agar plate (total 15 plates) and spread the culture using sterile spreader.
10. Incubate all BHI-agar culture plates at 37 °C aerobic incubator for 4–5 h (*see* **Note 4**).
11. Then, transfer all BHI-agar culture plates into anaerobic jar, create anaerobic atmosphere, and incubate at 37 °C incubator overnight (16–18 h).
12. Next day morning, remove plates from anaerobic jar and add 1 ml of sterile BHI + 10% glycerol broth to each plate. Scrape off culture from BHI plate using sterile spreader and transfer culture to 1.5 ml tube. The resulting culture suspension contains transposon mutants. Store mating mix/mutants at −80 °C for future use (*see* **Note 5**).

13. Remove one to two stocks from −80 °C and thaw on ice for 10–15 min. Then, transfer different volumes (10, 20, and 40 μl) of mating mix/mutants culture to BHI-agar/gentamycin(25 μg/ml)/erythromycin (10 μg/ml) plate and spread it with sterile spreader. Keep the plates in anaerobic jar, create the anaerobic atmosphere, and incubate at 37 °C for 2–3 days (*see* **Note 6**).
14. After 2–3 days, remove plates from anaerobic jar and count colonies in each plate (each colony is considered as unique mutant). Usually, 10–20 μl of culture may yield 250–500 colonies/mutant (*see* **Notes 6** and 7).
15. To create 50,000 mutant pool, calculate the number of plates required to obtain 50,000 colonies. Accordingly, prepare the plates and inoculate with mating mix/mutants. Keep plates in anaerobic jar, create anaerobic atmosphere, and incubate at 37 °C for 2–3 days (*see* **Note 8**).
16. Add 1 ml of LB + 20% glycerol broth to each plate and scrape off cells by spreader. Pool cells suspension from different plates (≅50,000 mutants). Resulting cell suspension is referred as transposon mutant library (TML1). Make 1 ml aliquots and store TML at −80 °C. Determine colony forming units/ml from these stocks (*see* **Note 9**).
17. For the biological reproducibility, create another TML library as described above from **steps 15** and **16** and name it as TML2 (*see* **Note 10**).

3.2 Sequencing Library Preparation and Sequencing

Following step facilitates construction of sequencing library and *en masse* identification mutated genes present in TML by next-generation sequencing (Fig. 1). To better understand the concept, please refer our paper Veeranagouda and Didier [12].

3.2.1 Genomic DNA (gDNA) Isolation, gDNA Shearing, and Adding C-Tail

This step fragments gDNA to 200–500 bp fragments (which is necessary for sequencing on Illumina sequencers) and adds C-tail to fragmented DNA. Addition of C-tail provides primer binding site for PCR.

1. For each TML (1 and 2), remove four tubes from −80 °C and thaw on ice for 10–15 min. Centrifuge culture tubes at 10,000 × *g* for 5 min, discard the supernatant, and resuspend the cell pellets in 1 ml of water (approximate wet weight of cells is equal to 350–500 mg) (*see* **Note 11**).
2. Using Quick-DNA™ Fungal/Bacterial Midiprep Kit, prepare the gDNA. Pool gDNA from four tubes into one tube.
3. For technical replicates (TR), split each pooled gDNA from each TML into two tubes and label them as TML1-TR1, TML1-RT2 and TML2-TR1, TML-TR2 for TML1 and

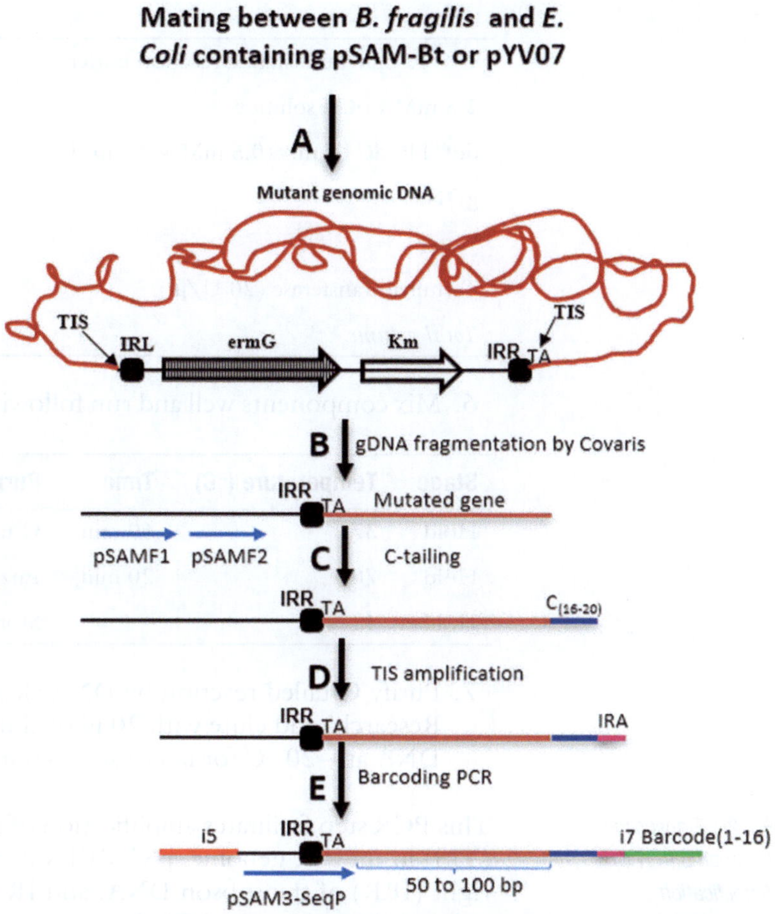

Fig. 1 Sequencing library construction using TML. (**a**) Mutation of *B. fragilis* using pSAM-Bt or pYV07 randomly inserted transposon DNA (region in Inverted Repeat Right and Left (IRL and IRR), which contains erythromycin (ermG) cassette (pSAM-Bt) and/or ermG and Kanamycin (Km) resistance cassette (pYV07)) at TA site in genome. Transposon insertion site (TIS) can be amplified using following method. (**b**) Mutant genomic DNA (gDNA) is fragmented to 200–500 bp using Covaris E220 focused ultrasonicator. (**c**) Addition of C-tail to fragmented gDNA using Terminal transferase. (**d**) Amplification of TIS using pSAMF1/IRA-PolyG primer. (**e**) Barcoding PCR using pSAMF2/Barcode (1–16) primers. Sequencing of amplicons using pSAM3-SeqP facilitates the identification of mutated gene

TML2, respectively. Quantify the gDNA by nanodrop and store at −80 °C for later use or continue to next step (*see* **Note 12**).

4. Resuspend 5 μg of gDNA in 50 μl of water and shear to 350 bp (average size) by using Covaris E220 focused ultrasonicator sheared gDNA (*see* **Note 13**).

5. For each sample, setup the C-tailing reaction in PCR tube as follows (*see* **Note 14**):

10× terminal transferase reaction buffer	6 µl
2.5 mM $CoCl_2$ solution	3 µl
ddCTP/dCTP mix (0.5 mM + 9.5 mM)	3 µl
gDNA	3 µg
Water	Up to 45 µl
Terminal transferase (20 U/µl)	3 µl
Total volume	*60 µl*

6. Mix components well and run following cycle on thermocycler:

Stage	Temperature (°C)	Time	Purpose
Hold	37	60 min	C-tailing by terminal transferase
Hold	70	20 min	Enzyme inactivation
Hold	4	–	Storage

7. Purify C-tailed reaction by DNA clean & concentrator (Zymo Research) and elute with 20 µl of elution buffer. Store C-tailed DNA at −20 °C for later use or continue to next step.

3.2.2 Transposon Insertion Site (TIS) Amplification

This PCR step facilitates amplification of transposon integrated site (TIS) in mutant genome. pSAMF1 will bind near inverted repeat right (IRR) of transposon DNA, and IRA PolyG will bind to "C" tail near TIS of mutated DNA.

1. For each TML replicate, set up the following reaction in 0.2 ml PCR tube:

pSAMF1 primer (30 µM)	3 µl
IRA polyG primer (30 µM)	3 µl
Sheared/C-tailed gDNA	15 µl
Water	54 µl
2X Phusion master mix	75 µl
Total volume	*150 µl*

2. Mix components well, split into three PCR tubes (50 µl each), and run following cycle on thermocycler:

Stage	Temperature (°C)	Time	Purpose
Hold	95	2 min	Initial denaturation
24 cycles	95	10 s	Denaturation

(continued)

Stage	Temperature (°C)	Time	Purpose
	60	30 s	Annealing
	72	60 s	Amplification
Hold	72	3 min	Final extension
Hold	4	–	Storage

3. Pool three reactions into one PCR tube and continue to next step.

3.2.3 Sample Barcoding PCR, Sequencing, and Analysis

The PCR step serves three purposes: (1) pSAMF2 primer binds downstream of pSAMF1 and hence preferentially enrich TIS containing DNA fragments, (2) use of barcodes primers facilitates unique labeling of samples, and (3) adds Illumina specific sequences (flow cell binding sequences and adapter sequence) to PCR products. Barcode primers are provided in Table 1. Use of barcode primers facilitates pooling of samples in single run. Use one barcode primer for each sample.

1. For each sample, set up following reaction in PCR tubes:

pSAMF2 primer (30 μM)	4 μl
Barcode R primer1–16 (30 μM)	4 μl (one/sample)
PCR mix from step 3.2.2.3	4 μl
Water	88 μl
2× Phusion master mix	100 μl
Total volume	*200 μl*

2. Mix components well, split into four PCR tubes (50 μl each), and run following cycle on thermocycler:

Stage	Temperature (°C)	Time	Purpose
Hold	95	2 min	Initial denaturation
14 cycles	95	10 s	Denaturation
	52	30 s	Annealing
	72	60 s	Amplification
Hold	72	5 min	Final extension
Hold	4	–	Storage

3. Pool four PCR reactions into single tube and purify PCR mix using QIA quick PCR purification kit. Elute bound DNA with 30 μl of water. Resulting PCR product is referred as sequencing library.

4. Run 1 μl of sequencing library on 2% agarose gel and determine the approximate size (usually 200–500 bp).
5. Quantify sequencing library (four TML: two biological and two technical TML) using Qubit HS kit and convert ng into nM concentration using following formula:

   Concentration in nM $= (x \times 1,000,000)/(y \times 660)$

   where

   x is the concentration of sequencing library as measured by Qubit in ng/μl,

   1,000,000 is the nM conversion factor,

   y is the average size of PCR product (as determined by 2% gel), and

   660 is the average weight of a DNA base pair.
6. Dilute each sequencing library to 10 nM (in 10 μl) and pool four sequencing libraries (10 μl X4 = 40 μl) into single tube and proceed to sequencing.
7. Depending on the sequencer used, dilute and denature pooled library. Load the sequencing library onto specific position on sequencing cartridge. Also, dilute sequencing primer (pSAM3-Seq) to appropriate concentration and load it on sequencing cartridge. Sequence library with following parameters (*see* **Note 15**):

Read 1	75–100 cycles
Index 1	6 cycles

8. After completion of sequencing, sort sequencing reads to respective samples by demultiplexing.
9. Resulting FASTQ files can be analyzed on Tufts University Galaxy server using TnSeq pipeline. Analysis workflow typically involves: (1) performing raw data QC, (2) clipping C-tail (use CCCCCCCCCCCCCCCCCCCCCCCCCC for clipping C-tail) and filtering reads less than 25 nucleotides, (3) mapping reads to *B. fragilis* 638R genome, (4) generating "hop count," (5) calculating number of unique insertions per gene and normalization to sequencing depth, (6) comparison between technical replicates and biological replicates, and (7) Determining essential genes (*see* **Note 16**).

4 Notes

1. *B. fragilis* 638R is commonly used and widely studied *B. fragilis* strain. It is devoid of any plasmid. *B. fragilis* 638R is resistant to rifampicin and gentamycin. It is sensitive to erythromycin (which is necessary for mutant selection). If you

are working with different *B. fragilis* strain, test its antibiotic resistance properties and use it accordingly. *E. coli* S-17 λpir-pSAM-Bt (Goodman et al) is mariner transposon vector. The pYV07(Veeranagouda et al) is derivative of pSAM-Bt. Both vectors can replicate in *E. coli* and don't have Ori (origin of replication) for *B. fragilis* and hence cannot replicate in *B. fragilis*. pSAM-Bt and pYV07 contain ampicillin and erythromycin resistant genes. Ampicillin gene is under the control of *E. coli* promotor and hence provide ampicillin resistance to *E. coli*, whereas erythromycin gene has *B. fragilis* promotor and hence expressed in *B. fragilis*. These two vectors are in *E. coli* S-17 λpir, which has pir and conjugation transfer genes (RP4/RK6) in its genome. When pSAM-Bt or pYV07 vector is mobilized into *B. fragilis* via biparental mating, DNA fragment within inverted repeat (IR) is integrated into random TA site in *B. fragilis* genome.

2. *E. coli* strains grow faster, while *B. fragilis* strain grows slowly. So, consider inoculating the cultures at different starting points. In general, first inoculate *B. fragilis* and continue incubation. After 2 h, inoculate *E. coli* strains.
3. The younger culture provides higher number of mutants. Hence, OD between 0.1 and 0.15 can yield higher number of mutants compared to cultures with OD between 0.2 and 0.3 OD. So, for efficient mutant generation, it is necessary to maintain OD between 0.1 and 1.5.
4. Incubating mating mix at 37 °C in aerobic condition promotes efficient mating. This condition is very optimal for *E. coli*. On the other hand, *B. fragilis* will not grow in aerobic atmosphere.
5. At this stage, mating mix contains *B. fragilis* transposon mutants. But keep in mind that this mix also contains wild-type *B. fragilis* (not mutated), mutated *B. fragilis*, and *E. coli* as well.
6. For isolation of transposon mutants and to determine number of transposon mutants/μl of stock culture, it is necessary to grow mating mix/mutants on dual antibiotic plate (BHI-agar/gentamycin (25 μg/ml)/erythromycin (10 μg/ml)). Gentamycin and erythromycin kill *E. coli* and wild-type *B. fragilis* strains. Only mutant *B. fragilis* strains, which receive transposon vector, gain resistance to erythromycin and grow on this dual antibiotic plate.
7. Each colony arises from unique mutant, and hence, each colony is considered as unique mutant. Make appropriate dilutions to obtain individual colonies on plate.
8. Most bacteria including *B. fragilis* approximately contain 5000 genes in genome. Creating 50,000 mutant pool can randomly provide 10 unique chances/gene for transposon-mediated

mutation. If resources are available, you may consider increasing number of mutants (100,000 mutants). Depending on number of colonies in mating mix, you may need 100 (500 colonies/plate) to 200 (250 colonies/plate) BHI plates for each TML.

9. TML can be stored at −80 °C for 10–12 months. Mutant survival rate may reduce for longer storage period. As and when required, determine CFU. This TML can be used for different purpose such as screening for drug-resistant mutants [13].
10. Creating two independent TML (biological replicates) facilitates systematic comparison between two pools and elimination of mutants present in one of TML.
11. Upon centrifugation, cell pellet may not completely settle. So carefully remove supernatant (avoid removing cells).
12. Technical replicates help to understand and eliminate the variations arising from library construction. So, for each TML, it is necessary to split gDNA into two parts and sequence it separately.
13. Compared to enzyme digestion, focused-ultrasonicators from Covaris provide precise fragmentation with very minimal sample loss. The gDNA fragmentation setting varies according to instrument type. Please follow manufacturer recommendations to obtain desired fragments (average 350 bp).
14. In this reaction, ddCTP and dCTP were used at 0.5 mM and 9.5 mM (1:19 ratio). At this ratio, approximately 20 "C" bases will be added to fragmented DNA by terminal transferase. Changing this ratio might affect the length of C-tail. Prepare ddCTP and dCTP mix at final concentrations of 0.5 mM and 9.5 mM (1:19), respectively, and store at −20 ° C. This mix can be used for up to 6 months.
15. Prepared NGS libraries can be sequenced Illumina sequencers such as NextSeq or NovaSeq. This custom library needs to be sequenced with custom primer (pSAM3-Seq). Hence, prepare sequencing primer concentration as per sequencer guidelines. 75–100 cycles are sufficient to capture TIS site. Hence, depending on the sequencers, use 75–100 cycles. Each sample requires minimum ten million reads. Since barcodes used in this library are 6 nt, sequencing Index 1 with 6 cycles facilitates sequencing of barcodes.
16. Analysis of FASTQ files requires basic bioinformatics knowledge. Any open source NGS tools can be used for the analysis of FASTQ files. For detailed analysis procedure, please refer ref. 7.

References

1. Wexler HM (2007) Bacteroides: the good, the bad, and the nitty-gritty. Clin Microbiol Rev 20 (4):593–621. https://doi.org/10.1128/CMR.00008-07
2. Hsiao EY, McBride SW, Hsien S et al (2013) Microbiota modulate behavioral and physiological abnormalities associated with neurodevelopmental disorders. Cell 155 (7):1451–1463. https://doi.org/10.1016/j.cell.2013.11.024
3. Wareham DW, Wilks M, Ahmed D et al (2005) Anaerobic sepsis due to multidrug-resistant Bacteroides fragilis: microbiological cure and clinical response with linezolid therapy. Clin Infect Dis 40:e67–e68. https://doi.org/10.1086/428623
4. Husain F, Veeranagouda Y, Hsi J et al (2013) Two multidrug-resistant clinical isolates of bacteroides fragilis carry a novel metronidazole resistance nim gene (nimJ). Antimicrob Agents Chemother 57:3767–3774. https://doi.org/10.1128/AAC.00386-13
5. Ghotaslou R, Bannazadeh Baghi H, Alizadeh N et al (2018) Mechanisms of Bacteroides fragilis resistance to metronidazole. Infect Genet Evol. https://doi.org/10.1016/j.meegid.2018.06.020
6. Merchan C, Parajuli S, Siegfried J et al (2016) Multidrug-resistant Bacteroides fragilis bacteremia in a US resident: an emerging challenge. Case Rep Infect Dis. https://doi.org/10.1155/2016/3607125
7. Veeranagouda Y, Husain F, Tenorio EL, Wexler HM (2014) Identification of genes required for the survival of B. fragilis using massive parallel sequencing of a saturated transposon mutant library. BMC Genomics. https://doi.org/10.1186/1471-2164-15-429
8. Privitera F, Dublanchet A, Sebald M (1979) Transfer of multiple antibiotic resistance between subspecies of bacteroides fragilis. J Infect Dis 139:97–101. https://doi.org/10.1093/infdis/139.1.97
9. Patrick S, Blakely GW, Houston S et al (2010) Twenty-eight divergent polysaccharide loci specifying within- and amongst-strain capsule diversity in three strains of Bacteroides fragilis. Microbiology 156:3255–3269. https://doi.org/10.1099/mic.0.042978-0
10. Goodman AL, McNulty NP, Zhao Y et al (2009) Identifying genetic determinants needed to establish a human gut symbiont in its habitat. Cell Host Microbe 6:279–289
11. Veeranagouda Y, Husain F, Wexler HM (2013) Transposon mutagenesis of Bacteroides fragilis using a mariner transposon vector. Anaerobe 22:126–129. https://doi.org/10.1016/j.anaerobe.2013.04.012
12. Veeranagouda Y, Didier M (2017) Transposon insertion site sequencing UNIT 21.35 (TIS-Seq): an efficient and high-throughput method for determining transposon insertion site(s) and their relative abundances in a PiggyBac transposon mutant pool by next-generation sequencing. Curr Protoc Mol Biol 21:35.1–35.11. https://doi.org/10.1002/cpmb.47
13. Veeranagouda Y, Husain F, Boente R et al (2014) Deficiency of the ferrous iron transporter FeoAB is linked with metronidazole resistance in Bacteroides fragilis. J Antimicrob Chemother 69:2634–2643. https://doi.org/10.1093/jac/dku219

Chapter 17

Repressible Promoter System to Study Essential Genes in Mycobacteria

Francesca Boldrin and Riccardo Manganelli

Abstract

Inducible gene expression systems represent powerful tools for studying essential gene function and for validation of drug targets in bacteria. Even if several regulated promoters have been characterized, only a few of them have been successfully used in *Mycobacteria*. Here we describe a successful mycobacterial gene regulation system based on the presence of two chromosomally encoded repressors: Pip and TetR, and a tunable promoter (P_{ptr}) that allows a tight regulation of gene expression.

Key words Mycobacteria, TetR/Pip-OFF, Essential gene, Gene expression, Conditional mutant, Repressible promoter

1 Introduction

Developing tools for conditional gene expression are of great importance to gene function research and drug target identification in bacteria [1]. Experimental control of target gene expression can be achieved by replacing the native promoter of the target gene with tightly regulated promoters whose activity can be chemically controlled adding a specific molecule (inducer) to the bacterial culture [1]. The TetR/Pip-OFF repressible promoter system here described, is based on two chromosomally encoded repressors: TetR and Pip, dependent on anhydrotetracycline (ATc) and pristinamycin, respectively, and P_{ptr}, a strong promoter repressible by Pip [2, 3]. The gene encoding the tetracycline-responsive repressor TetR is constitutively expressed, while the gene encoding the pristinamycin-responsive repressor Pip is expressed from a TetR-dependent promoter. In the absence of ATc, TetR represses *pip* expression and the target gene can be transcribed by the P_{ptr} promoter (Fig. 1a) [3]. However, in the presence of ATc, due to the release of its promoter by TetR, Pip is produced resulting in the

Ren Zhang (ed.), *Essential Genes and Genomes: Methods and Protocols*, Methods in Molecular Biology, vol. 2377,
https://doi.org/10.1007/978-1-0716-1720-5_17,

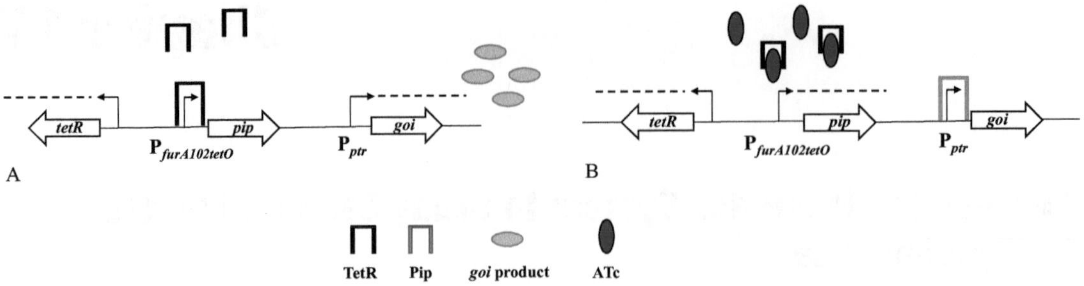

Fig. 1 Working model of the TetR/Pip-OFF system. In the absence of ATc, TetR binds to its operators present in the $P_{furA102tetO}$ promoter repressing *pip* transcription and thus allowing the expression of the gene of interest from P_{ptr} (**a**). In the presence of ATc, TetR is released from its operators upstream $P_{furA102tetO}$ allowing *pip* transcription which results in the repression of gene of interest (*goi*) repression (**b**). (Figure modified from [3])

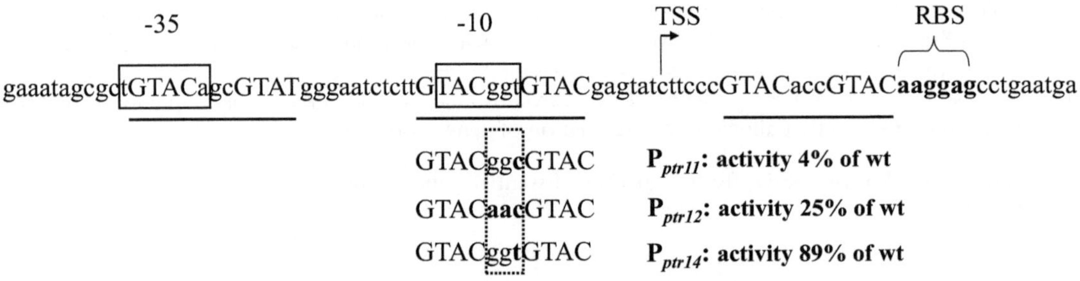

Fig. 2 P_{ptr} promoter regions and mutations introduced to modify its strength. Putative −10 and −35 consensus sequences are boxed, while the three Pip-binding regions are underlined. The transcriptional start site (TSS) is indicated by an arrow, the ribosome binding site (RBS) is in bold. Conserved repeats within the Pip-binding regions are in capital letters. The mutations introduced in the −10 consensus sequence to obtain P_{ptr11}, P_{ptr12}, and P_{ptr14} are shown below. (Figure modified from [2])

silencing of the P_{ptr}-expressed target gene (Fig. 1b). The main advantages of this model are its very high stringency, due to the presence of two repressors working in series, and its genetic stability, due to the integration of both repressor genes in the genome, making it suitable for long-term in vivo experiments [3].

Since the strength of the P_{ptr} promoter may represent a problem when the target gene is physiologically expressed at low levels, we recently constructed three P_{ptr} mutants with lower strength by site-directed mutagenesis: P_{ptr11}, P_{ptr12}, and P_{ptr14} (Fig. 2). The possibility to choose among promoters with different strength, strongly improved the performances of the TetR/Pip-OFF repressible system increasing the spectrum of manipulable target genes [2].

1.1 Strategies to Generate Conditional Mutants

1.1.1 Selection of the Promoter

One of the first issues to construct conditional mutants in which the expression of the target gene is effectively manipulated, is represented by the choice of the regulated promoter, which should guarantee a basal level of expression of the target gene comparable to that obtained by its physiologic promoter. At this purpose, the promoter of the target gene is cloned in front of a promoterless *egfp* gene in a reporter plasmid which is then introduced in *Mycobacterium tuberculosis*. The fluorescence produced by the resulting strain is than compared to that produced by *M. tuberculosis* strains containing a similar plasmid in which the *egfp* gene is transcribed from the different P_{ptr} mutants. The promoter giving the amount of fluorescence closest to the target gene physiologic promoter, is chosen to construct the conditional mutant.

1.1.2 Construction of the Conditional Mutant (First Approach: Promoter Replacement)

In the simplest approach, the physiological promoter of the target gene is replaced by insertional duplication with the P_{ptr} of choice. The first step of this approach is the construction of a suicide plasmid in which the first part of the gene (at least 500 bp) is cloned in frame with the P_{ptr} of choice. The resulting plasmid is then transformed in *M. tuberculosis* and transformants in which it has been integrated in the genome by insertional duplication are selected. Finally, an L5 mycobacteriophage-based integrative plasmid (pFRA61) is used to integrate the genes encoding the two repressors in the chromosome of one of these mutants placing target gene expression under the control of the TetR/Pip-OFF system (Fig. 3) (*see* **Note 1**).

This approach is surely the simplest and fastest, but can present some problems: (1) P_{ptr} insertion upstream of the target gene is obtained by insertional duplication, and consequently may be unstable; (2) the first part of the gene is duplicated and can still be transcribed and translated from its original promoter: in very small genes it can retain some activity and mask the effect of the repression of the target gene; (3) when the target gene is within an operon, the introduction of P_{ptr} will affect transcription of all the genes downstream the target gene (*see* **Note 2**).

1.1.3 Construction of the Conditional Mutant (Second Approach: Ectopic Expression)

To obviate to these problems, we suggest a second, alternative approach. During first step the entire ORF of the target gene is cloned downstream of the P_{ptr} of choice and the resulting construct is then introduced in the integrative plasmid carrying *tetR* and *pip* (pFRA61). This plasmid is then introduced in *M. tuberculosis* to generate a merodiploid strain in which the copy of the target gene under the control of the TetR/Pip-OFF system will be integrated in an ectopic locus of the chromosome.

Afterwards, the wild type copy of the target gene can be deleted, introducing an unmarked in-frame deletion using the favorite approach (Fig. 4).

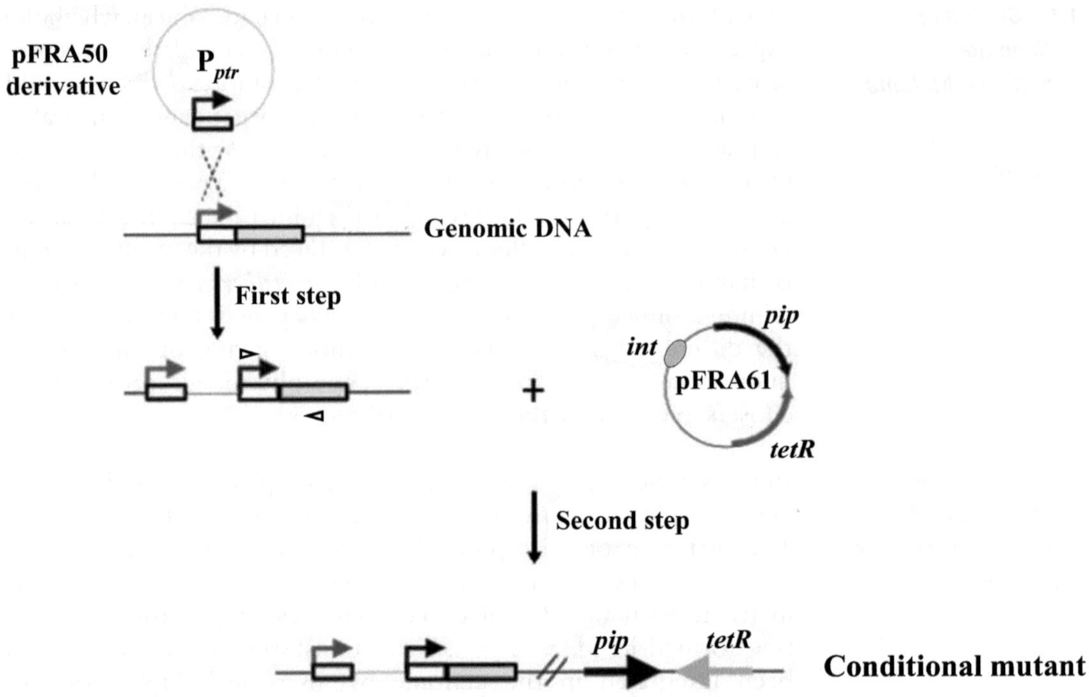

Fig. 3 Construction of conditional mutants, first approach: the first step consists on the replacement of the original promoter of the gene of interest by insertional duplication using a pFRA50 derivative containing the first portion of the gene of interest cloned downstream of the P_{ptr} of choice. The second step is represented by the pFRA61-mediated integration of *tetR* and *pip* at the mycobacteriophage L5 *attB* site. Black arrow: P_{ptr} promoter; gray arrow: promoter of the gene of interest; white boxes: part of the gene of interest used cloned in the suicide plasmid downstream P_{ptr} and used to drive the integration by insertional duplication; gray box: part of the gene of interest not included in the suicide plasmid; open arrowheads: PCR primers used to confirm the correct integration

1.2 Working Conditions

In the simplest study design, bacteria are grown in the absence of ATc, which is added at the time repression of the gene of interest is required. ATc concentration usually used is 50 ng/ml in *Mycobacterium smegmatis* and 500 ng/ml in *M. tuberculosis*. In long experiments, multiple addition of ATc may be required to obviate to its natural decay (*see* **Note 3**). In some case, the repression of the gene of interest shows its physiological effect only after several rounds of replication, since some time is required to titrate down the protein produced before the addition of ATc [4]. This is particularly true when P_{ptr} is stronger than the physiological promoter thus causing overexpression of the gene of interest.

Another interesting option of the TetR/Pip-OFF repressible system, is the possibility to relive the repression obtained adding ATc, by administrating a subinhibitory concentration of pristinamycin to the bacteria. This molecule, binding to Pip prevents its binding to the P_{ptr} operators releasing its expression in a dose-related manner [3].

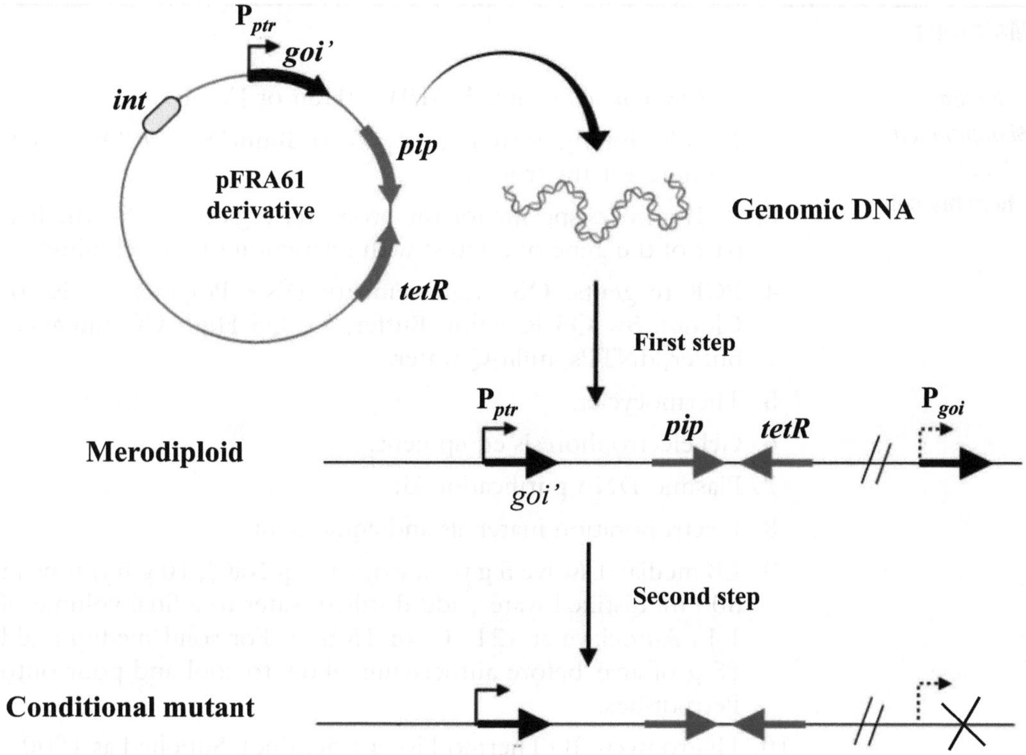

Fig. 4 Construction of conditional mutants, second approach: the first step consists on the integration of a copy of the gene of interest transcribed by P_{ptr} at the L5 mycobacteriophage *attB* site using a derivative of pFRA61 also carrying *tetR* and *pip*. The result is a merodiploid strain in which one copy of the gene of interest is transcribed from P_{ptr} and one is transcribed from its own physiological promoter. The second step consists on the deletion of the wt copy of the gene of interest to obtain the conditional mutant

1.3 Applications

The TetR/Pip OFF system has been successfully used in several mycobacterial species and in different setting as axenic cultures, and macrophage or mice infections, showing great versatility and robustness [4–11]. In these experiments, the role of several essential genes has been characterized allowing their validation as drug targets. It is worth noting that a simple variant of this system (not treated in this paper) is the Pip-ON inducible system [12]. Consequently, the gene of interest is not expressed unless pristinamycin is present. This system is less stringent of the TetR/Pip-OFF system, but allows a wider range of expression levels of the target gene and can be very useful to evaluate drug targets vulnerability [2] (*see* **Note 4**).

2 Materials

2.1 Cloning and Manipulation Procedures in Escherichia coli

1. *Escherichia coli* strains HB101, DH5α or JM110.
2. *E. coli* cloning system (e.g., Zero Blunt™ TOPO™ PCR Cloning Kit Invitrogen).
3. PCR primers specific for the promoter region and for the first part of the gene of interest with restriction site for cloning.
4. PCR reagents: Q5 High- Fidelity DNA Polymerase (EuroClone), 5× Q5 Reaction Buffer, 5× Q5 High GC Enhancer buffer, dNTPs, milli-Q water.
5. Thermocycler.
6. Gel electrophoresis equipment.
7. Plasmid DNA purification kit.
8. Electroporation materials and equipment.
9. LB media: dissolve 5 g yeast extract, 5 g NaCl, 10 g tryptone in 800 ml distilled water; add distilled water to a final volume of 1 L. Autoclave at 121 °C for 15 min. For solid medium add 15 g of agar before autoclaving; allow to cool and pour onto Petri dishes.
10. Hygromycin B (Thermo Fisher Scientific). Supplied as 1000× stock solution (50 mg/ml). Store at 4 °C.
11. Kanamycin sulfate (Sigma Aldrich). 2500× stock solution (50 mg/l): dissolve 500 mg in 10 ml milli-Q water and filter sterilize. Store at −20 °C.
12. Streptomycin sulfate (Sigma Aldrich). 1000× stock solution (20 mg/ml): dissolve 200 mg in 10 ml milli-Q water and filter sterilize. Store at −20 °C.

2.2 Mycobacterial Culture Media

1. ADN 10× stock solution: dissolve 20 g glucose, 50 g BSA (bovine serum albumin, Sigma Aldrich), and 8.5 g NaCl in 1 L Milli-Q water. Mix well and filter sterilize. Store at 4 °C.
2. Middlebrook 7H9 stock: dissolve 4.7 g 7H9 powder (Becton Dickinson, Sparks, MD) in 600 ml milli-Q water, add 4 ml glycerol 50% (Sigma Aldrich), and 500μl Tween-80 (Sigma Aldrich). Mix all the components and bring to 900 ml with milli-Q water. Autoclave at 121 °C for 10 min.
3. Middlebrook 7H9-ADN (for *M. tuberculosis*): add 100 ml ADN 10× to 900 ml of Middlebrook 7H9 stock.
4. Middlebrook 7H9 (for *M. smegmatis*): add 100 ml of milli-Q water to 900 ml of Middlebrook 7H9 stock.
5. 7H10 agar plates. Add 19 g 7H10 agar (Becton Dickinson, Sparks, MD), 10 ml glycerol 50%, and 500μl Tween-80 in 900 ml milli-Q water and autoclave 121 °C for 10 min. Cool melted agar to 50 °C and add either ADN or milli-Q water.

2.3 Mycobacterial DNA Extraction

1. 10 mM Tris–HCl pH 8: dissolve 1.22 g of Tris (Sigma) in 750 ml of deionized water. Adjust to pH 8 using concentrated HCl; fill to the final volume of 1 L with deionized water. Filter sterilize.
2. 1 mM EDTA: obtained by dilution from an EDTA 0.5 M stock: dissolve 186.1 g EDTA to 800 ml of H_2O; stir vigorously on a magnetic stirrer. Adjust to pH 8.0 with NaOH; sterilize by autoclaving, store at room temperature.
3. 0.1%NaN_3: dissolve 0.1 g of NaN_3 (Sigma) in 100 ml of deionized water. Filter with 0.45μm filter to remove particles.
4. Chelex 100 (Biorad).

2.4 Generation of Single-Stranded DNA by Alkali Denaturation

1. 10 M NaOH: dissolve 40 g of NaOH in 100 ml of water and autoclave at 121 °C for 15 min.
2. 0.5 M ethylenediaminetetraacetic acid (EDTA): dissolve 18.6 g of EDTA in 70 ml of milli-q water, adjust pH to 8 with NaOH, and bring volume to 100 ml with water. Autoclave at 121 °C for 15 min.
3. 0.2 mM EDTA/0.2 M NaOH solution: add 0.4 ml of 0.5 M EDTA and 20 ml of 10 M NaOH to 980 ml of sterile water. Prepare fresh just before use.
4. 3 M sodium acetate (NaOAc): dissolve 40.8 g of NaOAc in 70 ml of water, adjust H to 5.2 with acetic acid, and bring to 100 ml with water. Autoclave at 121 °C for 15 min.
5. 70% ethanol: add 70 ml of 100% ethanol to 30 ml of deionized water.

2.5 Electroporation of Mycobacteria

1. Preparation of electrocompetent cells have been previously described [5].
2. Electroporation apparatus with pulse controller (electroporator 2510 Eppendorf).
3. Electroporation cuvettes: 0.2 cm gap width (Sigma).

2.6 Fluorescence Assay

1. Nunclon 96-well Flat Bottom Black Polystyrol FluoroNunc microplates (Biosigma).
2. Infinite 200Pro Microplate Reader (Tecan Group, Mannendorf, Switzerland).

2.7 Plasmids

All plasmids are available from the authors.

2.7.1 pFRA50

Suicide vector derived from pBend, contains an Hyg cassette conferring hygromycin resistance in both *E. coli* and mycobacteria [13]. This plasmid contains P_{ptr} followed by a polylinker to allow the construction of the transcriptional fusion with the first portion of the gene of interest. The presence of a strong transcriptional

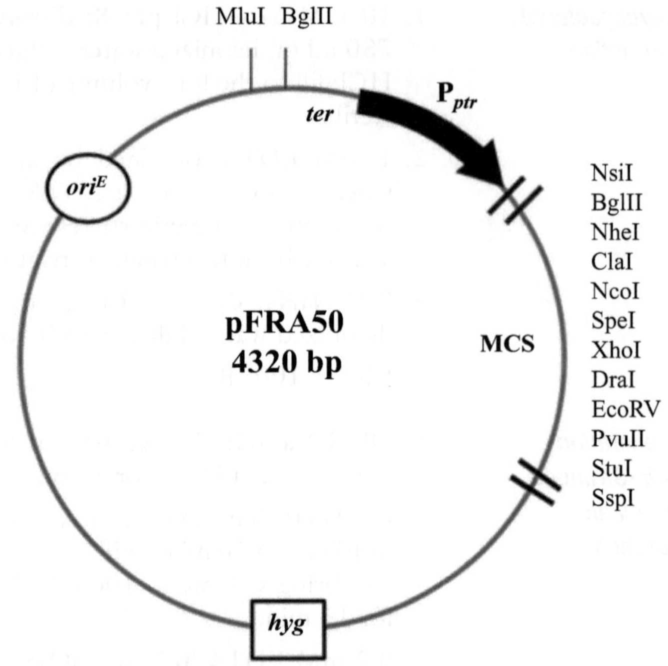

Fig. 5 Schematic representation of pFRA50 (4320 bp). This plasmid carries a hygromycin B (hyg) resistance cassette for selection in both *E. coli* and mycobacteria, an origin of replication for *E. coli* (*oriE*), and multiple cloning site (MCS). P_{ptr} is indicated by an arrow; *ter*- terminator sequence

terminator upstream P_{ptr} allows to minimize the effects of promoters located elsewhere in the plasmid or in regions flanking the chromosomal insertion site (Fig. 5).

2.7.2 pFRA50-Derivative Plasmids

pFRA114: pFRA50 derivative in which a promoterless *egfp* gene was cloned downstream of P_{ptr} (Fig. 6). pFRA169, pFRA170, and pFRA171 are pFRA114 derivatives in which P_{ptr} was replaced by P_{ptr11}, P_{ptr12}, or P_{ptr14}, respectively. The efficiency of these promoters respect to P_{ptr} is 4%, 25%, and 89%, respectively [2]. These plasmids maintain pFRA50 MCS starting from the NheI site (Fig. 7).

2.7.3 pFRA61

L5 mycobacteriophage-based integrative plasmid deriving from pMC1s, it carries a cassette conferring streptomycin resistance in both *E. coli* and mycobacteria [14]. Upon electroporation in mycobacteria, it is able to insert to the single L5 *attB* site present in all mycobacterial genomes. It carries the genes encoding TetR and Pip (Fig. 8).

2.8 Stock Solutions of Inducers

Anhydrotetracycline (ATc, Sigma). Stock solution (0.5 mg/ml): dissolve 50 mg in 10 ml milli-Q water filter sterilize and store at −20 °C. Keep in the dark since ATc is light sensitive.

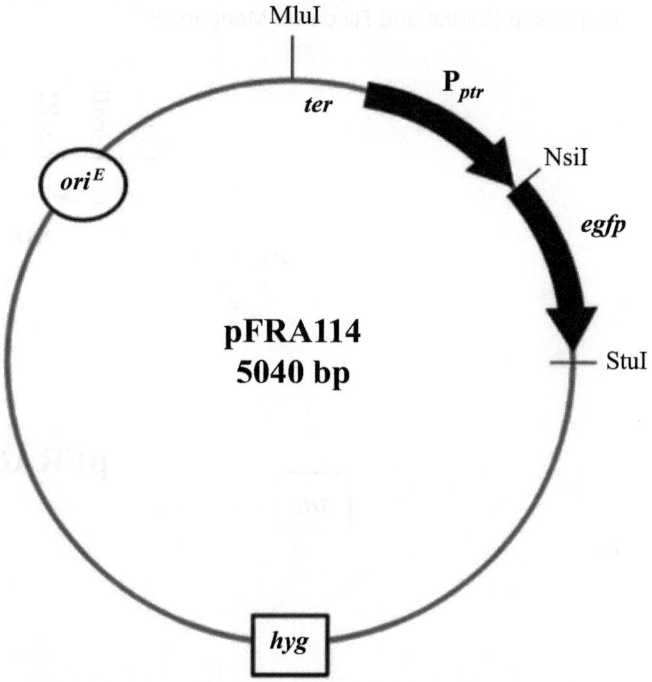

Fig. 6 Schematic representation of pFRA114 (5040 bp). It derives from pFRA50 and carries an *egfp* gene cloned downstream of P_{ptr} between the NsiI and the StuI sites

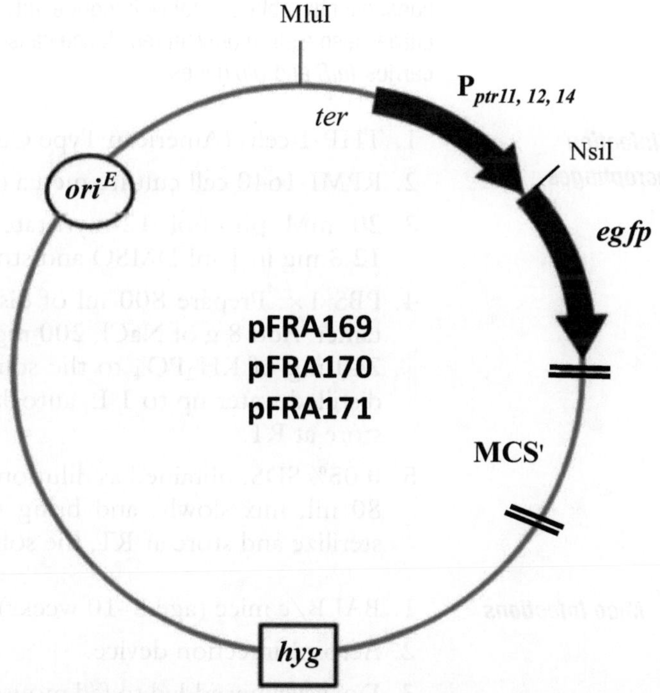

Fig. 7 Schematic representation of pFRA50-derivative plasmids in which the different promoter mutants were introduced upstream of an *egfp* promoter to obtain pFRA169 (P_{ptr11}), pFRA170 (P_{ptr12}), and pFRA171 (P_{ptr14}); MCS' derives from pFRA50 MCS starting from NheI

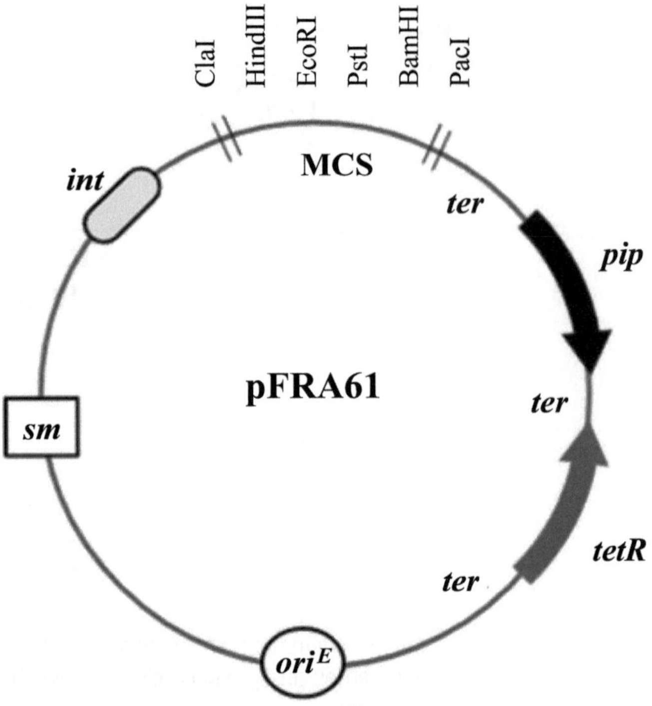

Fig. 8 Schematic model of the integrative plasmids pFRA61. This plasmid carries the mycobacteriophage L5 integrase gene and its *attP* sequence and is consequently able to stably integrate into the mycobacterial chromosome. It carries also a streptomycin resistance cassette, a multiple cloning site. Finally, it carries *tetR* and *pip* genes

2.9 Infection of Macrophages

1. THP-1 cells (American Type Culture Cultivation, ATCC).
2. RPMI-1640 cell culture media (Thermo Fisher Scientific).
3. 20 mM phorbol 12-myristate 13-acetate (PMA): dissolve 12.3 mg in 1 ml DMSO and store at −20 °C.
4. PBS 1×: Prepare 800 ml of distilled water in a suitable container. Add 8 g of NaCl, 200 mg of KCl, 1.44 g of Na_2HPO_4, 240 mg of KH_2PO_4 to the solution. Adjust to pH 7.4. Add distilled water up to 1 L, autoclave at 121 °C for 15 min, and store at RT.
5. 0.05% SDS: obtained as dilution of SDS 10%. Dissolve 10 g in 80 ml, mix slowly, and bring the volume to 100 ml. Filter sterilize and store at RT, the solution is stable for 6 months.

2.10 Mice Infections

1. BALB/c mice (age 8–10 weeks).
2. Aerosol-infection device.
3. Doxycycline: added to fed mouse chow at 2000 ppm (Research Diets).
4. *M. tuberculosis* frozen stock cultures 10^7 cfu/ml.
5. Middlebrook 7H10 plates with the appropriate antibiotics.

3 Methods

3.1 Construction of a Reporter Plasmid to Evaluate the Strength of the Promoter of the Gene of Interest

1. Clone the upstream region of the gene of interest containing its promoter, between the MluI and NsiI sites of pFRA114 to replace P_{ptr} (*see* **Note 5**).
2. Cut the promoter-*egfp* DNA fragment from the resulting plasmid using MluI and StuI and subclone it in the integrative plasmid pFRA61 using one of the restriction sites present in pFRA61 MCS.
3. Extract DNA (at least 1 μg) for bacterial transformation.
4. Electroporate in the mycobacterial strain of choice (*see* below).

3.2 Fluorescence Assay

1. Grow bacterial strains expressing eGFP from the different promoters to be compared, in 5 ml of Middlebrook 7H9 ADN, supplemented with appropriate antibiotics, until the $OD_{540} = 0.5$. Bacterial cultures of *M. smegmatis* or *M. tuberculosis* strains not expressing eGFP, are used as controls to determine bacterial autofluorescence.
2. Centrifuge the cells (5 min, 3000 × *g*) and resuspend the pellet in a volume of 0.1 ml of medium to obtain a concentration of about 10^9 cfu/ml.
3. Add the equivalent of 10^8 cfu in the microplates, and measure fluorescence values using excitation and emission wavelengths of 485 and 535 nm.

3.3 Construction of Conditional Mutants (First Approach, Fig. 3)

3.3.1 Preparation of Suicide Vector

1. Amplify the first portion of the gene of interest (about 600–800 bp).
2. Clone it downstream of the P_{ptr} promoter in frame with the ATG codon encoded in the NsiI site present in the correct pFRA114 derivative (*see* **Note 6**).
3. Electroporate the plasmid in the strain of choice (*see* below).

3.3.2 Mutant Selection

1. Plate the entire transformation onto 7H10 ADN plates containing hygromycin and incubate the plates at 37 °C until colonies appear.
2. Streak 3–4 colonies on a fresh hygromycin plate. After growth, resuspend bacterial cells in 300μl of 10 mM Tris–HCl pH 8, 1 mM EDTA, 0.1% NaN_3 containing 20% Chelex 100.
3. Incubate suspensions for 30 min at 56 °C, vortexing every 5 min. Finally, incubate at 100 °C for 12 min.
4. After spinning at 12,000 × *g* for 3 min at room temperature, collect the supernatant and store at 4 °C. Use this supernatant as template in PCR to confirm plasmid integration. At this

purpose use an upper primer able to anneal to P_{ptr}, and a lower primer able to anneal downstream of the DNA region used for the recombination (Fig. 3).

3.3.3 Introduction of the TetR/Pip OFF System

1. Electroporate the selected mutant strain with 1μg of the integrative plasmids pFRA61 and select the final conditional mutant on Middlebrook 7H10 plates added with hygromycin and streptomycin.

3.4 Construction of Conditional Mutants (Second Approach, Fig. 4)

1. Clone the entire gene of interest downstream of the P_{ptr} of choice in the appropriate pFRA114 derivative plasmid and then subclone the P_{ptr}-gene of interest fragment in one of the unique restriction sites in pFRA61 (ClaI, HindIII, EcoRI, PstI, BamHI, PacI).
2. Electroporate 1μg of the resulting plasmid in the mycobacterial strain of choice and select on Middlebrook 7H10 to obtain a merodiploid strain (*see* below).
3. Delete the wild type copy of the gene of interest using the method of choice [13–15].

3.5 Preparation of Single-Stranded Denatured Plasmid DNA

1. Precipitate 10μg of plasmid DNA with 1/10 vol of NaOAc and 2.5 vol of 100% ethanol at 4 °C for 20 min. Centrifuge at 10,000 × *g* for 15 min and remove supernatant.
2. Dry the DNA pellet and resuspend in 10μl of sterile deionized water; add 15μl of 0.2 mM EDTA/0.2 M NaOH to DNA, mix and incubate at 37 °C for 30 min.
3. Add 1/0 vol of 3 M NaOAc to denaturated DNA and precipitate the DNA with 2.5 vol of 100% ethanol at 4 °C for 20 min.
4. Centrifuge at 10,000 × *g* for 15 min to remove supernatant, and then wash the DNA pellet twice using 70% ethanol.
5. Dry the pellet and resuspend in 5–8μl of sterile deionized water prior to electroporation.

3.6 Electroporation

1. Place DNA in an electroporation cuvette and add 50μl of mycobacterial electrocompetent cells. When electroporating an integrative plasmid 1μg of DNA is enough. However, when electroporating a suicide plasmid that needs to integrate in the chromosome by homologous recombination, given the low efficiency of homologous recombination in mycobacteria, we suggest to perform several electroporations (minimum 8) using at least 10μg of plasmid DNA. Moreover, we suggest to perform at least half of them using single-stranded denaturated DNA.
2. Deliver a pulse using electroporation equipment; use the following settings: 2.5 kV, 1000 Ω, 25μF.

3. Transfer electroporated bacteria into 800μl 7H9 ADN and incubate at 37 °C (3 h for *M. smegmatis* or 24 h for *M. tuberculosis*) to allow expression of the drug-resistance gene.
4. Spread serial dilutions of the culture on Middlebrook 7H10 plates with the appropriate selection.
5. Incubate plates until colonies are visible (3–5 days for *M. smegmatis* or 2–5 weeks for *M. tuberculosis*).
6. Pick single colonies and culture in Middlebrook 7H9 ADN with the appropriate selection; Store at −80 °C after adding 10% glycerol.

3.7 Validation of Gene Essentiality with TetR/PipOFF System In Vitro

1. Grow bacteria in Middlebrok 7H9 containing different ATc concentrations (between 0 and 50 ng/ml for *M. smegmatis*, between 0 and 500 ng/ml for *M. tuberculosis*) (*see* **Notes** 7 and **8**).
2. To confirm the predicted essentiality of the gene, it is also possible to spread serial dilutions of bacteria in 7H10 in presence or absence of ATc (*see* **Note 9**).
3. To determine if gene depletion has a bacteriostatic or bactericidal effect, add a concentration of ATc able to stop bacterial replication. At different time points take an aliquot of the culture and plate serial dilutions on Middlebrook 7H10 plates without ATc.

3.8 Infection of Macrophages

1. Grow bacteria in exponential phase and prepare aliquots with a concentration of 3.7×10^4 cfu/ml.
2. Expand THP-1 cells in RPMI until their concentration reaches 7.5×10^5 cells/ml and then add 50 nM phorbol 12-myristate 13-acetate (PMA) to induce their differentiation into macrophages. Plate 7.5×10^4 cells in each well and incubate 24 h until they are differentiated.
3. Infect the cells by adding 100μl of bacterial suspension to reach a multiplicity of infection of 1:20 cfu/macrophage (*see* **Note 10**).
4. After 90 min of incubation at 37 °C, remove the RPMI and wash the cells twice using warm PBS (this step is necessary to remove extracellular bacteria).
5. Add warm RPMI with or without ATc (500 ng/ml) to cells infected with the conditional mutant or with the control strain (harboring the TetR/Pip-OFF system). Change medium with or without ATc every 48 h.
6. Every day (starting from T0) lyse the cells (two wells/strain) using 100μl of 0.05% SDS.
7. Dilute 50μl of the lysed cells suspension in 450μl of 7H9 and plate serial dilutions to determine viable counts.

3.9 Mice Infection

1. Infect BALB/c mice (age, 8–10 weeks) via low-dose aerosol with 10^2 cfu of either *M. tuberculosis* H37Rv or the mutant strain.
2. Feed different groups of mice with chow with or without 2000-ppm doxycycline (*see* **Note 11**).
3. Sacrifice four mice per group at day 0, week 2, week 4, week 8, week 12, and week 16 postinfection.
4. For CFU enumeration, plate appropriate dilutions of lung (spleen) homogenates on 7H10 ADN plates.

4 Notes

1. Even if it would be possible to transform the suicide plasmid in a recipient strain already carrying *tetR* and *pip*, this is not recommended since it can lead to the integration of the suicide plasmid on the integrative plasmid carrying *tetR* and *pip* due to shared *E. coli* sequences between the two plasmids.
2. Even if pFRA61 integration is very stable, it is still possible that in a small fraction of the bacteria it will excise from the chromosome with consequent loss of the possibility to regulate P_{ptr}. Moreover, we recently showed that *pip* can be inactivated by rare spontaneous mutations due to a single G insertion in correspondence of a GGGGG stretch, also causing the loss of the possibility to regulate P_{ptr}. To obviate to these problems, we recently developed more stable integrative vectors and a vector which enables the possibility to introduce a second copy of *pip* in the genome [16].
3. ATc is derivative of tetracycline with very low toxicity; in our experience, its activity decreases with time, so that in long experiments with *M. tuberculosis* it is required to add 500 ng/ml of fresh compound every 96 h.
4. The Pip-ON inducible system uses the same P_{ptr} promoters of the TetR/Pip-OFF repressible system. The only difference between the two systems is that in the Pip-ON system uses integrative vectors carrying only *pip* and not *tetR*.
5. At this purpose design, a lower primer in which the translational start site of your gene is replaced by the NsiI target sequence (ATGCAT).
6. In this case, the upper primer must be designed so that the translational start site of your gene is replaced by the NsiI target sequence (ATGCAT).
7. The use of different ATc concentrations allows decreasing gene transcription level in a dose-dependent manner to verify its impact on bacteria growth.

8. In some case, the repression of the gene of interest shows its physiological effect only after several rounds of replication, since some time is required to titrate down the protein produced before the addition of ATc. For this reason, if after 4–5 days of incubation with or without ATc, the conditional mutant does not show any difference, we recommend to dilute the old cultures 1:20 in fresh media (still with or without ATc). This operation may be repeated 3–4 times before some effect of ATc-mediated repression becomes visible [2, 17].
9. We recommended to use the highest concentration of ATc (50 ng/ml for *M. smegmatis* and 500 ng/ml for *M. tuberculosis*) in order to ensure its efficiency throughout the incubation time.
10. We suggest to calculate at least two wells/strain with ATc and two without ATc daily for a total of 8 days.
11. The chow with doxycycline can be administrated to mice before the infection or at different times after the infection to have information on the essentiality of the gene of interest in different phases of the disease [5].

References

1. Schnappinger D, Ehrt S (2014) Regulated expression systems for mycobacteria and their applications. Microbiol Spectr 2(1) MGM2-0018-2013. https://doi.org/10.1128/microbiolspec.MGM2-0018-2013
2. Boldrin F, Degiacomi G, Serafini A, Kolly GS, Ventura M, Sala C, Provvedi R, Palu G, Cole ST, Manganelli R (2018) Promoter mutagenesis for fine-tuning expression of essential genes in Mycobacterium tuberculosis. Microb Biotechnol 11(1):238–247. https://doi.org/10.1111/1751-7915.12875
3. Boldrin F, Casonato S, Dainese E, Sala C, Dhar N, Palu G, Riccardi G, Cole ST, Manganelli R (2010) Development of a repressible mycobacterial promoter system based on two transcriptional repressors. Nucleic Acids Res 38:e134
4. Kolly GS, Boldrin F, Sala C, Dhar N, Hartkoorn RC, Ventura M, Serafini A, McKinney JD, Manganelli R, Cole ST (2014) Assessing the essentiality of the decaprenyl-phospho-d-arabinofuranose pathway in Mycobacterium tuberculosis using conditional mutants. Mol Microbiol 92(1):194–211. https://doi.org/10.1111/mmi.12546
5. Boldrin F, Ventura M, Degiacomi G, Ravishankar S, Sala C, Svetlikova Z, Ambady A, Dhar N, Kordulakova J, Zhang M, Serafini A, Vishwas KG, Kolly GS, Kumar N, Palu G, Guerin ME, Mikusova K, Cole ST, Manganelli R (2014) The phosphatidyl-myo-inositol mannosyltransferase PimA is essential for Mycobacterium tuberculosis growth in vitro and in vivo. J Bacteriol 196(19):3441–3451. https://doi.org/10.1128/JB.01346-13
6. Degiacomi G, Benjak A, Madacki J, Boldrin F, Provvedi R, Palu G, Kordulakova J, Cole ST, Manganelli R (2017) Essentiality of mmpL3 and impact of its silencing on Mycobacterium tuberculosis gene expression. Sci Rep 7:43495. https://doi.org/10.1038/srep43495
7. Esposito M, Szadocka S, Degiacomi G, Orena BS, Mori G, Piano V, Boldrin F, Zemanova J, Huszar S, Barros D, Ekins S, Lelievre J, Manganelli R, Mattevi A, Pasca MR, Riccardi G, Ballell L, Mikusova K, Chiarelli LR (2017) A phenotypic-based target screening approach delivers new antitubercular CTP synthetase inhibitors. ACS Infect Dis 3 (6):428–437. https://doi.org/10.1021/acsinfecdis.7b00006
8. Serafini A, Boldrin F, Palu G, Manganelli R (2009) Characterization of a Mycobacterium tuberculosis ESX-3 conditional mutant: essentiality and rescue by iron and zinc. J Bacteriol 191(20):6340–6344
9. Ventura M, Rieck B, Boldrin F, Degiacomi G, Bellinzoni M, Barilone N, Alzaidi F, Alzari PM,

Manganelli R, O'Hare HM (2013) GarA is an essential regulator of metabolism in Mycobacterium tuberculosis. Mol Microbiol 90 (2):356–366. https://doi.org/10.1111/mmi.12368

10. Gola S, Munder T, Casonato S, Manganelli R, Vicente M (2015) The essential role of SepF in mycobacterial division. Mol Microbiol 97 (3):560–576. https://doi.org/10.1111/mmi.13050
11. Cortes M, Singh AK, Reyrat JM, Gaillard JL, Nassif X, Herrmann JL (2011) Conditional gene expression in Mycobacterium abscessus. PLoS One 6(12):e29306. https://doi.org/10.1371/journal.pone.0029306
12. Forti F, Crosta A, Ghisotti D (2009) Pristinamycin-inducible gene regulation in mycobacteria. J Biotechnol 140(3–4):270–277
13. Parish T, Stoker NG (2000) Use of a flexible cassette method to generate a double unmarked Mycobacterium tuberculosis tlyA plcABC mutant by gene replacement. Microbiology 146:1969–1975
14. Cascioferro A, Boldrin F, Serafini A, Provvedi R, Palu G, Manganelli R (2010) Xer site-specific recombination, an efficient tool to introduce unmarked deletions into mycobacteria. Appl Environ Microbiol 76 (15):5312–5316. https://doi.org/10.1128/AEM.00382-10
15. Bardarov S, Bardarov S Jr, Pavelka MS Jr, Sambandamurthy V, Larsen M, Tufariello J, Chan J, Hatfull G, Jacobs WR Jr (2002) Specialized transduction: an efficient method for generating marked and unmarked targeted gene disruptions in Mycobacterium tuberculosis, M. bovis BCG and M. smegmatis. Microbiology 148(Pt 10):3007–3017
16. Boldrin F, Anoosheh S, Serafini A, Cioetto Mazzabo L, Palu G, Provvedi R, Manganelli R (2019) Improving the stability of the TetR/Pip-OFF mycobacterial repressible promoter system. Sci Rep 9(1):5783. https://doi.org/10.1038/s41598-019-42319-2
17. Degiacomi G, Personne Y, Mondesert G, Ge X, Mandava CS, Hartkoorn RC, Boldrin F, Goel P, Peisker K, Benjak A, Barrio MB, Ventura M, Brown AC, Leblanc V, Bauer A, Sanyal S, Cole ST, Lagrange S, Parish T, Manganelli R (2016) Micrococcin P1 – a bactericidal thiopeptide active against Mycobacterium tuberculosis. Tuberculosis 100:95–101. https://doi.org/10.1016/j.tube.2016.07.011

Chapter 18

A Single-Cell RNA-Seq-Based Approach for Genome-Wide Identification of Cell Essential Genes

Yuqiu Lu and Shaolin Shi

Abstract

Identification of genes essential for structure, function, and survival of a cell type is critical for understanding of the underlying mechanisms. Unfortunately, there is no efficient way to identify such genes. Studies by single-cell RNA sequencing have shown that gene expressions of single cells of the same type are highly heterogeneous. We therefore speculate that the genes expressed in all individual cells of the same type are essential for the cell type, including the housekeeping genes and cell type-specific essential genes. Based on this rationale, we design a high-throughput approach to identify podocyte essential genes. In this approach, mouse podocytes are subjected to ultra-deep single-cell RNA-seq, and the genes expressed in all single podocytes are sorted out and considered as the candidates of podocyte essential genes. The essentiality of these genes for podocytes is assessed by bioinformatics, cross-species conserved expression, association with injury/disease, inclusion of known essential genes, and experimental validation. By comparison with the essential genes of other cell types, podocyte-specific essential genes can be distinguished. This approach applies to any cell types. In this chapter, we describe the approach and detailed methods.

Key words Essential genes, Single-cell RNA-seq, Podocyte, Genome-wide

1 Introduction

As one of the three intrinsic cell types in glomeruli, podocytes are highly differentiated with a large number of dendritic branches (foot processes); and together with glomerular basement membrane (GBM) and endothelial cells, they form the glomerular filtration barrier to filter blood to remove metabolic waste from the body. Podocytes are critical for glomerular structure and function, and their damage may lead to glomerular diseases such as focal segmental glomerulosclerosis (FSGS), membranous nephropathy (MN), diabetic nephropathy (DN), and minimal change disease (MCD), etc. At present, over a hundred genes have been shown to be essential for podocyte survival, structure, and function; and most podocyte essential genes have not been identified yet, which

Ren Zhang (ed.), *Essential Genes and Genomes: Methods and Protocols*, Methods in Molecular Biology, vol. 2377, https://doi.org/10.1007/978-1-0716-1720-5_18,

greatly limit our understanding of podocyte physiology and pathology.

In recent years, as a powerful tool for molecular biology research, single-cell RNA-seq is being widely used. Available single-cell RNA-seq studies have shown that gene expression in individual cells of the same type is highly heterogeneous [1–3], suggesting that only a small portion of genes detected by bulk sequencing of mixed population of a cell type are shared by all the individual cells while most of the genes are randomly expressed in individual cells. We speculate that genes expressed in all individual cells of the same type are essential for the cell type. These essential genes include housekeeping genes that are also expressed in other cell types to maintain cell survival and basic physiological activities; in addition, they also include cell type-specific essential genes that execute functions required for survival of the organism.

Based on single-cell RNA-seq and the above speculation, we design a high-throughput screening approach for identification of podocyte essential genes (Fig. 1). Through single-cell RNA-seq analysis of 20 mouse podocytes using the Fluidigm C1 system for single-cell isolation and cDNA synthesis, followed by sequencing library construction and sequencing, we have determined the expression of 335 genes in all of the 20 mouse podocytes using 0.5 RPKM as the expression cutoff. Among these genes, 239 genes were also found to be expressed in mesangial and/or endothelial cells, thus likely being housekeeping genes; while the remaining 92 genes were specifically expressed in podocytes compared with mesangial and endothelial cells [4, 5], and therefore the candidates of essential genes specific for podocytes. When gene expression cutoff is set 0.1 RPKM, 611 genes that were commonly expressed in all the 20 podocytes would be resulted in [6], significantly increasing the number of podocyte essential genes.

Several methods can be used to assess the essentiality of the genes, including bioinformatics analysis, cross-species conservation of gene expression, association between expression of the genes with podocyte injury and disease, inclusion of known essential genes, and, more importantly, in vitro and in vivo experimental validations. We have provided evidence to support that the 92 podocyte essential gene candidates are essential and mainly involved in the formation and stability of podocyte cytoskeletons.

2 Materials

1. Animals: 6–8 week C57BL/6 male mice, SPF grade, were purchased from the Model Animal Research Center of Nanjing University (http://www.nicemice.cn/).

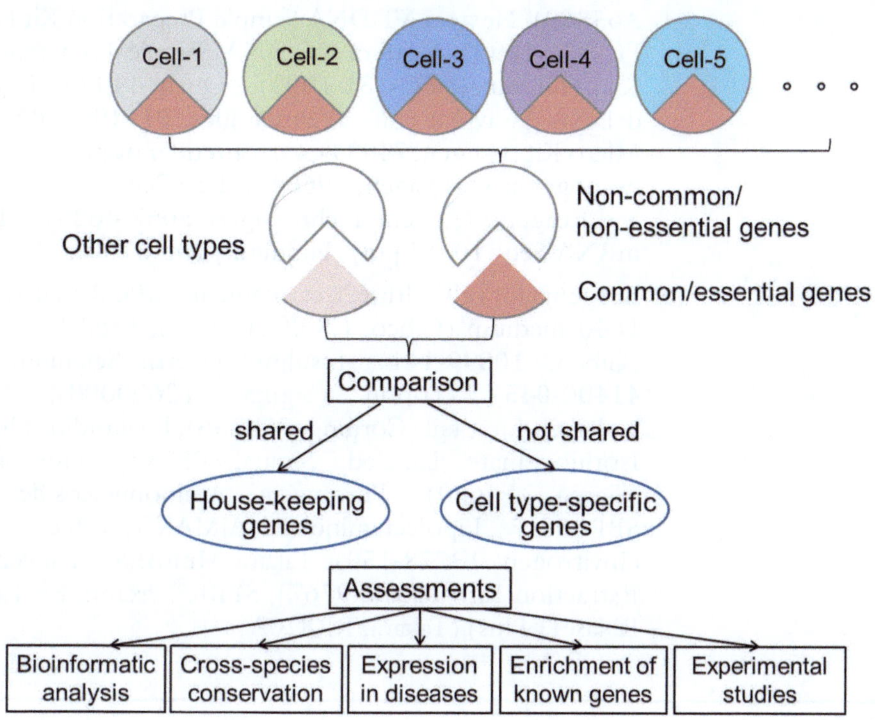

Fig. 1 The approach for genome-wide identification of genes essential for a cell type based on single-cell RNA-seq

2. The human immortalized podocyte line (HPC) was gift from Dr. Saleem M (University of Bristol, UK).
3. Reagents for mouse perfusion and glomerular cell suspension: Collagenase II (Sigma, C6885), Pronase E (Sigma, P6911), DNase I (Sigma, D5025), Dynabeads M450 tosylactivated (Invitrogen, 14013).
4. Reagents for Fluidigm C1 single-cell separation and cDNA synthesis system: Advantage® 2PCR kit (Clontech, 639207), SMARTer® Ultra low RNA kit for the Fluidigm C1 system (Clontech, 634834), dNTP (mix) (Biosky, D0056), EasyTaq buffer (Transgene, GK101-01), EasyTaq DNA polymerase (Transgene, AP111-03), 6× DNA loading buffer (Transgene, GH101-01X), 100-bp Plus DNA Ladder (Transgene, BM311-01), Quant-It Picogreen dsDNA Reagent and Kits (Invitrogen, P11496), SYBR® Premix Ex Taq II (Tli RNaseH Plus) (Takara, RR820A), PrimeScript™ RT Master Mix (Perfect Real Time) (Takara, RR036A), SMARTer® Ultra Low RNA Kit for Illumina Sequencing Advantage®2 PCR Kit (Clontech, 634936), Array Control™ RNA Spikes (Ambion, AM1780), the RNA Storage Solution (Ambion, AM7000).
5. Reagents for sequencing library construction and sequencing: Agencourt AMPure® XP (Agencourt Bioscience Corp,

A63880), Nextera XT DNA Sample Preparation Kit (Illumina, FC-131-1096), Nextera XT DNA Sample Preparation Index Kit (Illumina, FC-131-1002), Quant-ITTM PicoGreen® dsDNA Assay Kit (Life Technologies, P11496), RNeasy Plus Micro Kit (Qiagen, 74034), QIAshredder disposable cell-lysate homogenizers (Qiagen, 79654), High Sensitivity DNA Chips and Reagents (Agilent Technologies, 5067-4626), C1 IFC for mRNA Seq (10–17 μm) (Fluidigm, 100-5760).

6. Reagents for cell culture, treatment, and qPCR analysis: RPMI-1640 medium (Gibco, CS-0011-0022), Fetal Bovine Serum (Gibco, 10099-141), Insulin-Tansferrin-Selenium (Gibco, 41400-045), Tripsin (Sigma, T2600000), Corning® Collagen I, rat tail (Corning, 354236), Phalloidin (Fluorescein Isothiocyanate Labeled; Sigma, P5282), Fluoromount™ (Sigma, F4680), Puromycin Aminonucleoside (Sigma, SBR00017), Lipofectamine® RNAiMAX Transfection Reagent (Invitrogen, 13778-150), Takara MiniBEST Universal RNA Extraction Kit (Takara, 9767), SYBR® Premix Ex Taq II (Tli RNaseH Plus) (Takara, RR820A).

3 Methods

3.1 Isolation of Mouse Glomeruli Using Dynabeads and Magnetic Concentrator

1. Preparation of renal perfusion solution and digestive solution as follows:
   (a) 1× HBSS buffer: Ca^{2+} free, Mg^{2+} free, pH 7.2.
   (b) Digestive solution: Collagenase II (1 mg/ml), Pronase E (0.5 mg/ml), DNase I (50 U/ml), dissolved in HBSS.
   (c) Perfusion solution A: 5% Dynabeads M450 Tosylactivated in HBSS.
   (d) Perfusion solution B: Collagenase II (1 mg/ml), Pronase E (0.5 mg/ml), DNase I (50 U/ml), 5% Dynabeads M450 Tosylactivated, in HBSS.
2. Prewarm the prepared solutions at 37 °C until use.
3. Anesthetize a mouse by inhaling isoflurane until complete loss of consciousness, and fix it on an operation platform; open the abdominal cavity to fully expose the kidneys, renal artery, and abdominal aorta; and make a local blood circulation with abdominal aorta as the entrance and inferior vena cava as the exit.
4. Slowly push the following solutions into the entrance at abdominal aorta using a syringe: 5 ml HBSS, 2 ml perfusion solution A, and 2 ml perfusion solution B, respectively.

5. Once the perfusion is finished, excise both kidneys and place them on ice; remove the kidney capsule and cut the cortex into 1 mm^3 pieces.
6. Mix the sliced cortex tissue with 1 ml digestive solution in a 1.5 ml tube, and place the tube on a constant temperature oscillation incubator for digestion at 37 °C and 300 rpm for 10 min.
7. After digestion, transfer the tissue onto a 100- μM cell sieve which is inserted on a 50 ml Falcon centrifuge tubes, and gently squash and ground the tissue several times; rinse the dissociated tissue with HBSS (precooled at 4 °C) to make the glomeruli pass through the sieve. (Please note that tubuli also pass through the sieve.)
8. Centrifugate the glomeruli and tubuli suspension at 200 × *g* for 5 min at 4 °C, and discard the supernatant.
9. Perform magnetic concentration to purify glomeruli as follows: resuspend the pellet from centrifugation in 1.5 ml HBSS and transfer it to a 1.5 or 2.0 ml tube; place the tube on a magnetic concentrator and let it stand for 1 min to make the dynabeads-containing glomeruli move to the side of magnet and leave the tubuli in the solution; and then remove the solution with tubuli by pipetting.
10. Repeat above steps of purification for 2–3 times until the purity of glomeruli reaches >95% according to microscopy inspection.

3.2 Preparation of Mouse Glomerular Single-Cell Suspension

1. Add 1 ml of digestive solution to the obtained mouse glomeruli and place the sample on the constant temperature oscillation incubator for incubation at 37 °C and 1400 rpm for 30 min. During the digestion, use 1 ml pipet to gently pipette up and down for several times to facilitate digestion.
2. After digestion, place the sample on a magnetic concentrator and leave it for 1 min to remove the Dynabeads and undigested or incompletely digested glomeruli.
3. Transfer the digestive solution containing glomerular cells onto a 40-μm cell sieve on top of a 50 ml Falcon tube and rinse with HBSS.
4. Collect the filtrate and centrifuge at 4 °C and 500 × *g* for 5 min; discard the supernatant and add 300 μl HBSS containing 10% FBS to resuspend the cells.
5. Pipette 5 μl cell suspension to 40 μl HBSS in a tube, followed by addition of 5 μl 0.4% Trypan Blue; mix and add onto a slide to assess cell viability by microscopy, while leaving the remaining cell suspension on ice.

6. Once the viability of the cells is determined to be satisfactory (e.g., >95%), proceed with the Fluidigm C1 procedure.

3.3 Single-Cell Separation and Preparation of cDNA Samples

1. For the whole process of single-cell separation, cDNA synthesis and amplification, refer to the instruction in the Fluidigm's manual "Using the C1™ single cell auto prep system to generate mRNA from single cells and libraries for sequencing" (*see* **Note 1**).
2. The procedure is carried out in strict accordance with the instructions of Fluidigm C1 system. The main steps of single-cell capture and cDNA synthesis and amplification with Fluidigm C1 platform are depicted in Figs. 2 and 3 (*see* **Notes 2** and **3**)

3.4 Quality Assessment of cDNA Samples and Identification of Cell Type of cDNA

1. Quality assessment of cDNA samples: the obtained cDNA samples were analyzed for the concentration and size distribution using a 2100 Bioanalyzer (Aligent). For the detailed method, please refer to instrument and reagent instructions.
2. Determination of the cell type from which a cDNA sample derives: there are three intrinsic cell types in glomeruli, podocytes, mesangial cells, and endothelial cells. To determine the cell types from which a single-cell cDNA sample derives, qPCR analyses using podocyte markers, Nphs2 and Synpo; mesangial cell markers, Igfbp5, Gata3, and Fhl2; and endothelial cell markers, CD31, Tek, and Kdr, were performed. Primers are designed by software Oligo7. qPCR was carried out using the SYBR® Premix Ex Taq™ II (Tli RNaseH Plus) kit and following the manufacturer's instruction.

3.5 Construction of cDNA Library for Sequencing

1. Add fragmentation buffer to the cDNA samples to fragment the cDNA molecules.
2. Perform end repair with adenosine, followed by addition of adaptors that contain sequences of amplification primers and index (Fig. 3).
3. Use the Beckman Coulter (Beverly, USA) system to purify the cDNA and select the fragments with a size between 150 and 200 bp.
4. Add 3 μl USER Enzyme (Uracil-Specific Excision Reagent) (NEB, USA) to the cDNA fragment samples and incubate at 37 °C for 15 min, followed by 95 °C for 5 min.
5. Perform PCR amplification using Phusion High-Fidelity DNA polymerase, Universal PCR primers and Index (X) Primer.
6. Purify the PCR products using Ampure XP System to obtain the final cDNA library for sequencing.

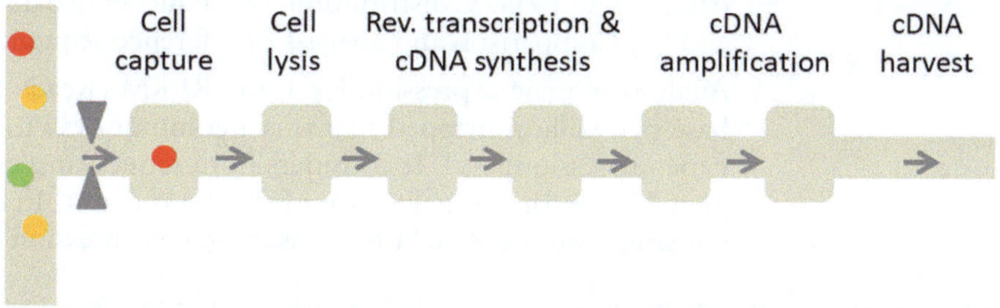

Fig. 2 Microfluidics of a chip for the Fluidigm C1 platform and the main steps of single-cell capture, cDNA synthesis, and amplification

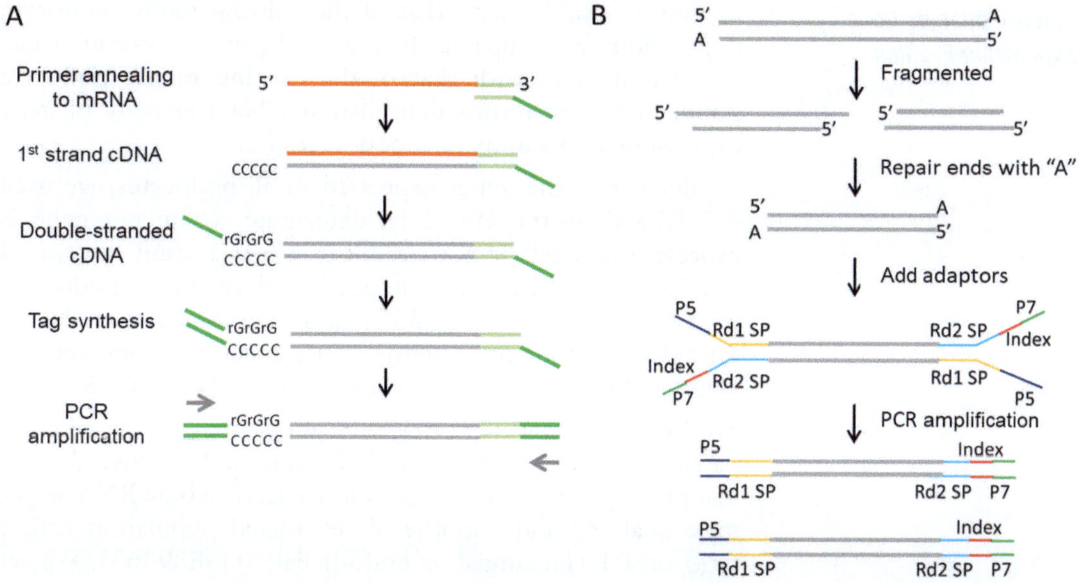

Fig. 3 cDNA synthesis/amplification using SMARTer® system (**a**) and construction of the sequencing library (**b**)

7. Finally, use Agilent 2100 Bioanalyzer to assess the quality of the library.

3.6 Sequencing

1. Enrich the cDNA samples (index-coded) on the cBot Cluster Generation System using reagent TruSeq PE Cluster Kit v3-cBot-HS (Illumina).
2. The cDNA library was sequenced by single-end reads of digital gene expression tag profiling (DGE) on the high-throughput sequencing platform (Illumina Hiseq 2000) with a length of 50 bp and a depth of 15 M.
3. Quality assessment of sequencing data: prior to further analysis, the quality of the original sequencing data is evaluated. These assessments include (1) error distribution of sequencing

data, (2) A/T/G/C distribution, (3) sequencing data filtering, and (4) Comparison and analysis of reference sequences.

4. Analysis of gene expression level: the RPKM (Reads Per Kilobase per Million mapped reads) is a common way to measure gene expression level. It considers both sequencing depth and gene length for normalization of the reads count. In our present study, we use RPKM to measure gene expression levels.

3.7 Identification of Podocyte Essential Gene Candidates Based on the Single-Podocyte Gene Expression Profiles

1. We firstly performed correlation analysis to demonstrate the gene expression heterogeneity between single podocytes as follows: (1) compared the gene expression profile between individual podocytes; (2) compared single podocyte gene expression profiles with that of the existing mouse podocyte population; (3) compared the averaged gene expression profile of 20 podocytes with that of the existing mouse podocyte population. The results demonstrated heterogeneity of gene expression across individual podocytes [4].
2. To determine the genes expressed in all podocytes, we used 0.5 RPKM as the cutoff to determine whether a gene is expressed in a cell or not (*see* **Note 4**). As a result, we sorted out 335 genes that were expressed in all of the 20 podocytes with a RPKM >0.5. These 335 genes were considered to be the candidates of podocyte essential genes, including housekeeping genes and podocyte-specific essential genes (*see* **Note 5**).
3. To distinguish the two categories of essential genes, we compared the mean RPKM of the 335 genes in 20 mouse glomerular podocytes with their expression levels in a bulk RNA-seq of mesangial cell and endothelial cell mixed population with a ratio of 1:1 (mesangial vs endothelial) (GSE89263). We set the following criterion for distinguishing housekeeping genes and podocyte-specific essential genes: if a gene has an expression level in podocytes that is ≥5-folds higher than that in the mixed mesangial/endothelial cells, it is then considered to be specifically expressed in podocytes and is thus a podocyte-specific essential gene candidate. With this criterion, we sorted out 92 podocyte-specific essential genes candidates. Accordingly, there were 239 genes whose expression in podocytes is <5-fold than that in mesangial and endothelial cells. They were considered as housekeeping genes candidates.

3.8 Assessment of Essentiality of the 92 Podocyte-Specific Essential Gene Candidates

1. Bioinformatics analysis: DAVID Bioinformatics Resources 6.7 (National Institute of Allergy and Infectious Diseases, NIH) was used to analyze 239 housekeeping gene candidates and 92 podocyte-specific essential gene candidates. These analyses included Gene Ontology, Functional category clusters, and KEGG pathways. The results showed that the housekeeping activities were indeed enriched in the 239 genes, e.g.,

mitochondrial energy metabolism and protein synthesis. Meanwhile, cytoskeleton-related genes were enriched in the 92 genes, which is characteristic of podocytes [4].

2. Assessment of evolutionary conservation of expression: cross-species conservation of gene expression reflects functional importance of a gene. We searched the Human Protein Atlas database (www.proteinatlas.org) to determine whether the 92 podocyte-specific essential gene candidates are also expressed in human podocytes to assess evolutionary conservation of their expression. We found that most of the 92 genes are expressed in human podocytes, demonstrating evolutionary conservation of their expression and suggesting functional importance in podocytes [4].
3. Expression down-regulation in podocyte diseases: Cell dysfunction or injury is often associated with deficiency in essential gene expression. We therefore investigated whether the 92 podocyte-specific essential gene candidates are down-regulated in expression by searching the Nephroseq, a database with glomerular gene expression in various podocyte diseases (www.nephroseq.org). In addition, the 92 genes were also down-regulated as shown by qPCR in cultured podocytes treated by purimycin aminonucleosides, a commonly-used podocyte injury model [4].
4. Inclusion of known podocyte essential genes: if our approach is feasible and effective, it would be expected that the 92 podocyte-specific essential gene candidates include those that are already known to be podocyte essential genes. Indeed, among the 92 genes we found ~30 genes that are known to be crucial for podocytes, supporting the validity of our approach for essential gene identification based on single-cell RNA seq.
5. Experimental validation: in vitro, the genes were knocked down by siRNA in cultured podocytes to determine whether the deficiency of these genes would cause podocyte injury. Several parameters were used to assess podocyte injury, e.g., qPCR that determined whether CD2AP and SYNPO expressions were down-regulated, which is characteristic of podocyte injury; in addition, F-actin staining using phalloidin was used to show the changes of F-actin stress fibers, including number reduction and relocation.

4 Notes

1. With Fluidigm C1 single-cell separation and cDNA synthesis system, it is important to choose an appropriate Fluidigm chip for capture of single cells according to the size of a cell type of

interest. We chose the one of 10–17 μm (1772×) that matches the size of glomerular cells.

2. After cell suspension is loaded to the Fluidigm C1 instrument and the step of single-cell capturing is completed, it is important to examine each well of the chip and record the capturing efficiency and quality using "no cell", "one cell", and "multiple cells". Once cDNA synthesis and amplification are completed, keep the cDNA samples from the wells with only one cell, and discard the others.
3. In addition to Fluidigm C1 platform by which we conducted the single-cell RNA-seq of podocytes, there are a couple of additional single-cell RNA-seq methods that have been developed later, including Drop-seq, inDROP, microwell-seq, SPLit-seq, etc. These new technologies are more efficient concerning the number of cell analyzed and the cost per cell. They are particularly useful in profiling the cell types and subtypes in a tissue or organ. However, the number of genes detected in each cell could be as low as hundreds or one or two thousand by deeper sequencing. In contrast, Fluidigm C1 platform allows independent sequencing library construction for each single cell, maximizing the sequencing depth and resulting in maximal number of genes detected. For example, we detected >6000 genes on average in each podocyte. The availability of maximal number of detected genes is extremely crucial for our approach to identify podocyte essential genes because it minimizes the failure in gene expression detection.
4. Choice of gene expression cutoff: this is very important and directly affects the number of identified podocyte essential gene candidates. When using 0.5 RPKM, we found 335 genes; and when using 0.1 RPKM, 611 genes were obtained. Studies have shown that there may be ~2000 essential genes for a typical cell type [7, 8]; therefore, we may further lower the RPKM cutoff to obtain more genes.
5. In our approach, we defined a podocyte essential gene by detection of its expression in every podocyte that was subjected to single-cell RNA-seq. However, this criterion may be too stringent as the technical variations during the entire process of single-cell RNA-seq could result in failure of detection of a gene in a cell. Therefore, permitting one or few cells to have 0 RPKM in determining essential genes would increase the total number of podocyte essential gene candidates. In fact, we did find several known essential genes that had 0 RPKM in one or few single cells in our previous studies [4, 5].

Acknowledgments

This work was supported by the grants from the National Natural Science Foundation of China (81800631), Shanghai Sailing Program (18YF1421300), and Tongji Hospital Research Program (TJ1716) to Lu Y.; and the National Natural Science Foundation of China (81670653, 81970619) to Shi S.

References

1. Marinov GK, Williams BA, McCue K, Schroth GP, Gertz J et al (2014) From single-cell to cell-pool transcriptomes: stochasticity in gene expression and RNA splicing. Genome Res 24:496–510
2. Raser JM, O'Shea EK (2005) Noise in gene expression: origins, consequences, and control. Science 309:2010–2013
3. Miyamoto DT, Zheng Y, Wittner BS, Lee RJ, Zhu H et al (2015) RNA-Seq of single prostate CTCs implicates noncanonical Wnt signaling in antiandrogen resistance. Science 349:1351–1356
4. Lu Y, Ye Y, Bao W, Yang Q, Wang J et al (2017) Genome-wide identification of genes essential for podocyte cytoskeletons based on single-cell RNA sequencing. Kidney Int 92:1119–1129
5. Lu Y, Ye Y, Yang Q, Shi S (2017) Single-cell RNA-sequence analysis of mouse glomerular mesangial cells uncovers mesangial cell essential genes. Kidney Int 92:504–513
6. Zhang S, Yang Q, Liu Z, Shi S (2020) POFUT1 is dispensible for podocyte structure, function and survival. Am J Transl Res 12(5):2212–2224
7. Wang T, Birsoy K, Hughes NW, Krupczak KM, Post Y et al (2015) Identification and characterization of essential genes in the human genome. Science 350:1096–1101
8. Mair B, Tomic J, Masud SN, Tonge P, Weiss A et al (2019) Essential gene profiles for human pluripotent stem cells identify uncharacterized genes and substrate dependencies. Cell Rep 27:599–615

Chapter 19

Identification of Essential Genes in *Caenorhabditis elegans* with Lethal Mutations Maintained by Genetic Balancers

Shicheng Yu, Chaoran Zheng, and Jeffrey Shih-Chieh Chu

Abstract

Genetic balancer systems, which allow effective capture and maintenance of lethal mutations stably, play an important role in identifying essential genes. Whole-genome sequencing (WGS) followed by bioinformatics analysis, combined with genetic mapping data analysis, allows for an efficient and economical means of identifying genomic mutations in essential genes. Using this approach, we successfully identified 104 essential genes on ChrI, ChrIII, and ChrV in *C. elegans*. In this report, we described a protocol that sequences the genome of prebalanced *Caenorhabditis elegans* (*C. elegans*) strains to carry lethal mutations and identifies candidate causal mutations and candidate essential genes using a robust bioinformatics procedure.

Key words Essential gene, Balancer, Duplication, Translocation, Illumina sequencing

1 Introduction

Caenorhabditis elegans, a small, free-living, and self-fertilizing hermaphroditic nematode, is a powerful tool for eukaryotic genetic research [1]. The ease of handling and the ability to generate hundreds and thousands of mutant strains make *C. elegans* an excellent genetic tool to decipher the functions of important genes. Thousands of strains exhibiting various phenotypes have been generated using chemical, irradiation, transposable elements, and CRISPR methodologies over the last 40–50 years. One class of genes remain challenging to study, and it is genes that are essential for survival. The challenge arises from the fact that strains carrying knock-out mutations in essential genes are difficult to maintain. These strains will not generate viable progeny, and the lethal alleles are lost over subsequent generations. Essential genes are large groups of genes in *C. elegans* that occupy approximately 25% of all the genes [2–4] and have been shown to have considerable

Shicheng Yu and Chaoran Zheng contributed equally with all other contributors.

Ren Zhang (ed.), *Essential Genes and Genomes: Methods and Protocols*, Methods in Molecular Biology, vol. 2377, https://doi.org/10.1007/978-1-0716-1720-5_19, © Springer Science+Business Media, LLC, part of Springer Nature 2022

homology with humans [5, 6]. *C. elegans* has been an excellent model for studying essential genes. With its ease of handling and with many of its genes are homologous to those related to human diseases [7], *C. elegans* has been valuable in identifying essential genes, including *let-60/Ras* [8] and *let-740/dcr-1* [9, 10].

The classical methods of identifying essential genes are the use of random mutagenesis. Itaya used *E.coli* gene engineering system to discover six genes necessary for the survival of *Bacillus subtilis* [11]. Subsequent systematic research using chemical mutagens, such as ethylnitrosourea (ENU) in zebrafish [12], P-element insertion in *Drosophila melanogaster* [13], and transposon mutagenesis in *Mycoplasma genitalium* [14] has greatly expanded the identification and understanding of essential genes. In *C. elegans*, random mutagenesis using Ethyl methanesulfonate (EMS), ENU, and ultraviolet/trimethylpsoralen (UV/TMP) [15] is the most common method to obtain mutant strains. Through the vigorous effort of the *C. elegans* Knock-out Consortium and the Million Mutation Project (MMP), 13,760 loss-of-function alleles were obtained in 20,514 protein-coding genes [16]. However, random mutagenesis methods for identifying candidate genes required numerous genetic mapping and PCR experiments to identify the causal mutation, which can take months or years. The MMP has since employed the use of high-throughput sequencing to identify the genomic variations that were generated in the mutagenesis process. However, mutants obtained in this effort were largely non-lethal [16]. Genome-wide attempts in identifying essential genes in *C. elegans* have been carried out using other various methods. RNAi knock-down has estimated about 3500 genes (~17%) of the *C. elegans* genome were essential [4, 7, 17]. Although RNAi technique allows the identification of genetic phenotypes by applying a genome-wide targeted approach, it can only down-regulate gene expression and is incapable of maintaining the phenotype over more extended periods [4, 18]. Taken together, an effective and high-throughput method is needed to identify essential genes.

Genetic balancer systems, which are either extrachromosomal DNA or balanced translocations that carry a wild allele and thereby rescuing the lethal phenotype [1], are effective systems where one can easily capture and maintain lethal mutations stably over generations. With the use of duplications, reciprocal translocations, and inversions, the researchers have balanced over 70% of *C. elegans* genome [1, 19]. Duplication balancers can provide a third wild-type allele to an otherwise homozygous mutation. Duplication balancers also rarely cross-over with normal chromosomes and thus allow stable maintenance of lethal mutations [1]. Overall, 640 essential genes in *C. elegans* have been identified using balancers [20–23], but fewer than 90 essential genes (let-*) have been identified with precise chromosomal location [17, 24, 25]. With the use of powerful methods including genetic mapping, high-throughput sequencing, and bioinformatics analyses, we were able to identify the location of causal mutations in *C. elegans* strains carrying lethal alleles balanced by genetic balancers [7]. The high-throughput sequencing-based method increases the efficiency and reduces the cost of a large-scale essential gene identification [7, 26]. We detail a protocol below to demonstrate how one can extract genomic DNA from balanced lethal strains, sequence its genome, and analyze for the causal mutation.

2 Materials

2.1 Strain Maintenance and Sequencing

2.1.1 The Maintenance of Each C. elegans Strain

1. *Caenorhabditis elegans* KR235 [*dpy-5 (e61)*, +, *unc-13 (e450)/ dpy-5(e61)*, *unc-15(e73)*, +; *sDp2*].
2. *C. elegans* strains carrying lethal mutations balanced by sDp2.
3. *Escherichia coli* OP50.
4. EMS: 12 mM.
5. 2-in. plates.
6. M9 (Na_2HPO_4 6 g, KH_2PO_4 3 g, NaCl 5 g, $MgSO_4$ 0.2 g, add Milli-Q to 1 L).
7. Milli-Q.
8. NGM (3 g NaCl, 17 g agar, 2.5 g peptone, and 975 ml H_2O; Autoclave for 50 min; Cool to 55 °C; Add 1 ml 1 M $CaCl_2$, 1 ml 5 mg/ml cholesterol in ethanol, 1 ml 1 M $MgSO_4$, and 25 ml 1 M KPO_4 buffer; Fill 2/3 of plates with the NGM solution; Cooling at room temperature).

2.1.2 Genomic DNA Extraction and Sequencing

1. Sodium dodecyl sulfate (SDS).
2. Chloroform.
3. RNase A.
4. Proteinase K.
5. LoTE buffer (3 mM Tris–HCl, pH 7.5; 0.2 mM EDTA).
6. 100% ethanol.
7. Phenol.
8. Elution buffer (LoTE buffer: 7.5 M ammonium acetate = 5:1).
9. DNeasy Blood & Tissue Kit (QIAGEN, 69504).
10. Qubit dsDNA High Sensitivity Assay Kit.
11. 1 kb DNA ladder (Fermentas).
12. QIAquick Gel Extraction Kit (QIAGEN, 28704).
13. Milli-Q.

Table 1
Software involved in this study

Software	Function	Website	Ref.
BLASTP	Blast	http://blast.ncbi.nlm.nih.gov/Blast.cgi	[27]
BWA (Burrows-Wheeler Aligner)	Alignment	https://sourceforge.net/projects/bio-bwa/files/	[28]
SAMtools	Alignment, call SNV, analyzing the sequencing depth	https://sourceforge.net/	[29]
Picard tools	Remove duplicates	https://sourceforge.net/projects/picard/	[30]
GATK (Genome Analysis Toolkit)	Re-alignment, detecting the SNVs and indels	https://gatk.broadinstitute.org/hc/en-us	[31]
VarScan	Detecting the SNVs and indels	http://varscan.sourceforge.net/	[32]
IGV (Integrative Genomics Viewer)	View the SNP and indel	http://software.broadinstitute.org/software/igv/	[33, 34]
ANNOVAR	Variation annotation	http://annovar.openbioinformatics.org/	[35]
PANTHER database	Gene ontology (GO) analysis and PANTHER Protein Class	http://pantherdb.org/	[36]
Primer3	Design primer	https://sourceforge.net/projects/primer3/	[37, 38]
FastQC	Quality control tool of sequence data	http://www.bioinformatics.babraham.ac.uk/projects/fastqc/	[39]
Trimmomatic	Read trimming tool for sequencing data	http://www.usadellab.org/cms/?page=trimmomatic	[40]
bam2fastq	Convert the BAM file to a FASTQ file	https://gsl.hudsonalpha.org/information/software/bam2fastq	[41]

2.2 Database Used in the Protocol

The sequencing data used in our publications [26] are available from the NCBI Sequence Read Archive (accession numbers: SRR6739866–SRR6739996). One can also use any sequencing output in BAM/FASTQ format for the analysis protocol we describe below.

2.3 Software Used in the Protocol

See Table 1.

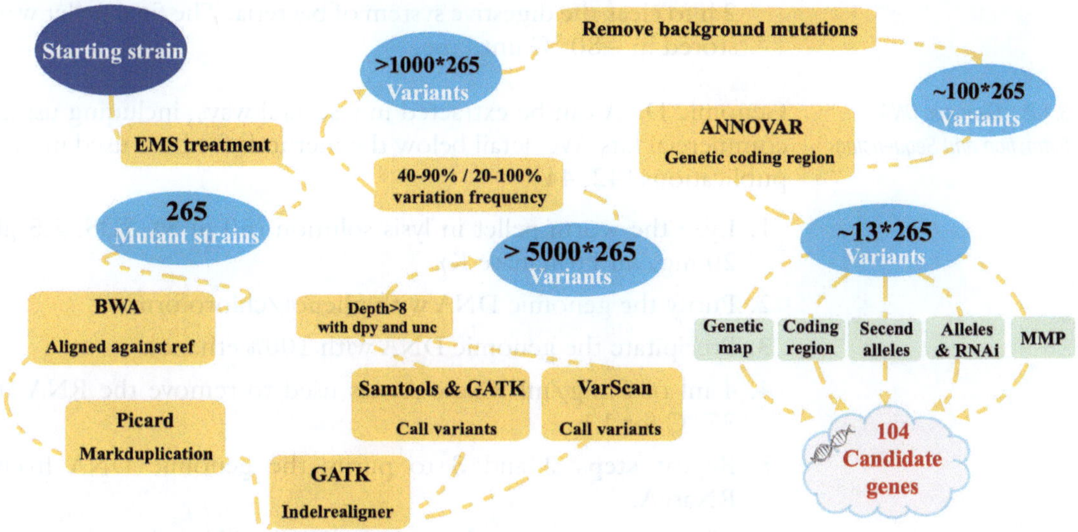

Fig. 1 Overview of the mutation identification procedure. Numbers in blue ovals are averages with data from [26, 42]

3 Methods

The methods presented here describe (1) The maintenance of *C. elgeans* lethal strains balanced by sDp2 and its genomic DNA sequencing, (2) identification of mutation sites, and (3) identification of high-confidence candidate essential genes. The method is illustrated by the examples in duplications and translocations balancer systems (Fig. 1).

3.1 Maintaining *C. elegans* Strains and Whole-Genome Sequencing

3.1.1 Maintaining *C. elegans* Strains Carrying Lethal Mutations

1. *C. elegans* could be maintained on NGM plates streaked with *E. coli* OP50 [43]. Strains carrying lethal mutations balanced by *sDp2,* which is a duplication balancer covering the first 7 Mb of ChrI, were generated from EMS mutagenesis experiments on KR235 (*see* Subheading 2 for genotype) (*see* **Note 1**). *sDp2* provides a rescuing phenotype for *dpy-5 (e61)* and thus animals with *sDp2* show Unc phenotype while animals without *sDp2* show Dpy Unc phenotype. Maintenance of the lethal strain is by selecting three Unc hermaphrodites on a new plate every 5 days.
2. To collect the worms for genomic DNA extraction, we cultured each strain on 20 2-in. plates at 20 °C until confluent. The worms were washed off the plate using M9 buffer and collected in 15-ml Falcon tubes. The worms were pelleted by centrifugation at 400 × *g* for 1 min. The pellet was washed with 12 ml of M9 buffer followed by centrifugation three times. The worm pellet was incubated at room temperature on a rocker for

2 h to clear the digestive system of bacteria. The final pellet was stored in −80 °C until use.

3.1.2 Genomic DNA Extraction and Sequencing

Genomic DNA can be extracted in a several ways, including using commercial kits. We detail below the methods we have used in our publications [42, 44]:

1. Lyze the worm pellet in lysis solution (50 μl 5% SDS, 2.5 μl 20 mg/ml Proteinase K).
2. Purify the genomic DNA with phenol/chloroform.
3. Precipitate the genomic DNA with 100% ethanol.
4. 4 ml of 5 mg/ml RNase A was used to remove the RNA at 37 °C for 1 h.
5. Repeat **steps 2** and **3** to purify the genomic DNA from RNaseA.
6. Library construction from the purified DNA follows Illumina paired-end library preparation protocol and sequenced on the Illumina platform.

3.2 Mutation Identification

3.2.1 Download the Reference Genome and the Sequencing Data

1. Obtain reference genome from WormBase (https://wormbase.org/). For this protocol, we have used the WS246 version of the *C. elegans* genome.
2. Download the BAM format sequencing data from the NCBI Sequence Read Archive (accession numbers: SRR6739866–SRR6739996). In order to align the sequenced reads to the latest version of reference genome, the BAM format file should be converted to a FASTQ format file with bam2fastq. If fastq files have been downloaded, this step can be skipped.

```
> bam2fastq --aligned input.bam -o output.fq
```

3. Clean the FASTQ reads using Trimmomatic. The expected output files will contain two paired files and two unpaired files with four continuous bases of reads with average quality below 30 removed.

```
> trimmomatic PE SRR6739866_1.fq SRR6739866_2.fq
SRR6739866_1.fastq unpaired_1.fq SRR6739866_2.fastq un-
paired_2.fq SLIDINGWINDOW:4:30
```

4. Quality check (QC) sequencing data using FastQC and ensure the proportion of reads with Q20 is >90% and Q30 is >85%.

```
> fastqc -f fastq -o result/ SRR6739866_1.fastq
SRR6739866_2.fastq
```

3.2.2 Aligning the FASTQ Reads to the C. elegans *Reference Genome*

1. Before aligning reads to the reference genome, an index for the reference genome must be constructed. The index construction is dependent on the alignment tool and each tool will generally have a function to construct a fasta index. In our case, we will use bwa to align the reads and thus we will use bwa to construct the reference genome index.

```
> bwa index c_elegans.PRJNA13758.WS246.genomic.fa
```

2. Align FASTQ data to the reference genome. The "mem" means maximal exact matches algorithm. Add "-M" to mark the shorter split hits, which is important to improve compatibility with Picard. The "@RG" line, acts as a sample tag, is required for the GATC analysis. The output includes SAM format file with the aligned reads mapped to the reference genome.

```
> bwa mem -M -R '@RG\tID:My_ID\tPL:illumina\tPU:
D272YACXX.2\tLB:My_Library\tSM:My_Sample\tCN:BCCAGSC' ~/ref/
c_elegans.PRJNA13758.WS246.genomic.fa SRR6739866_1.fastq
SRR6739866_2.fastq > ~/H_all_out/SRR6739866/SRR6739866.sam
```

3.2.3 Realignment Around Indels

The older GATK will require manual realignment of reads around indels. However, recent releases of GATK workflow will no longer require manual realignment around indels. Please also refer to the latest best practices found on the GATK website (https://gatk.broadinstitute.org/hc/en-us).

1. Convert SAM format to BAM format that allows faster data access and a smaller footprint.

```
> samtools view -bS SRR6739866.sam -o SRR6739866.bam
```

2. Sort the reads in BAM files by alignment coordinate.

```
> java -Xmx4g -jar ~/picard/picard-tools-1.96/SortSam.jar
\
INPUT= SRR6739866.bam \
OUTPUT= SRR6739866.sort.bam \
SORT_ORDER=coordinate
VALIDATION_STRINGENCY=SILENT
```

3. Mark duplications. Mark the duplicate reads to the metrics format file. This step removes reads that may arise from PCR duplications.

```
> java -Xmx4g -jar ~/picard/picard-tools-1.96/MarkDupli-
cates.jar
   MAX_FILE_HANDLES_FOR_READ_ENDS_MAP=8000
   INPUT= SRR6739866.sort.bam
   OUTPUT= SRR6739866.dedup.bam
   METRICS_FILE= SRR6739866.dedup.metrics
```

4. Build index for the BAM file and generate BAI files.

```
> java -jar ~/picard/picard-tools-1.96/BuildBamIndex.jar
INPUT= SRR6739866.dedup.bam
```

5. Create a target intervals file.

```
> java -jar ~/gatk/GenomeAnalysisTK-2.3-0-g9593e74/Geno-
meAnalysisTK.jar
   -T RealignerTargetCreator
   -R ~/ref/c_elegans.PRJNA13758.WS246.genomic.fa
   -I SRR6739866.dedup.bam
   -o target_intervals.list
```

6. Realign the deduplicated reads using the target_intervals.list. Output includes the BAM format file of the realigned reads.

```
> java -jar ~/gatk/GenomeAnalysisTK-2.3-0-g9593e74/Geno-
meAnalysisTK.jar
   -T IndelRealigner
   -R ~/ref/c_elegans.PRJNA13758.WS246.genomic.fa
   -I SRR6739866.dedup.bam
   -targetIntervals target_intervals.list
   -o SRR6739866.realigned_reads.bam
```

7. Build index for the BAM file and generate BAI files.

```
> java -jar ~/picard/picard-tools-1.96/BuildBamIndex.jar
INPUT= SRR6739866.realigned_reads.bam
```

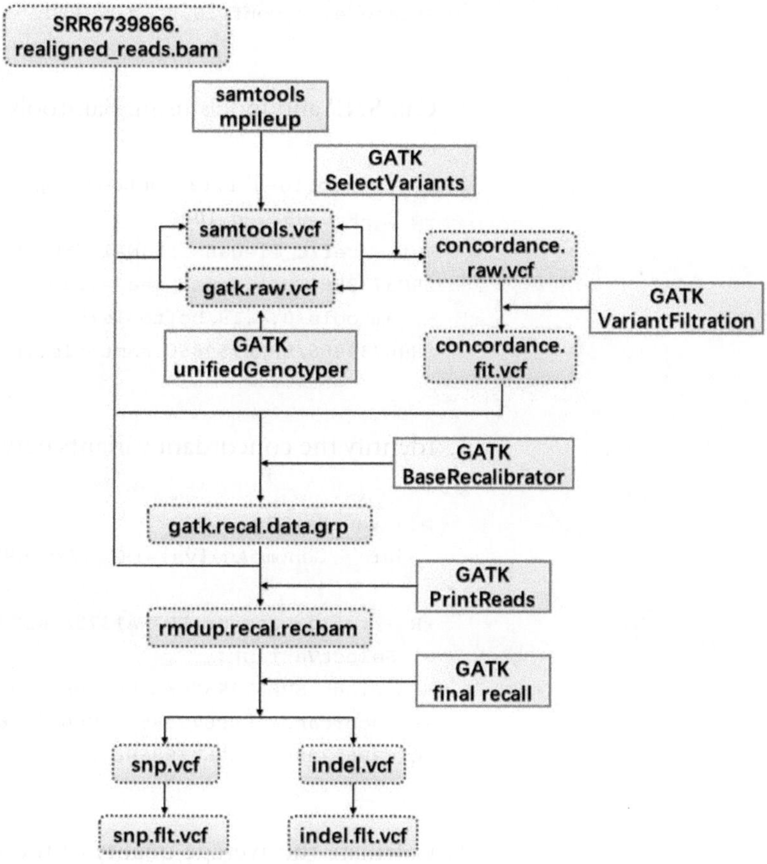

Fig. 2 Overview of variation calling

3.2.4 Variation Calling

The steps below shows an older GATK workflow that was used to generate the data presented in the protocol [31]. GATK is constantly being updated and we do recommend following the latest best practices. The following is our flow chart and command line (Fig. 2).

1. Call SNP and indels using GATK UnifiedGenotyper.

```
> java -Xmx10g
  -jar ~/gatk/GenomeAnalysisTK-2.3-0-g9593e74/GenomeAnaly-
sisTK.jar
  -R ~/ref/c_elegans.PRJNA13758.WS246.genomic.fa
  -T UnifiedGenotyper
  -I SRR6739866.realigned_reads.bam
  -o SRR6739866/SRR6739866.gatk.raw.vcf
  -dcov 200 -A QualByDepth -A FisherStrand
  -A AlleleBalance -A Coverage
  -A MappingQualityZero -A TandemRepeatAnnotator
  -baq CALCULATE_AS_NECESSARY -stand_call_conf 30.0
```

```
-stand_emit_conf 10.0 -glm BOTH -rf BadCigar
```

2. Call SNP and indels using Samtools and bcftools.

```
> ~/samtools-0.1.19/samtools mpileup
-m 2 -F 0.002 -d 1000
-uf ~/ref/c_elegans.PRJNA13758.WS246.genomic.fa
SRR6739866.realigned_reads.bam |
~/samtools-0.1.19/bcftools/bcftools view -cvg ->
SRR6739866/SRR6739866.samtools.vcf
```

3. Identify the concordant variants between GATK and Samtools.

```
> java -Xmx10g
 -jar ~/GenomeAnalysisTK-2.3-0-g9593e74/GenomeAnalysisTK.
jar
-R ~/ref/c_elegans.PRJNA13758.WS246.genomic.fa
-T SelectVariants
--variant SRR6739866/SRR6739866.gatk.raw.vcf
--concordance SRR6739866/SRR6739866.samtools.vcf
-o SRR6739866/SRR6739866.gatk.concordance.raw.vcf
```

4. Calculate the average quality of the variants.

```
> MEANQUAL=\$(awk '{ if (\$1 !~ /#/) { total += \$6; count+
+ } } END { print total/count }' SRR6739866/SRR6739866.gatk.
concordance.raw.vcf)
```

5. Filter for high-quality variants with QD, ReadPosRankSum, and FS scores.

```
> java -Xmx10g
 -jar ~/GenomeAnalysisTK-2.3-0-g9593e74/GenomeAnalysisTK.
jar
-R ~/ref/c_elegans.PRJNA13758.WS246.genomic.fa -T Variant-
Filtration
--filterExpression 'QD < 20.0 || ReadPosRankSum < -8.0 ||
FS > 10.0 || QUAL < \$MEANQUAL'
--filterName LowQualFilter
--missingValuesInExpressionsShouldEvaluateAsFailing
--variant SRR6739866/SRR6739866.gatk.concordance.raw.vcf
--logging_level ERROR
-o SRR6739866/SRR6739866.gatk.concordance.flt.vcf
```

6. Select the high-quality variants to a VCF format file.

```
> grep -v 'Filter' SRR6739866/SRR6739866.gatk.concordance.
flt.vcf > SRR6739866/SRR6739866.concordance.flt.knowsites.vcf
```

7. Identify the base quality scores by BaseRecalibrator model and genertae the GRP format file.

```
> java -Xmx10g
  -jar ~/GenomeAnalysisTK-2.3-0-g9593e74/GenomeAnalysisTK.
jar
  -R ~/ref/c_elegans.PRJNA13758.WS246.genomic.fa
  -T BaseRecalibrator
  -I SRR6739866/SRR6739866.rmdup.realn.bam
   -knownSites SRR6739866/SRR6739866.concordance.flt.know-
sites.vcf
  -o SRR6739866/SRR6739866.rmdup.realn.bam.recal.data.grp
```

8. The BQSR was used to filter the high-quality base. However, the BQSR was replaced by ApplyBQSR in the new version of GATK. The SRR6739866.rmdup.realn.recal.bam file is the final well-aligned file of the protocol. The starting strain and the other strains were processed with the same pipeline.

```
> java -Xmx10g
  -jar ~/GenomeAnalysisTK-2.3-0-g9593e74/GenomeAnalysisTK.
jar
  -R ~/ref/c_elegans.PRJNA13758.WS246.genomic.fa
  -T PrintReads
  -I SRR6739866/SRR6739866.rmdup.realn.bam
 -BQSR SRR6739866/SRR6739866.rmdup.realn.bam.recal.data.grp
  -o SRR6739866/SRR6739866.rmdup.realn.recal.bam
```

9. Based on recalibrated base quality scores, call indels using GATK.

```
  > java -Xms5g -Xmx5g -Djava.io.tmpdir=$outDir ~/gatk/
GenomeAnalysisTK-2.3-0-g9593e74/GenomeAnalysisTK.jar \
      -T UnifiedGenotyper \
      -I SRR6739866/SRR6739866.rmdup.realn.recal.bam \
      -R ~/ref/c_elegans.PRJNA13758.WS246.genomic.fa \
      -o SRR6739866.indel.vcf $optTR \
      -dcov 200 \
```

```
-A QualByDepth \
-A FisherStrand \
-A AlleleBalance \
-A Coverage \
-A MappingQualityZero \
-A TandemRepeatAnnotator \
-baq CALCULATE_AS_NECESSARY \
-stand_call_conf 30.0 \
-stand_emit_conf 10.0 \
-glm INDEL \
-et NO_ET \
-K $GATKKey $opt_P
```

10. Filter for high-quality indels based on QD, ReadPosRankSum, and FS scores.

```
>java -Xms5g -Xmx5g -Djava.io.tmpdir=$outDir ~/gatk/
GenomeAnalysisTK-2.3-0-g9593e74/GenomeAnalysisTK.jar \
-T VariantFiltration \
-R ~/ref/c_elegans.PRJNA13758.WS246.genomic.fa \
-o SRR6739866.indel.flt.vcf \
-V SRR6739866.indel.vcf \
--filterExpression "QD < 2.0 || ReadPosRankSum < -20.0
|| FS > 200.0" \
--filterName "StandardFilter" \
-et NO_ET \
-K $GATKKey
```

11. Calling SNPs with UnifiedGenotyper.

```
> java -Xms5g -Xmx5g -Djava.io.tmpdir=$outDir ~/gatk/
GenomeAnalysisTK-2.3-0-g9593e74/GenomeAnalysisTK.jar \
-T UnifiedGenotyper \
-I SRR6739866/SRR6739866.rmdup.realn.recal.bam \
-R ~/ref/c_elegans.PRJNA13758.WS246.genomic.fa \
-o SRR6739866.snp.vcf $optTR \
-dcov 200 \
-A QualByDepth \
-A FisherStrand \
-A AlleleBalance \
-A Coverage \
-A MappingQualityZero \
-A TandemRepeatAnnotator \
-baq CALCULATE_AS_NECESSARY \
-stand_call_conf 30.0 \
```

```
-stand_emit_conf 10.0 \
-glm SNP \
-et NO_ET \
-K $GATKKey $opt_P
```

12. Filter for high-quality SNPs using QD, MQ, FS, Haplotype-Score, MRankSum, and ReadPosRanksum scores.

```
> java -Xms5g -Xmx5g -Djava.io.tmpdir=$outDir ~/gatk/
GenomeAnalysisTK-2.3-0-g9593e74/GenomeAnalysisTK.jar \
  -T VariantFiltration \
  -R ~/ref/c_elegans.PRJNA13758.WS246.genomic.fa \
  -o SRR6739866.snp.flt.vcf \
  -V SRR6739866.snp.vcf \
  --filterExpression "QD < 2.0 || MQ < 40.0 || FS > 60.0
|| HaplotypeScore > 13.0 || MQRankSum < -12.5 || ReadPosRank-
Sum < -8.0" \
  --filterName "StandardFilter" \
  -et NO_ET \
  -K $GATKKey
```

3.2.5 Identify High Confidence Variations

Aside from SNP quality filtering, we employed a series of filters to remove lower confidence SNPs further.

1. We examine mutations ratios in *dpy-5* and *unc-13* alleles and should follow the expected pattern such that *dpy-5 (e61)* should exhibit ~66% variant ratio due to the rescuing allele on sDp2 and *unc-13* (e450) should exhibit a homozygous variant. Strains with unexpected *dpy-5* and *unc-13* variant ratios are not analyzed further.
2. We expect the lethal mutation to exhibit a ~66% variant ratio, and thus we only examined SNV sites that show the variant ratio between 40% and 90%.
3. We filtered away all variants identified in KR235, which indicates these variants were parental (*see* **Note 2**).
4. We further filtered SNVs that with than eight coverage and were not supported by reads from both directions. IGV was used to check the mutation ratio and coverage of variations (*see* Fig. 3).
5. Select SNVs that fall within the genetically map zones.

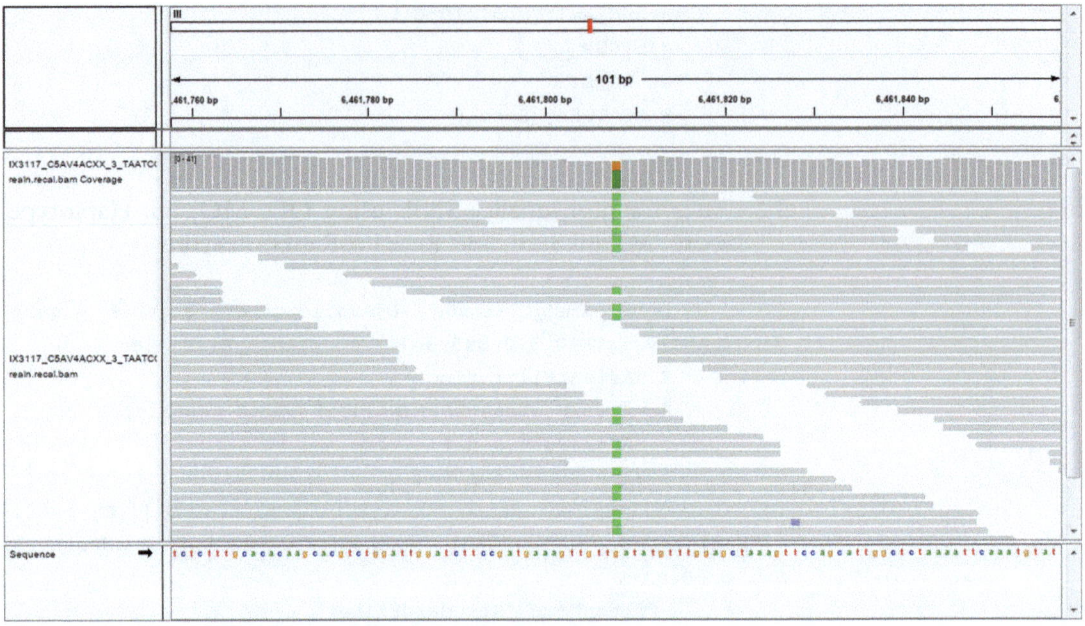

Fig. 3 Using IGV for visualization of Whole-genome sequencing data. BAM files and Fasta files (for WormBase) were uploaded to generate tracts, in which the mutation locus becomes a green square inside the gray arrow

Removed the mutations of each strain that are not in its corresponding genetic map area (left of Chromosome I) of the lethal allele [45, 46]. To allow for potential errors in genetic mapping data, we look for variants that fall within two genetic map zones (*see* **Note 3**).

6. Detrimental mutations, such as splicing or nonsense mutations, from MMP were also removed [16] (*see* **Note 4**).

3.2.6 Identify the High-Confidence Candidate Essential Genes

To identify high-confidence essential gene candidates, we first examined the effects of the variants (*see* Fig. 2). Variants that cause detrimental mutations (such as nonsense, splice-site, and start-codon mutations) will have higher confidence. Variations that fall in intergenic, introns, or those that causing synonymous mutations will be considered last. We annotated the variants using Annovar.

1. Download and install the software from http://www.openbioinformatics.org/annovar/.
2. Annotate variations with location information. The output file includes the genomic location, mutant base, and mutation frequency of variants.

```
> ~/Variation_Annotation/convert2annovar.pl -format
vcf4old 4090SRR6739866.varScan.snp.vcf > variant.avinput
```

3. Annotate variations with mutation types. The output files include the genomic location, mutation type, and genomic changes.

```
> ~/Variation_Annotation/annotate_variation.pl -buildver
WormBase_c_elegans.PRJNA13758.WS246.annotations -geneanno
variant.avinput /data/elegans_lethal/anno/SRR6739866
```

4. To provide secondary support for each candidate gene, we mined WormBase to see if the candidate gene shows lethal phenotype in RNAi experiments or have other characterized alleles that show lethal phenotype.

3.2.7 Confirmation Candidate Essential Gene by PCR of Second Alleles

C. elegans has an extensive genetic library that has been tested by complementation tests. Alleles that complement each other are thought to be on different genes and alleles that fail to complement are thought to be on the same gene. Many essential genes have more than one alleles available. To experimentally confirm our candidate genes, we would amplify the candidate gene region in a second allelic strain and perform Sanger sequencing to identify mutations.

1. We designed primers with staggered orientation using Primer3 [37, 38] such that the primers can cover all exons.
2. PCR amplicons were purified using MinElute PCR Purification Kit (Qiagen) followed by Sanger sequencing.
3. We analyzed the sequencing results in both directions with SeqMan (DNASTAR, https://www.dnastar.com/) (*see* Fig. 4).
4. To confirm a mutation, we would expect to see a clear double peak or a frameshift mutation.

3.3 Confirmation by the Complementation Testing

In the case where a candidate gene is a previously known essential gene, we could use complementation test as described in [22]. Specifically, we took our lethal-carrying strain for complementation testing with another, a strain with known essential gene.

1. Mate 1 hermaphrodites *(let-x dpy-5 unc-13/let-x dpy-5 unc-13; sDp2)* per plate to wild-type males.
2. Crosse F1 males *(let-x dpy-5 unc-13/ + + +)* to 1 hermaphrodite per plate of the known essential gene (*let-y dpy-5 unc-13/let-y dpy-5 unc-13; sDp2*).
3. Screen for the phenotype of the Dpy Unc progeny. If the Dpy Unc animals are viable, this indicates the two *let* alleles complement each other and thus are not on the same gene (i.e., the two strains provide a rescuing alleles to each other). If the Dpy

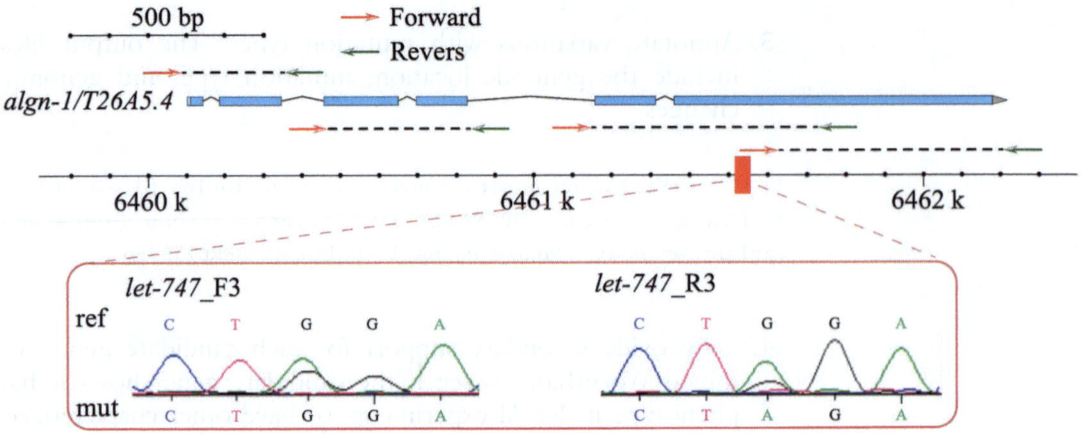

Fig. 4 Using SeqMan (DNASTAR) for visualization of mutation locus [42]. The sequencing map shows two peaks at the third base site

Unc animals show lethal phenotype, then the two alleles fail to complement and indicate the two alleles are on the same gene.

4 Notes

1. We do not detail how the lethal strains were generated in this protocol. The schematic for how lethal strains were generated can be found in [7].
2. The difference between the starting strains and the reference genome was not subjected to mutagenesis.
3. After a lot of three-factor mapping and complementation test of duplications and deficiencies, researchers mapped gene into 60 zones in the left half part of Chromosome I [22], 14 zones in the middle part of Chromosome III [21], and 22 zones in the left half part of Chromosome V [47].
4. In the MMP project, researchers identified 800,000 different mutations in 2007 strains [16]. The genetic mutations identified by MMP are principally not corresponding to essential genes for they cannot maintain homozygous lethal mutations, such as nonsense mutations. We downloaded the variations of each sequenced mutant from the MMP (http://genome.sfu.ca/mmp/).

Acknowledgments

We thank the book editor for his encouragement and advice. This work was supported by the Guangdong Postdoctoral Scientific

Research Foundation of China (Grant No. O0390302). Shicheng Yu and Chaoran Zheng contributed equally to this work.

References

1. Edgley ML, Baillie DL, Riddle DL et al (2006) Genetic balancers. WormBook:1–32
2. Johnsen RC, Baillie DL (1997) Mutation. In: Riddle DL, Blumenthal T, Meyer BJ, Priess JR (eds) C. elegans II. Cold Spring Harbor Laboratory Press, Cold Spring Harbor, NY
3. Ramani AK, Chuluunbaatar T, Verster AJ et al (2012) The majority of animal genes are required for wild-type fitness. Cell 148:792–802
4. Kamath RS, Fraser AG, Dong Y et al (2003) Systematic functional analysis of the Caenorhabditis elegans genome using RNAi. Nature 421:231–237
5. Yuan J, Shaham S, Ledoux S et al (1993) The C. elegans cell death gene ced-3 encodes a protein similar to mammalian interleukin-1 beta-converting enzyme. Cell 75:641–652
6. Shaye DD, Greenwald I (2011) OrthoList: a compendium of C. elegans genes with human orthologs. PLoS One 6:e20085
7. Chu JS, Chua SY, Wong K et al (2014) High-throughput capturing and characterization of mutations in essential genes of Caenorhabditis elegans. BMC Genomics 15:361
8. Han M, Sternberg PW (1990) let-60, a gene that specifies cell fates during C. elegans vulval induction, encodes a ras protein. Cell 63:921–931
9. Grishok A, Pasquinelli AE, Conte D et al (2001) Genes and mechanisms related to RNA interference regulate expression of the small temporal RNAs that control C. elegans developmental timing. Cell 106:23–34
10. Hill DA, Ivanovich J, Priest JR et al (2009) DICER1 mutations in familial pleuropulmonary blastoma. Science 325:965
11. Itaya M (1995) An estimation of minimal genome size required for life. FEBS Lett 362:257–260
12. Haffter P, Granato M, Brand M et al (1996) The identification of genes with unique and essential functions in the development of the zebrafish, Danio rerio. Development 123:1–36
13. Salzberg A, Prokopenko SN, He Y et al (1997) P-element insertion alleles of essential genes on the third chromosome of Drosophila melanogaster: mutations affecting embryonic PNS development. Genetics 147:1723–1741
14. Hutchison CA, Peterson SN, Gill SR et al (1999) Global transposon mutagenesis and a minimal Mycoplasma genome. Science 286:2165–2169
15. Flibotte S, Edgley ML, Chaudhry I et al (2010) Whole-genome profiling of mutagenesis in Caenorhabditis elegans. Genetics 185:431–441
16. Thompson O, Edgley M, Strasbourger P et al (2013) The million mutation project: a new approach to genetics in Caenorhabditis elegans. Genome Res 23:1749–1762
17. Yook K, Harris TW, Bieri T et al (2012) WormBase 2012: more genomes, more data, new website. Nucleic Acids Res 40:D735–D741
18. Kamath RS, Martinez-Campos M, Zipperlen P et al (2001) Effectiveness of specific RNA-mediated interference through ingested double-stranded RNA in Caenorhabditis elegans. Genome Biol 2:RESEARCH0002
19. Rose AM, Baillie DL, Curran J (1984) Meiotic pairing behavior of two free duplications of linkage group I in Caenorhabditis elegans. Mol Gen Genet 195:52–56
20. Adames KA, Gawne J, Wicky C et al (1998) Mapping a telomere using the translocation eT1(III;V) in Caenorhabditis elegans. Genetics 150:1059–1066
21. Stewart HI, O'neil NJ, Janke DL et al (1998) Lethal mutations defining 112 complementation groups in a 4.5 Mb sequenced region of Caenorhabditis elegans chromosome III. Mol Gen Genet 260:280–288
22. Johnsen RC, Jones SJ, Rose AM (2000) Mutational accessibility of essential genes on chromosome I (left) in Caenorhabditis elegans. Mol Gen Genet 263:239–252
23. Edgley ML, Riddle DL (2001) LG II balancer chromosomes in Caenorhabditis elegans: mT1 (II;III) and the mIn1 set of dominantly and recessively marked inversions. Mol Gen Genomics 266:385–395
24. Grove C, Cain S, Chen WJ et al (2018) Using WormBase: a genome biology resource for Caenorhabditis elegans and related nematodes. Methods Mol Biol 1757:399–470
25. Chen N, Harris TW, Antoshechkin I et al (2005) WormBase: a comprehensive data resource for Caenorhabditis biology and genomics. Nucleic Acids Res 33:D383–D389

26. Yu S, Zheng C, Zhou F et al (2018) Genomic identification and functional analysis of essential genes in Caenorhabditis elegans. BMC Genomics 19:871
27. Altschul SF, Madden TL, Schaffer AA et al (1997) Gapped BLAST and PSI-BLAST: a new generation of protein database search programs. Nucleic Acids Res 25:3389–3402
28. Li H, Durbin R (2009) Fast and accurate short read alignment with Burrows–Wheeler transform. Bioinformatics 25:1754–1760
29. Li H, Handsaker B, Wysoker A et al (2009) The sequence alignment/map format and SAMtools. Bioinformatics 25:2078–2079
30. Institute B (2019) Picard toolkit. In: Broad Institute. http://broadinstitute.github.io/picard/
31. Mckenna A, Hanna M, Banks E et al (2010) The genome analysis toolkit: a MapReduce framework for analyzing next-generation DNA sequencing data. Genome Res 20:1297–1303
32. Koboldt DC, Zhang Q, Larson DE et al (2012) VarScan 2: somatic mutation and copy number alteration discovery in cancer by exome sequencing. Genome Res 22:568–576
33. Robinson JT, Thorvaldsdottir H, Winckler W et al (2011) Integrative genomics viewer. Nat Biotechnol 29:24–26
34. Thorvaldsdottir H, Robinson JT, Mesirov JP (2013) Integrative Genomics Viewer (IGV): high-performance genomics data visualization and exploration. Brief Bioinform 14:178–192
35. Wang K, Li M, Hakonarson H (2010) ANNOVAR: functional annotation of genetic variants from high-throughput sequencing data. Nucleic Acids Res 38:e164
36. Mi H, Muruganujan A, Casagrande JT et al (2013) Large-scale gene function analysis with the PANTHER classification system. Nat Protoc 8:1551–1566
37. Koressaar T, Remm M (2007) Enhancements and modifications of primer design program Primer3. Bioinformatics 23:1289–1291
38. Untergasser A, Cutcutache I, Koressaar T et al (2012) Primer3—new capabilities and interfaces. Nucleic Acids Res 40:e115
39. Andrews S, Gilley J, Coleman MP (2010) Difference Tracker: ImageJ plugins for fully automated analysis of multiple axonal transport parameters. J Neurosci Methods 193:281–287
40. Bolger AM, Lohse M, Usadel B (2014) Trimmomatic: a flexible trimmer for Illumina sequence data. Bioinformatics 30:2114–2120
41. Discovery H (2010) Bam2fastq. In: HudsonAlpha Discovery. https://gslweb.discoveryls.com/information/software/bam2fastq
42. Qin Z, Johnsen R, Yu S et al (2018) Genomic identification and functional characterization of essential genes in Caenorhabditis elegans. G3 (Bethesda) 8:981–997
43. Brenner S (1974) The genetics of Caenorhabditis elegans. Genetics 77:71–94
44. Chu JS, Johnsen RC, Chua SY et al (2012) Allelic ratios and the mutational landscape reveal biologically significant heterozygous SNVs. Genetics 190:1225–1233
45. Rosenbluth RE, Rogalski TM, Johnsen RC et al (1988) Genomic organization in Caenorhabditis elegans: deficiency mapping on linkage group V(left). Genet Res 52:105–118
46. Johnsen RC, Baillie DL (1991) Genetic analysis of a major segment [LGV(left)] of the genome of Caenorhabditis elegans. Genetics 129:735–752
47. Rosenbluth RE, Baillie DL (1981) The genetic analysis of a reciprocal translocation, eT1(III; V), in Caenorhabditis elegans. Genetics 99:415–428

Chapter 20

A Screening Method to Identify Essential Yeast Genes for Responses Towards Spermine

Mustapha Aouida and Dindial Ramotar

Abstract

We exploited the yeast *DAmP* mutant collection to identify essential genes that play a role in polyamine resistance. Herein, we described in details the methodology to obtain these genes. This approach is applicable for screening many nontoxic and toxic drugs.

Key words Essential gene, Polyamine resistance, Yeast, Spermine

1 Introduction

The budding yeast *Saccharomyces cerevisiae* has been used to delineate many molecular pathways that are conserved and involved in cancer development [1–3]. To unravel these complex pathways, many investigators took advantage of the various *S. cerevisiae* mutant collections to perform genome-wide screens to identify gene deletions that resulted in specific phenotypes [1–3]. Our group has capitalized on this approach for drug discovery, mainly to understand the physiological processes responsible for resistance to anticancer drugs, as well as other compounds including polyamines [1, 4, 5].

Polyamines such as spermine are involved in regulating many biological functions including the stabilization of DNA and mRNA [6, 7]. High concentration of spermine is toxic to cells and the toxicity could be due to the degradation of spermine by polyamine oxidases, which produce reactive aldehydes including acrolein that can damage cellular macromolecules [8]. One of the factors involved in the homeostasis of polyamines in vivo is the regulation of influx by uptake transporter at the level of plasma membrane [6].

We have conducted a genome-wide screen with the polyamine analogue (bleomycin-A5) and identified Agp2, previously reported to be a L-carnitine transporter with a new function in the uptake of

Ren Zhang (ed.), *Essential Genes and Genomes: Methods and Protocols*, Methods in Molecular Biology, vol. 2377,
https://doi.org/10.1007/978-1-0716-1720-5_20,

polyamines into yeast cells [4, 9]. Several observations indicated that Agp2 might not directly transport polyamines and instead functions as a nontransporting transceptor, i.e., a transporter which has no ability to transport ligands, but acts as a sensor to regulate the expression of substrate-specific permeases as in the case of Ssy1 [9, 10]. Ssy1 senses amino acid availability by direct interaction with extracellular amino acids and triggers the expression of several downstream target genes, that encode amino acid permeases such as *AGP1*, *BAP2*, and *BAP3* [10, 11].

We investigate whether Agp2 possesses regulatory function since two uptake transporters, Dur3 and Sam3, were reported to be major high-affinity permeases for polyamine uptake in yeast [12]. We found that the *agp2Δ* single mutant and the *dur3Δ sam3Δ* double mutant displayed similar phenotypes towards polyamines, in contrast to the *dur3Δ* or *sam3Δ* single mutant [9]. Further analysis revealed that cells lacking a functional *AGP2* gene down-regulated nearly 172 genes as compared to the parent [9]. Amongst the down-regulated genes, many encode membrane transporters and these include *DUR3*, *SAM3*, *HNM1*, *TPO2*, and *HXT3* [9]. The data prompted us to test whether Agp2 conveys a signal to a transcription factor. Since no such factor was identified from the haploid mutant nonessential gene collection [1], we check a collection of essential genes that when down-regulated would confer resistance to polyamines [13].

In this book chapter, we described an alternative approach to screen the yeast DAmP essential collections in anticipation of finding a transcription factor that responds to spermine. Figure 1 provides an overview of the assay procedure in the context of typical drug screening. Using this screening method, we identified nearly 115 essential genes that confer hyper-resistance to spermine [13]. Some of these genes encode transcription factors, and others play a role in nucleic acid metabolisms.

2 Materials

2.1 Strains

1. Use the yeast wild type strain BY4741 as a control.
2. Use the Decreased Abundance by mRNA Perturbation (*DAmP*) Yeast Library (Dharmacon, Horizon Discovery Group Company, USA) [14].

   The Decreased Abundance by mRNA Perturbation (*DAmP*) Yeast Library is a collection of 842 *Saccharomyces cerevisiae* haploid strains containing compromised alleles of essential genes. This collection also contains the controls that serve as wild-type. The DAmP collection library was created by disruption of the 3′ UTR (untranslated region) with an

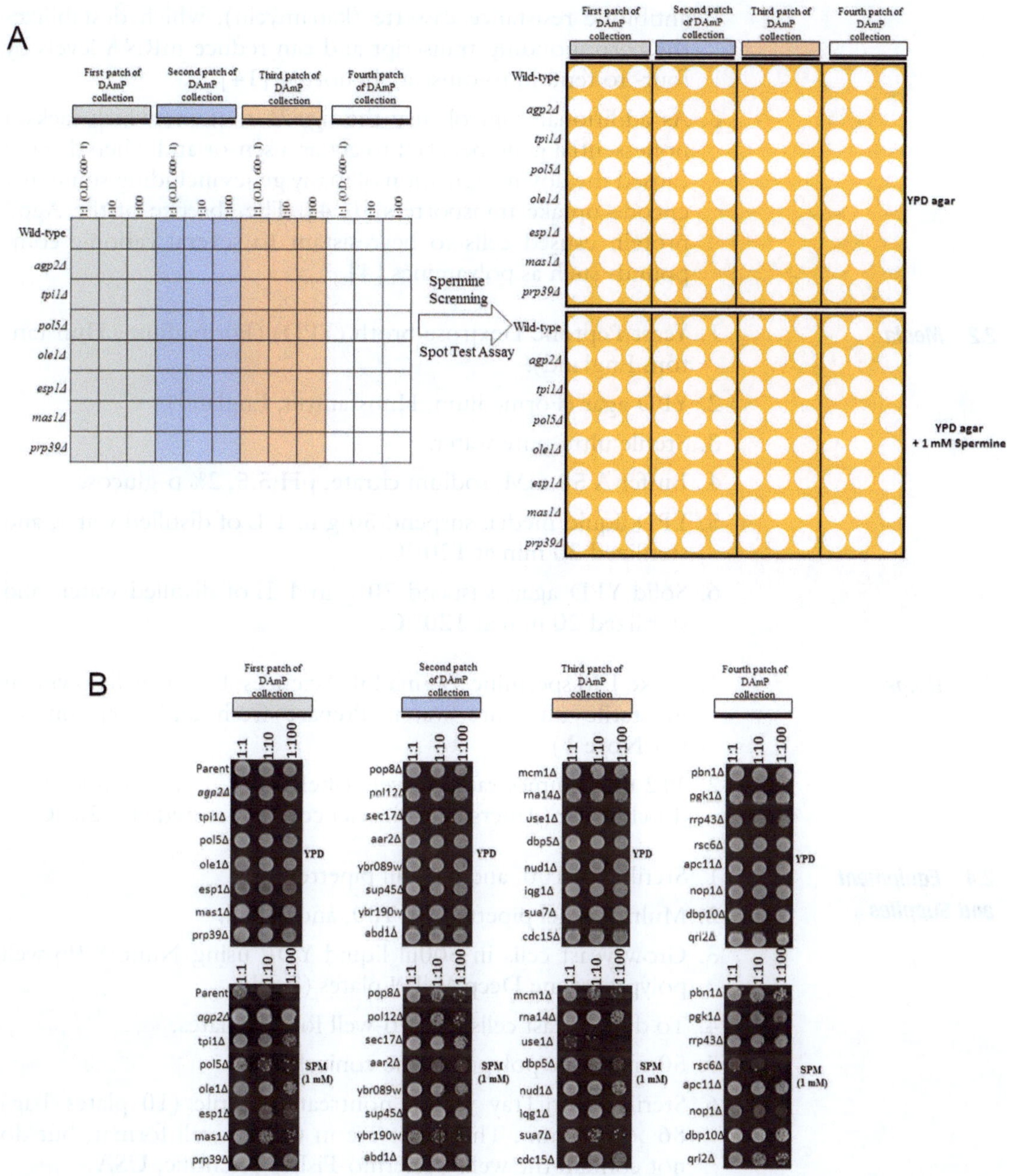

Fig. 1 Overview of the screening method used during this study. (**a**) A simple outline of the steps to screen the essential mutant collection (DAmP) towards spermine. (**b**) Spot test analysis of some of *DAmP* yeast strains displaying resistance to spermine. Cells were grown overnight in liquid YPD, serially diluted, and spotted onto YPD solid media with increasing concentrations of spermine. The plates were photographed after 2 days of incubation at 30 °C

antibiotic resistance cassette (kanamycin), which destabilizes the corresponding transcript and can reduce mRNA levels by four- to tenfold to cause a phenotype [14].

3. As additional control, use the *agp2Δ* mutant, which lacks a nonessential gene believed to act as a sensor and when deleted caused the down-regulation of many genes including some that encode uptake transporters [1, 4]. The absence of the Agp2 protein caused cells to be resistant to several cationic compounds such as polyamines [4].

2.2 Media

1. Yeast Peptone Dextrose broth (YPD) (Formedium, Hunstanton, England).
2. YPD agar (Formedium, Hunstanton, England).
3. Sterile ultra-pure water.
4. Buffer A 50 mM, sodium citrate, pH 5.5, 2% D-glucose.
5. YPD liquid media, suspend 50 g in 1 L of distilled water, and sterilized 20 min at 120 °C.
6. Solid YPD agar, suspend 70 g in 1 L of distilled water, and sterilized 20 min at 120 °C.

2.3 Drugs

1. Make 1 M spermine (Sigma Life Sciences, USA) by dissolving it in sterile ultrapure water. Prepare fresh and keep on ice (*see* **Note 1**).
2. 112 mCi/mmol radioactively labeled [^{14}C]-Spermidine trihydrochloride (Amersham Biosciences) and stored at −20 °C.

2.4 Equipment and Supplies

1. Sterile 10, 100, and 1000μl pipette tips.
2. Multichannel pipettes 10, 100, and 300μl.
3. Grow yeast cells in 500μl liquid YDP using Nunc™ 96-well polypropylene DeepWell™ plates (2 ml).
4. To dilute yeast cells, use 96-well format plates.
5. 50-ml sterile polypropylene conical tube.
6. Sterile OmniTray w/lid, nontreated sterile (10 plates/bag) 86 × 128 mm. The plates are in the 96-well format, but do not contain the wells (Thermo-Fisher Scientific, USA).
7. Orbital shaker incubator.
8. Incubator.
9. Vortex.
10. Spectrophotometer.
11. Alpha major gel documentation system.

3 Methods

3.1 Inoculation, Adjustments of Cell Density, and Dilution of the Cells for Spot Test

1. Using a multichannel pipette, inoculate 10μl of the thawed yeast *DAmP* strain collection stored at −80 °C in 96-well plate into the Nunc™ 96-well polypropylene DeepWell™ plates containing 500μl of YPD liquid. Incubate the inoculums at 30 °C overnight shaking at 200 rpm in the orbital incubator shaker.
2. Next day, measure the optical density of the overnight culture. Take 100μl of each strain and add it to 900μl of YPD and monitor the OD at 600 nm using the spectrophotometer. Normalize the OD 600 to 1.0 for each strain and place in column 1 (e.g., well A1) (*see* **Notes 2** and **3**). Dilute all strains in column 1 by 1:10 and place in column 2. From column 2 dilute another 1:10 and place in column 3 using the same 96-well plate (*see* Fig. 1a).

3.2 Preparation of Spermine-Containing Plates

1. Melt YPD agar using microwave, or use directly after autoclaving, and keep at 55 °C for at least 1 h.
2. Prepare a fresh aliquot of 1 M spermine in sterile distill water.
3. Make 1.0 mM spermine containing plates by adding 40μl of 1 M spermine to 40 ml of the melted YPD agar (kept at 55 °C, *see* **Note 4**) in a 50-ml capped falcon tube, gently invert a few times to ensure proper mixing of the spermine with the liquid YPD agar before pouring the entire content directly into the OmniTray. Prepare control plates in the same manner without spermine. Leave the YPD agar plates to solidify, then place the plates in an incubator with the lids removed, and allow them to dry at 37 °C for 1 h. Drying is to allow evaporation of droplets of water on the surface of the YPD agar to prevent spreading of the cells during the spotting assay (*see* **Note 5**).

3.3 Spot Test Assay

The spot test assay can be done as previously described [15].

1. Briefly, from the dilution plates prepared as above (Subheading 3.1, **step 2**), use a multichannel pipette to spot 4μl of each diluted strain onto plates without and with 1 mM spermine (Fig. 1a, b) (*see* **Note 6**).
2. Allow the wet spots to air dry (approximately 15 min).
3. Transfer the plates to a 30 °C incubator to allow the strains to grow for 48 h.
4. Photograph the plates after 48 h of growth, using a gel-dock with a black background (alpha major gel documentation system).

3.4 Uptake Assay

1. Prior to spermidine uptake assay, grow wild type, *agp2Δ*, and the selected essential mutant strains in YPD to midlogarithmic phase OD (600 nm) = 0.6.
2. Centrifuge yeast cells 1 min at 6000 rpm (g-force 3381).

3. Wash the cells three times in 5 ml of uptake buffer A.
4. Resuspend the cells in 100μl of the same buffer at 2×10^7 cells/ml.
5. Add 10μM of [^{14}C] spermidine at the indicated specific activity (as above) to initiate the uptake assay, and then incubate the cells in a water bath at 30 °C with shaking (*see* **Note 7**).
6. Stop the reaction at predetermine intervals (min) by adding 1 ml of ice-cold uptake buffer.
7. Wash the cells three times with 5 ml of uptake buffer to remove the free-labeled spermidine and then resuspend the cells in 100μl of this buffer.
8. Add 5 ml of scintillation mixture (Amersham Biosciences) to each sample and determine the retained radioactivity by liquid scintillation spectrometry.

3.5 Data Analyses

1. Compile all the data in a single computer file. All the strains should grow to same density on the YPD agar plates without spermine unless they are slow growers (Fig. 1b) [13].
2. Strains showing diminished growth as compared to the parent on YPD agar plates with 1 mM spermine are considered sensitive.
3. Strains that showed comparable growth as the *agp2Δ* mutant on YPD agar plates with 1 mM spermine are considered resistant.
4. Categorize the spermine resistant strains by the gene function and place them in Table 1 of ref. 13. These genes are considered to encode essential proteins that when defective cause the cells to be resistant towards spermine.
5. Select few spermine-resistant strains to test for the uptake of labeled spermidine [13]. This step is necessary to demonstrate that the resistant phenotype is related to a defect in spermine uptake.

4 Notes

1. Use spermine that is freshly prepared, as it is unstable. Previously, we made stocks of 1 M spermine and stored at −20 °C, but discovered that the potency of spermine significantly decreases the following day.
2. Adjust cell OD to 1.0 to ensure that slow-growing cells are spotted to nearly the same density as the controls.
3. For large number of strains, you can use a plate reader to determine the OD.

4. Add spermine to melted YPD agar that is kept at 55 °C. Adding spermine to melted agar at higher temperature can lead to its decomposition.
5. Plates must be thoroughly dried, as droplets of water can cause spotted cells to diffuse and run into each other leading to erroneous data.
6. Spot cells only once.
7. Use label spermidine instead of spermine as the latter is unstable.

Acknowledgments

This work was supported by the College of Health and Life Sciences at Hamad Bin Khalifa University to M.A. and D.R.

References

1. Aouida M, Page N, Leduc A, Peter M, Ramotar D (2004) A genome-wide screen in *Saccharomyces cerevisiae* reveals altered transport as a mechanism of resistance to the anticancer drug bleomycin. Cancer Res 64:1102–1109
2. Parsons AB, Brost RL, Ding H, Li Z, Zhang C, Sheikh B, Brown GW, Kane PM, Hughes TR, Boone C (2004) Integration of chemical-genetic and genetic interaction data links bio active compounds to cellular target pathways. Nat Biotechnol 22:62–69
3. Wu HI, Brown JA, Dorie MJ, Lazzeroni L, Brown JM (2004) Genome-wide identification of genes conferring resistance to the anticancer agents cisplatin, oxaliplatin, and mitomycin C. Cancer Res 64:3940–3948
4. Aouida M, Leduc A, Poulin R, Ramotar D (2005) AGP2 encodes the major permease for high-affinity polyamine import in *Saccharomyces cerevisiae.* J Biol Chem 280:24267–24276
5. Aouida M, Poulin R, Ramotar D (2010) The human carnitine transporter SLC22A16 mediates high affinity uptake of the anticancer polyamine analogue bleomycin-A5. J Biol Chem 285:6275–6284
6. Gerner EW, Meyskens FL Jr (2004) Polyamines and cancer: old molecules, new understanding. Nat Rev Cancer 4:781–792
7. Marton LJ, Pegg AE (1995) Polyamines as targets for therapeutic intervention. Annu Rev Pharmacol Toxicol 35:55–91
8. Pegg AE (2013) Toxicity of polyamines and their metabolic products. Chem Res Toxicol 26:1782–1800
9. Aouida M, Rubio-Texeira M, Thevelein JM, Poulin R, Ramotar D (2013) Agp2, a member of the yeast amino acid permease family, positively regulates polyamine transport at the transcriptional level. PLoS One 8:e65717
10. Gaber RF, Ottow K, Andersen HA, Kielland-Brandt MC (1998) Constitutive and hyperresponsive signaling by mutant forms of *Saccharomyces cerevisiae* amino acid sensor Ssy1. Eukaryot Cell 2:922–929
11. Didion T, Regenberg B, Jorgensen MU, Kielland-Brandt MC, Andersen HA (1998) The permease homologue Ssy1p controls the expression of amino acid and peptide transporter genes in *Saccharomyces cerevisiae.* Mol Microbiol 27:643–650
12. Uemura T, Kashiwagi K, Igarashi K (2007) Polyamine uptake by DUR3 and SAM3 in *Saccharomyces cerevisiae.* J Biol Chem 282:7733–7741
13. Aouida M, Ramotar D (2018) Identification of essential yeast genes involved in polyamine resistance. Gene 677:361–369
14. Breslow DK, Cameron DM, Collins SR, Schuldiner M, Stewart-Ornstein J, Newman HW, Braun S, Madhani HD, Krogan NJ, Weissman JS (2008) A comprehensive strategy enabling high-resolution functional analysis of the yeast genome. Nat Methods 5:711–718
15. Leduc A, He CH, Ramotar D (2003) Disruption of the *Saccharomyces cerevisiae* cell-wall pathway gene SLG1 causes hypersensitivity to the antitumor drug bleomycin. Mol Gen Genomics 269:78–89

Chapter 21

An Effective Preprocessing Method for High-Quality Pan-Genome Analysis of *Bacillus subtilis* and *Escherichia coli*

Hao Wu, Zhi-Kai Yang, Tong Yang, Dan Wang, Hao Luo, and Feng Gao

Abstract

Bacillus subtilis and *Escherichia coli*, as widely used microbial species, are of great significance in studying microbial community relationships, adaptive evolution in various niches, engineering cell factories that produce specific products, and designing genome reduction. The pan-genome analysis is an effective method for studying the characteristics and functions of genes among and within species. Many research directions and conclusions usually depend on accurate gene identification and reliable pan-genome results. However, there currently lack enough studies showing how to achieve high-quality pan-genome results between or within certain species. This chapter will take *Bacillus subtilis* as an example to introduce a stepwise manner for improving the quality of the pan-genome by gradually removing confounding strains step-by-step, and ultimately obtaining a reliable high-quality pan-genome landscape of *Bacillus subtilis*, which could be used as a quality control protocol in pan-genome analysis pipeline. Finally, we suggest further improving the pan-genome analysis results of *Escherichia coli* to prove the feasibility and credibility of the quality control protocol for obtaining high-quality pan-genome landscape.

Key words Pan-genome analysis, Phylogenetic relationship, Artificial genome, Essential gene, Quality control protocol

1 Introduction

"Pan-genome" refers to the global gene repertoire of a species, which ideally covers all the genes from all known and unknown strains of this species [1]. Horizontal gene transfer (HGT) has led to varying degrees of diversity in the gene content among different strains from the same species. Fortunately, the pan-genome analysis provides an excellent way to explore not only the evolution of species but also the changes in gene content [2]. Pan-genome studies define a set of "core genes" shared by all strains, "accessory genes" shared by two or more strains, and "unique genes" that are specific to a certain strain [3, 4]. Pan-genome analyses have been applied widely from bacteria to plant and animal [1]. *Bacillus*

Ren Zhang (ed.), *Essential Genes and Genomes: Methods and Protocols*, Methods in Molecular Biology, vol. 2377,
https://doi.org/10.1007/978-1-0716-1720-5_21,

subtilis and *Escherichia coli* are currently widely used microbial species, and have important applications and research in many fields [5–8]. Bacterial pan-genomic research has strong practical applications and can be used for the selection of antibacterial targets and candidate vaccines [9]. Because the protein products of essential genes constitute the main target of antibiotic drugs, essential genes are not only a key genetic component indispensable for survival, but also have great significance for the development of antibiotics [10]. At the same time, accurate core gene identification and analysis can guide the selection of candidate vaccines. These candidate vaccines often have low sequence mutations in core gene sets, which increase the chances of antigen recognition postimmunization [11]. Pan-genome studies of fungal pathogens can be used to track relationships between isolates of different virulence and identify new genes involved in infection and host response [1]. The pan-genome analysis of plants can be used to identify disease resistance genes lost during domestication and improvement [12]. The sequence map of the human pan-genome built by Li et al. identified ~5 Mb of novel DNA sequences that were either individual- or population-specific. And this group estimated that ~19–40 Mb of novel sequence information was contained in a complete human pan-genome, which was not identified in the extant reference genome [13]. Moreover, as Sherman et al. mentioned, vast amounts of variation in the population play an important role in understanding and analyzing the genetic instructions that make us human by pan-genome analysis [14].

The impact of the selection of various strains on the results of pan-genome analyses is poorly understood. Furthermore, there is currently no definitive protocol for obtaining a high-quality pan--genome landscape for a given species or isolate sets. We found that the involvement of incorrectly classified strains, phylogenetically distinct strains, chimeric strains, engineered genome-reduced strains, strains with abnormally numerous unique genes or large proportion of pseudogenes, and multiple nearly clonal laboratory strains can significantly impact the performance and quality of pan-genome analyses of *B. subtilis.* Furthermore, redundant genomes could impact the judgment of true pan-genome state of a species (open *vs* close) [15]. This chapter will introduce a quality control protocol for obtaining representatively high-quality pan--genome landscape of certain species by removing the confounding strains in pan-genome analysis data sets.

2 Materials

The complete and chromosome-level genome sequence of *B. subtilis* strains collected from different geographic locations and isolation sources were retrieved from NCBI for pan-genome analyses in Ref. [16] as well as this chapter. Since the data and

Table 1
The programs and parameters used in this chapter

Software	Version	Setting
Prokka	1.13.3	–
InterProScan	5.31–70.0	-t -iprlookup -goterms –f html –f tsv –dp –pa –dra -b
OrthoFinder	2.2.7	-t 48–a 48
Roary	3.12.0	-p 48
PGAP	1.2.1	--cluster –method GF - pangenome
BLASTN/BLASTX	2.7.1+	-outfmt 6 -evalue 1e-5 -num_threads 48
[a]AnnotationForge	1.24.0	–
[a]clusterProfiler	3.10.1	–
[a]DOSE	3.8.2	–
[a]pheatmap	1.0.12	–
Eggnog-mapper	1.0.3	-m diamond –d bact --cpu 48--translate
Diamond	0.8.22	–
PYANI	0.2.9	-m ANIb –g –v
RNAmmer	1.2	-S bac -m ssu
[a]ggtree	1.14.6	–
[a]GOSemSim	2.8.2	–

[a]R package used in this chapter

processes for pan-genome analysis are based on known sequencing data and tools, there is no experimental process content, so the material section basically introduces the programs used (Table 1) and analysis processes applied in this chapter (Fig. 1).

3 Methods

3.1 Sequencing Genome Data

1. Collecting and organizing strains data for pan-genome analysis.
2. Based on the information of strain sequences, first removing the chimeric strain from the data set, if a known chimeric strain exists.
3. Calculating and removing strains with N base content greater than 1% within genomes. The N base content of the sequence can be implemented or calculated by online tools or custom scripts.
4. The above statistics of chimeric strains and strains with large N base content are confusing strains, which need to be removed

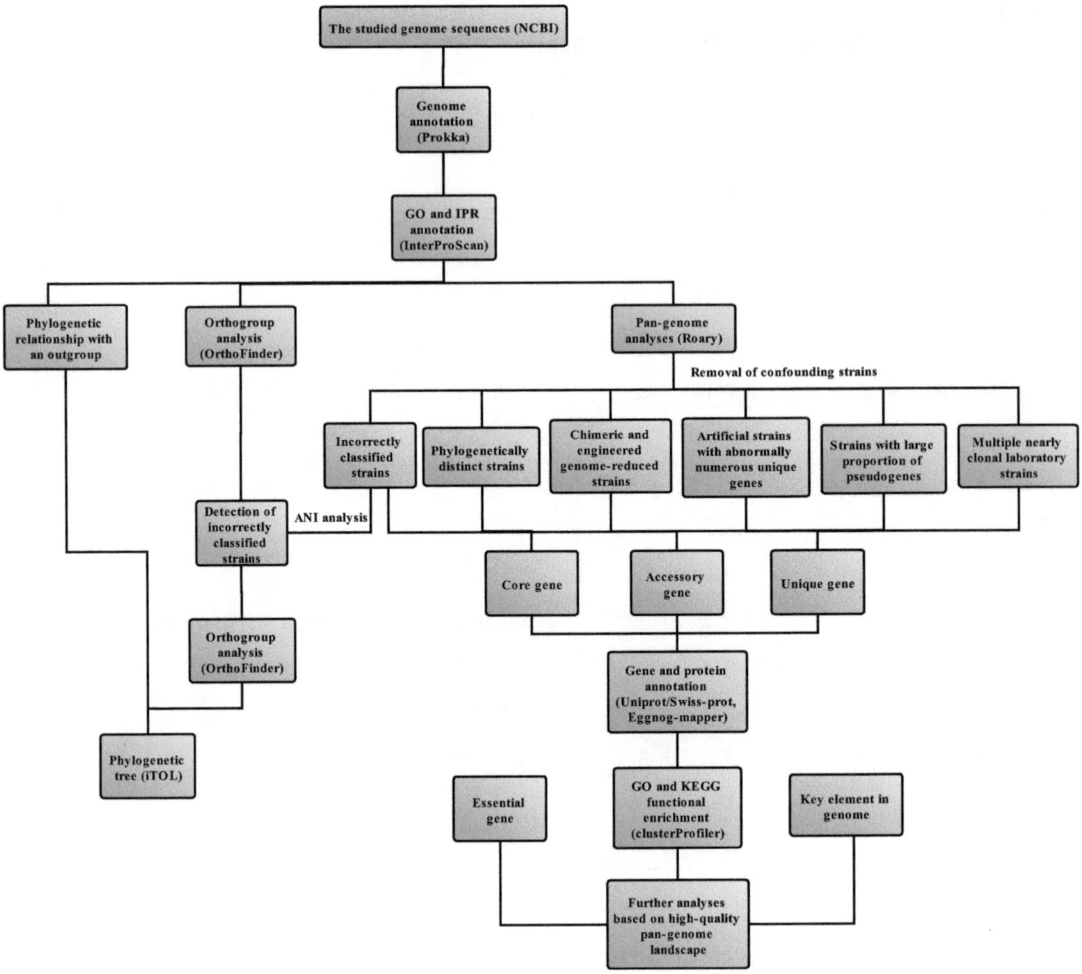

Fig. 1 The whole analysis process and bioinformatics tools used in this chapter

in pan-genome analysis, as they could cause biases in pan-genome analysis results (*see* **Note 1**).

3.2 Genome Annotation

To carry out the subsequent genome research and ensure the consistency and reliability of the annotation and gene prediction of the genomes. The program Prokaryotic Genome Annotation System (Prokka) [17] and InterProScan [18] are used to predict and annotate genes of each strain. Compiling a Shell script to batch processes of each strain, which is shown as follows (*see* **Note 2**):

```
for i in ls '/path/to/input_sequence'
do
mkdir -p /path/to/output/${i%.*}
 prokka /path/to/input_sequence/${i} --force --outdir /path/
to/output/${i%.*} --prefix ${i%.*} --addgenes --gcode 11 --
```

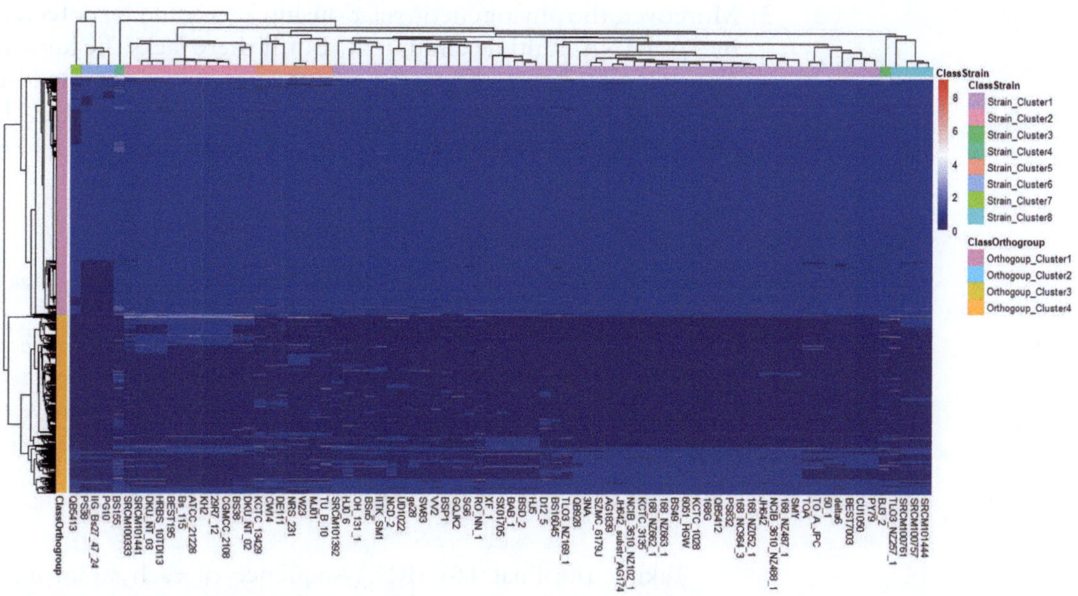

Fig. 2 The heatmap and dendrogram show the classified strain clusters and orthogroup clusters based on the OrthoFinder analysis results. The color tends to the gene count in the exited strains, where the greater the number, the closer it is to red

```
cpus 48
 /path/to/interproscan.sh -i /path/to/input_sequence/${i} -d /
path/to/output -t -iprlookup -goterms -f html -f tsv -dp -pa -
dra -b -cpu 48
 done
```

3.3 Phylogenetic Relationship Analysis

1. The gene protein sequence file with suffix of faa of each strain obtained by prokka is organized into the same folder "input_-faa". The program OrthoFinder [19] then is used to analyze the homology relationship and phylogenetic relationship among strains. Running the command line as follows:

```
$nohup orthifinder -f /path/to/input_faa/ -t 48 -a 48 &
```

   The homology relationship obtained from OthoFinder could show that each orthogroup exists in each strain or not and the gene count in the exited strains, which will contribute to detect phylogenetic relationship within strains based on the single copy orthogroups (Fig. 2).

2. Dragging the tree file obtained by orthofinder analysis into iTol (https://itol.embl.de/) to obtain the visualized phylogenetic relationship among strains.

3. Moreover, the phylogenetic relationship also could be detected by 16s rRNA multiple alignment. But if there lacks the known 16s rRNA sequence of studied strains, it could be obtained by the program RNAmmer [20]. The corresponding SHELL script is shown below:

```
for filename in 'ls /path/to/input_sequence'
do
mkdir -p /path/to/output/${filename%.*} && cd /path/to/output
${filename%.*}
/path/to/rnammer -S bac -m ssu -f /path/to/input_sequence
${filename%.*}.fasta -h ${filename%.*}.hmmreport -xml ${file-
name%.*}.xml -gff ${filename%.*}.gff2 /path/to/input_se-
quence/${filename}
done
```

Taking the final 16s rRNA sequence of each strain into MEGA5 [21] for plotting the evolutionary relationship tree as below (Fig. 3).

4. Making a statistics of abnormally phylogenetically distinct strains, and strains that are far from the other strains in evolutionary distance in phylogenetic trees.

3.4 Average Nucleotide Identity Analysis

1. Due to assigned names and classifications of several species in RefSeq have not been verified. Therefore, it is necessary to confirm the truly homology relationship and classification of strains within the genus through Average nucleotide identity (ANI) analysis. The PYANI package (v0.2.9; https://github.com/widdowquinn/pyani) [22] is carried out to calculate the ANI value between two strains. The fna files of each strain annotated by prokka are organized into the same folder "input_fna". ANI based on the BLAST algorithm (ANIb) was performed using average_nucleotide_identity.py script of the PYANI module. If MUMmer comparison is used, set the "-m" parameter to "ANIm". Running the command line as follows:

```
$nohup average_nucleotide_identity.py -i path/to/input/in-
put_fna -o /path/to/output -m ANIb -g -v &
```

2. For the obtained "ANIb_percentage_identity.tab" file, which contains the ANI values among strains, we bring this file into R, and use the pheatmap package to draw the ANI heatmap for showing the distribution of ANI values among strains.

```
library(pheatmap)
```

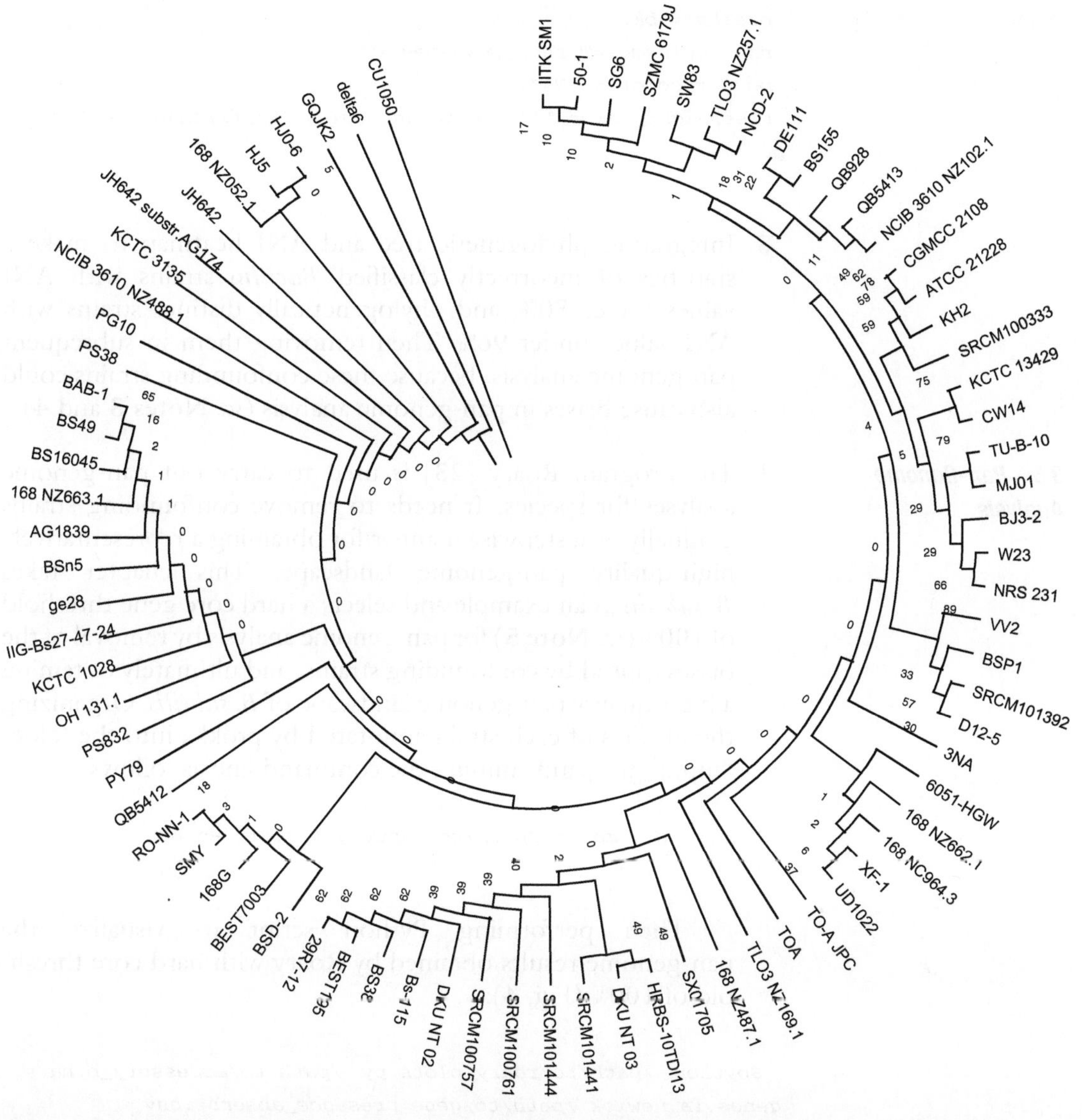

Fig. 3 The phylogenetic relationship tree of 79 *B. subtilis* strains obtained by MEGA5 with 16s rRNA sequences of each strain, showing the rich evolutionary relationship within studied strains

```
df<-read.table("ANIb_percentage_identity.tab", row.names =
1, header = T, as.is = T)
colnames(df) <- rownames(df)
dim(df)
min(df)
range(df)
bk <- c(seq(0.75,1, by=0.01))
pheatmap(df, cellwidth=8, cellheight=8, fontsize=6.5, angle_-
col=315, color=colorRampPalette(colors=c("#1B5A9C","#4F9B-
C7","#59A1CA","#D2E6F0","#DAE9F2","#FDDDCB","#FCD7C2","#D-
F765E","#DD7059","#B82531"))(25),
```

```
breaks = bk,
show_colnames=T,show_rownames =T,
border_color = "NA",
treeheight_row = "0",treeheight_col = "0",filename="ANI_heat.
pdf")
```

3. Integrating phylogenetic tree and ANI heatmap to make a statistics of incorrectly classified *Bacillus* strains with ANI values under 80%, and phylogenetically distinct strains with ANI values under 96%. Then removing them in subsequent pan-genome analysis, because these confounding strains could also cause biases in pan-genome analysis (*see* **Notes 3** and **4**).

3.5 Pan-Genome Analysis

1. The program Roary [23] is used to carry out pan-genome analyses for species. It needs to remove confounding strains gradually by a stepwise manner for obtaining a representatively high-quality pan-genome landscape. This chapter takes *B. subtilis* as an example and selects a hard core gene threshold of 100% (*see* **Note 5**) for pan-genome analysis by removal of the biases caused by confounding strains, and ultimately obtaining a high-quality pan-genome landscape of *B. subtilis.* Organizing the gff files of each strain annotated by prokka into the folder "input_gff", and running the command line as follows:

```
$Roary -e --mafft /path/to/input_gff/*.gff -p 48
```

Then performing Python script to visualize the pan-genome results obtained by Roary with hard core threshold of 100% (Fig. 4):

```
$python /path/to/roary_plots.py /path/to/accessory_binary_-
genes.fa.newick /path/to/gene_presence_absence.csv
```

2. In order to explore the effects of these confounding strains on the *B. subtilis* pan-genome analyses results, we need to perform pan-genome analyses on the data set including these strains and the data set excluding them. In the results obtained by Roary, the file "summary_statistics.txt" is a summary statistic result for the classified genes of pan-genome with hard core threshold. Importantly, the file "gene_presence_absence.csv" sorts out the relations of the annotated genes in the corresponding homologous genes of each strain. The partial python script of integrating the files "pan_genome_reference.fa" and "gene_-presence_absence.csv" to extract the nucleotide and protein sequences of core, accessory, and unique genes are as follows:

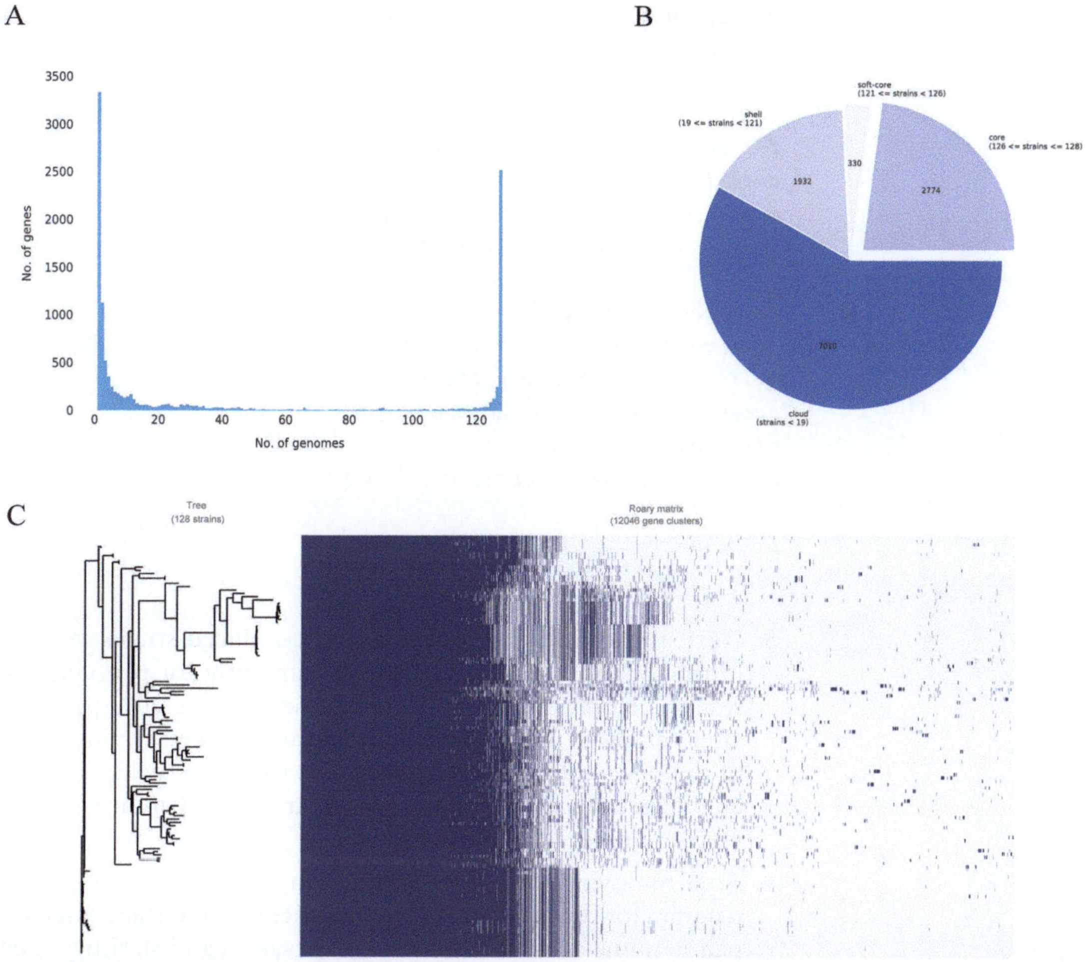

Fig. 4 The visualization plots of pan-genome results obtained by Roary with hard core threshold of 100%. (**a**) The pan-genome frequency plot shows the frequency of genes versus the number of genomes. (**b**) The pie chart shows the breakdown of genes and the number of isolate they are present in. (**c**) The pan-genome analysis tree compared to a matrix with the presence and absence of core and accessory genes

```
import pandas as pd
from Bio import SeqIO

file1 = "gene_presence_absence.csv"
file2 = "pan_genome_reference.fa"

with open(file1, 'r') as f1:
data1 = pd.read_csv(f1)
core_list = list(data1['Gene'])[0:2822] #order of core gene
in file "gene_presence_absence.csv"
 accessory_list = list(data1['Gene'])[2822:6689] #order of
accessory gene in file "gene_presence_absence.csv"
 specific_list = list(data1['Gene'])[6689:8812] #order of
unique gene in file "gene_presence_absence.csv"
```

```
core_result = []
accessory_result = []
specific_result = []

for record in SeqIO.parse(file2, 'fasta'):
description = record.description
num = description.split(' ')[0]
gene = description.split(' ')[-1]

if gene in core_list:
core_result.append(num + '\n')
elif gene in accessory_list:
accessory_result.append(num + '\n')
elif gene in specific_list:
specific_result.append(num + '\n')
else:
print(num)
```

3. Next, the known engineered genome-reduced strains existing in the data set are analyzed in the same way for pan-genome results before and after including them. Generally speaking, after removing the engineered genome-reduced strains, most of the newly obtained core genes are the genes deleted in the engineered genome-reduced strains from the parent strain to varying degrees (*see* **Note 6**).
4. After removing the above confounding strains, we plot a petal diagram for the number of unique genes of each strain based on pan-genome analysis results for observation and statistics. The R script for plot petal diagram is shown as follows partially:

```
par(bty = "n", ann = F, xaxt = "n", yaxt = "n",mar = c
(0,2,0,2))
plot(c(0,10.5),c(0,10.5),type="n")
N <- length(strain)
deg <- 360/n
res <- lapply(1:n, function(t){
draw.ellipse(x = 5 + 2*cos((start + deg * (t - 1)) * pi / 180),
y = 5 + 2*sin((start + deg * (t - 1)) * pi / 180),
col = ellipse_col[t],
border = border_col,
a = a, b = b, angle = deg * (t - 1),lwd = 3)
text(x = 5 + 3.9 * cos((start + deg * (t - 1)) * pi / 180),
y = 5 + 3.9 * sin((start + deg * (t - 1)) * pi / 180),
value[t], cex = 1.1
)

if (deg * (t - 1) < 180 && deg * (t - 1) > 0) {
```

```
text(x = 5 + 4.1 * cos((start + deg * (t - 1)) * pi / 180),
y = 5 + 4.1 * sin((start + deg * (t - 1)) * pi / 180),
strain[t],
srt = deg * (t - 1) - start,
adj = 1,
cex = 1.4,
font = 1
)

} else {
text(x = 5 + 4.1 * cos((start + deg * (t - 1)) * pi / 180),
y = 5 + 4.1 * sin((start + deg * (t - 1)) * pi / 180),
strain[t],
srt = deg * (t - 1) + start,
adj = 0,
cex = 1.4,
font = 1
)
}
```

5. Combined with petal diagram, the artificial strains with abnormally numerous unique genes were counted, and these strains are removed from the pan-genome analysis. Moreover, through the characteristics of GC content distribution to study the transformation process of specific regions of these artificial strains and their impact on pan-genome analysis [24] (*see* **Note 7**).
6. Next, counting the number and proportion of pseudogenes contained in each strain based on the information in GenBank. Strains with pseudogene ratio above 10% should be removed in the pan-genome analyses (*see* **Note 8**).
7. Finally, given the information of known multiple nearly clonal laboratory strains, these strains need to be removed from the data set for pan-genome analysis as well (*see* **Note 9**).

3.6 Gene Functional Enrichment Analysis

1. After removing incorrectly classified strains, phylogenetically distinct strains, chimeric strains, engineered genome-reduced strains, strains with abnormally numerous unique genes or large proportion of pseudogenes, and multiple nearly clonal laboratory strains. We think that the consequent pan-genome analysis results after removal of these confounding strains are representatively high-quality pan-genome landscape of *B. subtilis*.
2. Obtaining the sequences and names of the final core genes, accessory genes, and unique genes as the code provided in Subheading 3.5, **step 2**, and then carrying out the COG

enrichment analysis. Firstly, downloading the latest COG database protein sequences (ftp://ftp.ncbi.nih.gov/pub/COG/COG2014/data), next making COG alignment database and aligning each gene of core, accessory, and unique gene sets to the established COG database as follows:

```
$makeblastdb -in /path/to/prot2003-2014.fa -dbtype prot -par-
se_seqids -out /path/to/COG_db
$blastx -query /path/to/core_gene.fasta -out /path/to/out -db
/path/to/COG_db -outfmt 6 -evalue 1e-10 -num_threads 48
```

Integrating the alignment results of the COG database with the files "cog2003–2014.csv", "fun2003–2014.tab", and "cognames2003–2014.tab" to count the COG category classification of each gene through a python script as follows partially:

```
import sys
import re

cat_class={}
with open("cognames2003-2014.tab") as f:
for line in f:
line = line.split('\t')
cat_class[line[0]]=line[1]
cat_class["None"]="None"

num_class={}
with open('cog2003-2014.csv') as f:
for line in f:
item = line.split(",")
num = item[0]
cog = item[6]
num_class[num] = cog

class_des = {}
with open('fun2003-2014.tab') as f:
for line in f:
line = line.strip('\n')
if re.search(r'\[',line):
data = re.split(r'(\s\[)|(\]\s)',line)
class_des[data[3]] = data[6]
else:
continue

for i in cat_class.values():
if i not in class_des.keys():
```

```
class_des[i]="None"
```

3. Obtaining the sequences and names of the final core, accessory, and unique gene as Subheading 3.6, **step 2** did, then aligning each gene to the DEG database (http://www.essentialgene.org/) [25] as follows:

```
$makeblastdb -in /path/to/degaa-p.dat -dbtype prot -parse_-
seqids -out /path/to/DEG_db
$blastx -query /path/to/core_gene.fasta -out /path/to/out -db
/path/to/DEG_db -outfmt 6 -evalue 1e-10 -num_threads 48
```

   Counting the statistical comparison results that how many core, accessory, and unique genes are matched to the essential gene data of *B. subtilis.* There are 271 essential genes of *B. subtilis* (DEG10010001-DEG10010271, prefixed with DEG1001 in DEG database).

4. Next, carrying out the functional enrichment analysis for three types of genes (core, accessory, and unique) obtained from high-quality pan-genomes analysis results.
   (a) Three types of genes are aligned to the UniProt/Swiss-Prot (reviewed) database (ftp://ftp.uniprot.org/pub/databases/uniprot/current_release/knowledgebase/complete/uniprot_sprot.fasta.gz) to obtain the matched UniProtKB ID of each gene; next, the above utilized UniProtKB ID was aligned against idmapping.tb.gz (ftp://ftp.pir.georgetown.edu/databases/idmapping/idmapping.tb.gz) to obtain the GO ID of each gene. The method of making aligned database and aligning sequences is similar to that provided in Subheading 3.6, **steps 2** and **3**.
   (b) Using the online Kyoto Encyclopedia of Genes and Genomes (KEGG) Automatic Annotation Server (KAAS) (https://www.genome.jp/kegg/kaas/) tool in the bi-directional best hit mode to annotate the KEGG ORTHOLOGY number of each gene and merged them into one file.
   (c) Using the local version eggnog-mapper comprehensively to annotate genes with the aligning program DIAMOND as follows:

```
$.python /path/to/emapper.py -m diamond -i /path/to/core_-
gene.fasta -d bact -o /path/to/out --data_dir /path/to/emap-
per_database
```

(d) Checking whether there is a known OrgDB annotation package corresponding to the research species in Bioconductor (http://bioconductor.org/packages/3.9/data/annotation/). If it exists, you can directly apply it for functional enrichment analysis. Otherwise, it needs to establish a custom OrgDB package for functional enrichment analysis. After the *B. subtilis* OrgDb package org.BSubtilis.eg.db is constructed, the functional enrichment analysis code in R is shown as follows partially:

```
library(tidyverse)
library(stringr)
library(clusterProfiler)
library(DOSE)
library(enrichplot)
library(topGO)
library(org.BSubtilis.eg.db)
library(purrr)
library(AnnotationForge)

core_list <- read_csv("core_gene.txt", col_names = FALSE)
core_list <- as.character(core_list$X1)

core_go <- enrichGO(gene = core_list,
keyType = "GID",
OrgDb = org.BSubtilis.eg.db,
ont = "ALL",
pvalueCutoff = 0.01,
qvalueCutoff = 0.01,
pAdjustMethod = "BH",
readable = TRUE)
core_go_results<-as.data.frame(core_go)
 write.csv(core_go_results,file = "GO_result/core_GO_result.
csv")

dim(core_go)
 barplot(core_go, showCategory=20, x = "GeneRatio",drop =
TRUE)
 dotplot(core_go, showCategory=20)
 p <- dotplot(core_go, split="ONTOLOGY") + facet_grid(ONTOL-
OGY~., scales = "free")
 p + scale_y_discrete(labels = function(y) str_wrap(y, width =
50))

pathway2gene <- AnnotationDbi::select(org.BSubtilis.eg.db,
```

```
keys = keys(org.BSubtilis.eg.db),
columns = c("Pathway","Ko")) %>%
na.omit() %>%
dplyr::select(Pathway,GID)
load("kegg_info.RData")

core_kegg <- enricher(core_list,
TERM2GENE = pathway2gene,
TERM2NAME = pathway2name,
pvalueCutoff = 0.1,
qvalueCutoff = 0.1,
pAdjustMethod = "BH",
minGSSize = 1)
core_kegg_results <- as.data.frame(core_kegg)
dim(core_kegg_results)
 write.csv(core_kegg_results,file = "KEGG_result/core_KEG-
G_result.csv")

kp <- barplot(core_kegg, showCategory = 20,
font.size = 10,x = "GeneRatio")
kp + scale_x_discrete(labels = function(x) str_wrap(x, width
= 50))
```

5. GO and KEGG functional enrichment analyses demonstrated that the three types of gene sets (core, accessory, and unique genes) had distinctly different functions in BP (Biological Process), CC (Cellular Component), and MF (Molecular Function) (Fig. 5) (*see* **Note 10**).

3.7 Pan-Genome State

New genes play an important role in judging the pan-genome state of species (open vs close). We carried out PGAP [28] with the parameter "– cluster -pangenome -method GF" and PanGP [29] with the distance guide (DG) algorithm to detect the pan-genome state of *B. subtilis.*

```
 perl /path/to/PGAP-1.2.1/PGAP.pl -strains name1+name2+...
+nameN(connect strains' name with "+") -input /path/to/input/
strain_data -output /path/to/output --cluster --thread 48 --
evalue 0.00001 -pangenome -method GF
```

Finally taking the file "Orthologs_Cluster.txt" into local version PanGP with DG mode to get the pan-genome profile and new gene curve.

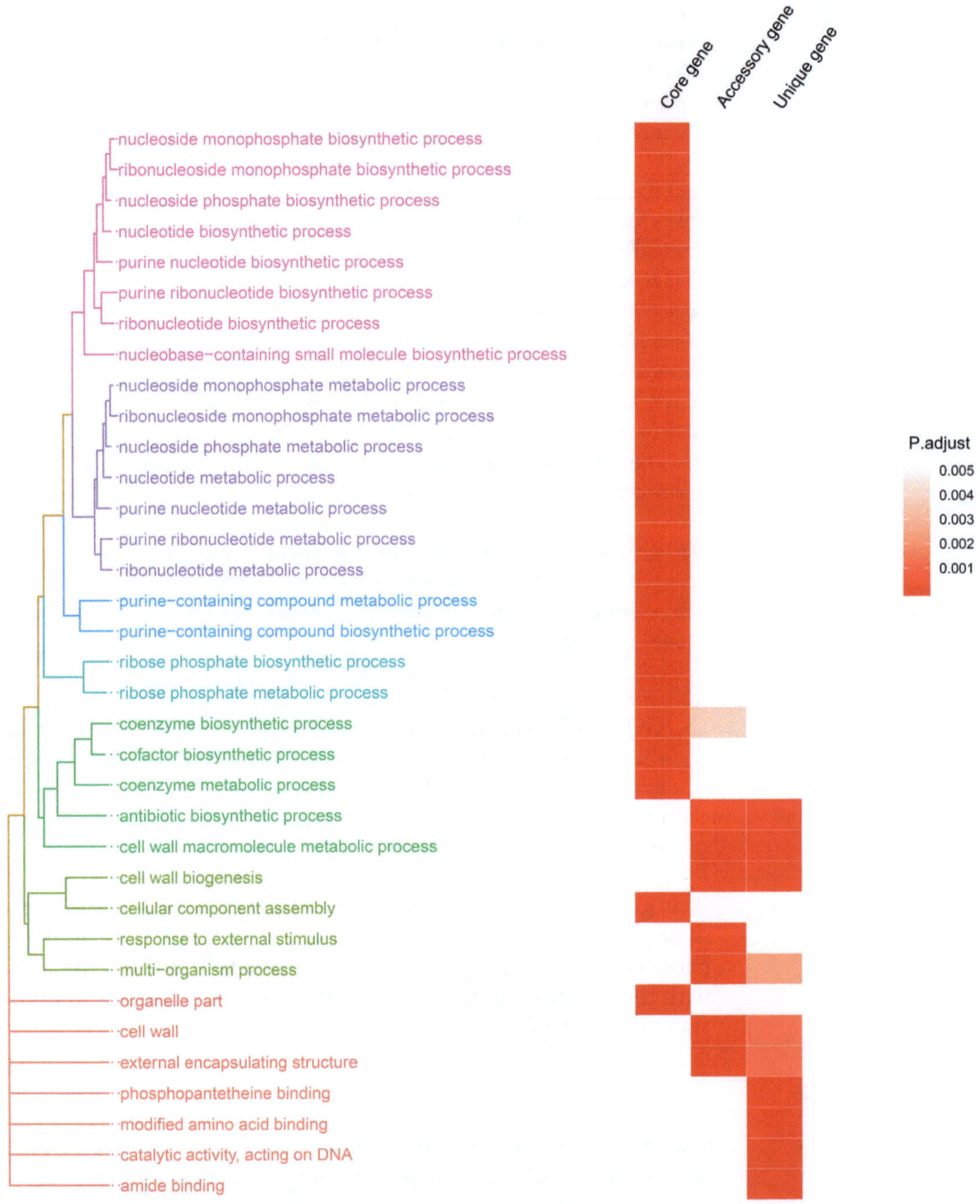

Fig. 5 The left tree of heatmap conveys the relationship of GO terms and grouped by ggtree [26] package with calculating similarity using GOSemSim [27] package. Consequently, we focus on the terms with *q*-value $<10^{-8}$, and limited final list of genes to examine term sets that annotate fewer than 300 genes and more than 100 genes. The color of heatmap corresponds to the *p*.adjust value (*q*value) in GO terms. The lower the value is, the redder is the box Column names correspond to core gene, accessory gene, and unique gene derived from pan-genome analysis results, which could better reflect the performance and quality of landscape of *B. subtilis* pan-genome

3.8 Pan-Genome of *E. coli*

Currently, the pan-genome of *E. coli* has been extensively studied based on different number or sources of strains or by different methods. Remarkably, the study by Yang et al. should be the most representative one in the existing *E. coli* pan-genomic studies, which constructed currently the best quality of *E. coli* pan-genome based on 491 strains with complete genomes [30]. On the basis of this pan-genome, Yang et al. further revealed that accessory genes played an important role in maintaining the harmony between *E. coli* and the human host, also found that core genes provided some important evidences for the reduction of *E. coli* genome. However, as discussed above, we found that some confounding strains were present in sequenced *B. subtilis* strains, which obviously impacted the result of its pan-genome analysis. After excluding these confounding strains, we obtained a high-quality pan-genome of *B. subtilis*. So, we wonder if there are also such confounding strains in 491 *E. coli* strains analyzed by Yang et al. If so, we suggest removing these strains from pan-genome analysis, and the analysis method is similar to that of *B. subtilis*.

4 Notes

1. Among these *B. subtilis* strains, we removed strains, whose N base content was greater than 1% of the genomic size, and we should remove the chimeric genome BEST7613 with a genome size of 7.6 Mb in *B. subtilis* species, which harbors the complete bacterial genomes of Synechocystis PCC6803 and *B. subtilis* strain 168 in one cell. Since most genes from Synechocystis strain PCC6803 would be considered as unique genes from strain BEST7613; consequently, leading to inaccurate results from pan-genome analyses.
2. It should be noted that, in batch processing of strain sequences through shell scripts, it is necessary to avoid parallel processing. Prokka and InterProScan will prompt errors due to insufficient memory during parallel processing.
3. The number of core genes drastically increases following the removal of abnormally and phylogenetically distinct strains. Inclusion of these strains in pan-genome analyses would lead to the identification of few core genes and what we would consider to be a low-quality pan-genome landscape of *B. subtilis*.
4. Phylogenetically distinct strains might cause an incorrect identification of unique genes in pan-genome analyses, and they should be removed to obtain an accurate identification of unique genes among strain sets without biases from distant evolution or artificial defects.

5. The hard core threshold of 100% obtained from Roary pan-genome pipeline was used as the threshold value of core genes. A hard core threshold (99% ≤ strains ≤ 100%) was adopted instead of a soft core threshold (95% ≤ strains ≤ 99%) mainly for the following reasons:
   (a) Several studies rely on high-quality pan-genome results, such as rational genome designs in synthetic biology and metabolic engineering. These studies require a precise identification of core, accessory, and unique genes, i.e., genes present in all strains included.
   (b) In the pan-genome analyses, the data sets that we collected are all strains with complete or chromosome assembly, because they have high assembly quality with credible annotations.
   (c) When the number of studied strains in a pan-genome analysis is less than 100, a hard core threshold of 100% is used to classify core genes with Roary acquiescently; otherwise a hard core threshold of 99% is adopted. Furthermore, as the core threshold is implemented more stringently, the number of core essential genes decreases, while the number of accessory essential genes increases, which means some core essential genes will be identified as accessory essential genes.

   Therefore, in order to maintain the consistency of the core gene thresholds, no matter the number of studied strains is (more or less than 100), we chose 100% as the hard core threshold to identify the core genes that were shared by all strains.
6. Some engineered genome-reduced strains have been produced for *B. subtilis*, including delta6 (Δ6), IIG-Bs27-47-24, PG10, and PS38. The inclusion of these genome-reduced strains could reduce the core gene set of *B. subtilis* and could suggest that the functional capacity of these four artificial strains might be impaired to some degree. Therefore, these engineered genome-reduced strains should be removed in the pan-genome analyses.
7. Strain QB5413 has the most number of unique genes which is more numerous than that in other *B. subtilis* strains in the petal diagram. It is surprising to find a long, straight line in the GC profile of strain QB5413, indicating that the GC content is steady between the genomic positions 3,230,184–4,217,957 bp in the QB5413 genome obtained by GC profile with web server GC-Profile, which might result in the erroneous performance of the pan-genome analysis. Therefore, the artificially engineered strain QB5413 that special strain with abnormally numerous unique genes should be considered carefully in the process of pan-genome analyses.

8. BS155 and D12-5 possess the largest proportion of pseudogenes (37.96% and 11.32%) among the *B. subtilis* strains. Unique genes derived from the pan-genome analysis that included BS155 and D12-5 are incorrectly identified due to a large proportion of pseudogenes. Therefore, strains with a large proportion of pseudogenes should be examined more carefully and excluded in the process of pan-genome analyses for an improved representation of the unique genes and for the improvement of pan-genome landscape quality.
9. When multiple clones of a strain are present in the data set, these strain-specific genes (unique genes) would be shared by two or more strains, so they would be identified as "accessory genes", which will cause bias in the pan-genome analysis results. Therefore, these multiple clonal strains should be removed from pan-genome analysis.
10. The functional enrichment of ultimately high-quality classified core genes, accessory genes, and unique genes would be useful to study the symbiotic regulation of microorganisms and hosts, loss and acquisition of genes in HGT, acquisition and formation of genomic islands, and mechanisms of pan-genome evolution.

Acknowledgments

The authors would like to thank Prof. Chun-Ting Zhang for the invaluable assistance and inspiring discussions. The present work was supported in part by National Key Research and Development Program of China (grant number 2018YFA0903700) and National Natural Science Foundation of China (Grant numbers 21621004 and 31571358).

References

1. Golicz AA, Bayer PE, Bhalla PL et al (2020) Pangenomics comes of age: from bacteria to plant and animal applications. Trends Genet 36:132–145
2. Vogan AA, Higgs PG (2011) The advantages and disadvantages of horizontal gene transfer and the emergence of the first species. Biol Direct 6:1
3. Medini D, Donati C, Tettelin H et al (2005) The microbial pan-genome. Curr Opin Genet Dev 15:589–594
4. Vernikos G, Medini D, Riley DR et al (2015) Ten years of pan-genome analyses. Curr Opin Microbiol 23:148–154
5. Earl AM, Losick R, Kolter R (2008) Ecology and genomics of *Bacillus subtilis*. Trends Microbiol 16:269–275
6. Higgins D, Dworkin J (2012) Recent progress in *Bacillus subtilis* sporulation. FEMS Microbiol Rev 36:131–148
7. Golicz AA, Bayer PE, Bhalla PL et al (2020) Pan-genomics comes of age: from bacteria to plant and animal applications. Trends Genet 36:132–145
8. Jafari A, Aslani MM, Bouzari S (2012) *Escherichia coli*: a brief review of diarrheagenic pathotypes and their role in diarrheal diseases in Iran. Iran J Microbiol 4:102–117

9. Tettelin H, Masignani V, Cieslewicz MJ et al (2005) Genome analysis of multiple pathogenic isolates of *Streptococcus agalactiae*: implications for the microbial "pan-genome". Proc Natl Acad Sci U S A 102:13950–13955
10. Poulsen BE et al (2019) Defining the core essential genome of *Pseudomonas aeruginosa*. Proc Natl Acad Sci U S A 116:10072–10080
11. Davies MR et al (2019) Atlas of group A streptococcal vaccine candidates compiled using large-scale comparative genomics. Nat Genet 51:1035–1043
12. Bayer PE et al (2019) Variation in abundance of predicted resistance genes in the *Brassica oleracea* pan-genome. Plant Biotechnol J 17:789–800
13. Li R, Li Y, Zheng H et al (2010) Building the sequence map of the human pan-genome. Nat Biotechnol 28:57–63
14. Sherman RM, Salzberg SL (2020) Pan-genomics in the human genome era. Nat Rev Genet 21:243–254
15. Chan AP, Sutton G, DePew J et al (2015) A novel method of consensus pan-chromosome assembly and large-scale comparative analysis reveal the highly flexible pan-genome of *Acinetobacter baumannii*. Genome Biol 16:143
16. Wu H, Wang D, Gao F (2021) Toward a high-quality pan-genome landscape of *Bacillus subtilis* by removal of confounding strains. Brief Bioinform 22:1951–1971
17. Seemann T (2014) Prokka: rapid prokaryotic genome annotation. Bioinformatics 30:2068–2069
18. Jones P, Binns D, Chang HY et al (2014) InterProScan 5: genome-scale protein function classification. Bioinformatics 30:1236–1240
19. Emms DM, Kelly S (2015) OrthoFinder: solving fundamental biases in whole genome comparisons dramatically improves orthogroup inference accuracy. Genome Biol 16:157
20. Lagesen K, Hallin P, Rødland EA et al (2007) RNAmmer: consistent and rapid annotation of ribosomal RNA genes. Nucleic Acids Res 35:3100–3108
21. Tamura K, Peterson D, Peterson N et al (2011) MEGA5: molecular evolutionary genetics analysis using maximum likelihood, evolutionary distance, and maximum parsimony methods. Mol Biol Evol 28:2731–2739
22. Pritchard L, Glover RH, Humphris S et al (2015) Genomics and taxonomy in diagnostics for food security: soft-rotting enterobacterial plant pathogens. Anal Methods 8(1):12–24
23. Page AJ, Cummins CA, Hunt M et al (2015) Roary: rapid large-scale prokaryote pan-genome analysis. Bioinformatics 31:3691–3693
24. Gao F, Zhang C-T (2006) GC-Profile: a web-based tool for visualizing and analyzing the variation of GC content in genomic sequences. Nucleic Acids Res 34:W686–W691
25. Luo H, Lin Y, Liu T et al (2021) DEG 15, an update of the database of essential genes that includes built-in analysis tools. Nucleic Acids Res 49:D677–D686
26. Yu G, Smith DK, Zhu H et al (2017) ggtree: an r package for visualization and annotation of phylogenetic trees with their covariates and other associated data. Methods Ecol Evol 8:28–36
27. Yu G, Li F, Qin Y et al (2010) GOSemSim: an R package for measuring semantic similarity among GO terms and gene products. Bioinformatics 26:976–978
28. Zhao Y, Wu J, Yang J et al (2012) PGAP: pan-genomes analysis pipeline. Bioinformatics 28:416–418
29. Zhao Y, Jia X, Yang J et al (2014) PanGP: a tool for quickly analyzing bacterial pan-genome profile. Bioinformatics 30:1297–1299
30. Yang ZK, Luo H, Zhang Y et al (2019) Pan-genomic analysis provides novel insights into the association of *E. coli* with human host and its minimal genome. Bioinformatics 35:1987–1991

Chapter 22

Analysis of Gene Essentiality from TnSeq Data Using Transit

Thomas R. Ioerger

Abstract

TnSeq, or sequencing of transposon insertion libraries, has proven to be a valuable method for probing the functions of genes in a wide range of bacteria. TnSeq has found many applications for studying genes involved in core functions (such as cell division or metabolism), stress response, virulence, etc., as well as to identify potential drug targets. Two of the most commonly used transposons in practice are Himar1, which inserts randomly at TA dinucleotides, and Tn5, which can insert more broadly throughout the genome. These insertions cause putative gene function disruption, and clones with insertions in genes that cannot tolerate disruption (in a given condition) are eliminated from the population. Deep sequencing can be used to efficiently profile the surviving members, with insertions in genes that can be inferred to be non-essential. Data from TnSeq experiments (i.e. transposon insertion counts at specific genomic locations) is inherently noisy, making rigorous statistical analysis (e.g. quantifying significance) challenging. In this chapter, we describe Transit, a Python-based software package for analyzing TnSeq data that combines a variety of data processing tools, quality assessment methods, and analytical algorithms for identifying essential (or conditionally essential) genes.

Key words Essentiality, Transposon mutagenesis, TnSeq, Statistical analysis, Software

1 Introduction

TnSeq, or sequencing of transposon insertion libraries, has proven to be a valuable method for probing the functions of genes in bacteria [1–3]. Libraries are generated by random insertion of a transposon throughout a genome, creating a pool of mutants. DNA from bacteria grown in a particular condition is extracted, the junctions between transposon and adjacent genomic region are amplified by PCR, the samples are sequenced by high-throughput sequencing, and insertions at individual sites are counted. The pattern of insertion counts can be used to infer the difference between essential (ES) and non-essential (NE) regions; regions that tolerate insertions are generally considered non-essential, while regions lacking insertions are taken as evidence of essentiality. In addition to the binary distinction of essential vs non-essential, some genes can exhibit a partial reduction of insertion counts,

Ren Zhang (ed.), *Essential Genes and Genomes: Methods and Protocols*, Methods in Molecular Biology, vol. 2377,
https://doi.org/10.1007/978-1-0716-1720-5_22, © Springer Science+Business Media, LLC, part of Springer Nature 2022

which can be interpreted as a growth defect (GD) caused by disruption of the gene, and indeed quantitative changes in insertion counts in has been shown to reflect changes in fitness *Escherichia coli* [1]. There are several variations of sample preparation protocols currently in use, going by names such as TraDIS [4], HITS [5], and InSeq [6], which differ based on the transposon used, method of DNA fragmentation, and transposon junction amplification. More recently, Barseq [7] has been introduced to exploit bar coding to afford a very high level of multiplexing on next-gen sequencers, allowing the efficient analysis of hundreds of samples in parallel.

TnSeq has found many applications in many different bacteria, including *E. coli*, *Mycobacteria*, *Pseudomonas*, *Vibrio*, *Haemophilus*, *Salmonella*, etc., to study genes involved in metabolism, stress response, virulence, etc. (*see* **Note 1**). TnSeq gives a very different readout on gene function from RNA-seq (transcriptomics), as essentiality of a gene is orthogonal to (i.e. not necessarily correlated with) the level of expression [8]; one can change without the other. Gene essentiality is especially useful for drug discovery efforts, such as interpreting mechanisms of resistance through mutations found in resistant mutants and identifying vulnerable targets and pathways that are critical for survival in vivo (during infection) [9]. Databases of essential genes, such as DEG [10], have been assembled from curated collections of published results of TnSeq analyses, along with inferences of essential genes extended to other species by homology.

Two of the most commonly used transposons in practice are Himar1 [11, 12] and Tn5 [4, 13]. While Himar1, a member of the mariner family, is restricted to inserting at TA dinucleotides [14], Tn5 can insert effectively at any site in the genome, which has consequences for statistical analysis. The magnitude of counts at an individual site is highly dependent on the library, as insertion into the chromosome is a stochastic process, which is a major factor contributing to variability. Although the magnitude of insertion counts can vary significantly between adjacent TA sites in non-essential regions, no strong sequence-dependent bias has yet been found for insertion of various transposons (*see* **Note 2**). Thus, insertion counts are usually treated as a random variable, and analysis methods rely on averaging over multiple TA sites in a gene to make a statistical assessment of essentiality. Even a GC-rich organism like *M. tuberculosis* has 3 or more TA sites for most genes (median is 13 TA sites; only 2.7% of genes have less than 3 TA sites), an analysis of genes with only 1–2 TA sites is typically not reliable (analysis of short genes becomes highly dependent on the degree of saturation), and of course, genes with 0 TA sites are unanalyzable by TnSeq. However, intergenic regions and ncRNAs are often short enough to have only 0–2 TA sites and are difficult to analyze using TnSeq. Many of the analytical methods for Himar1 can be extended to Tn5 data, provided that the saturation is high enough and that assuming the locations and magnitudes of insertions can effectively be treated as random [2]. It is important to note that essential genes, which typically lack insertions throughout the body of the ORF, have often been observed to tolerate insertions at the N- and C-termini, as well as in linkers between domains, and can even contain non-essential domains, which also pose challenges for rigorous identification of essentials (*see* **Note 3**).

Data from TnSeq experiments (i.e. insertion counts) is inherently noisy, making rigorous statistical analysis challenging. Variability can come from a variety of sources, including representation (abundance) in the library, stochastic differences between identically treated samples (plates, cultures, and animals) and the sequencing process. Reducing raw counts to template counts using barcodes can help ameliorate PCR jackpotting [15]. Loss of diversity (especially, in animal infections) and amplification of fitness differences due to differences in growth time (generations) can also lead to significant artifacts such as skewing of read-count distributions [16]. Larger genes with more TA sites ease this because the impact of isolated deviations (e.g. missing or outlier counts) can be mitigated by averaging over multiple observations. Conversely, higher noise and lower saturation can make statistical calls for smaller genes uncertain. Hence, normalization and rigorous statistical analysis are critical to identifying significant genes (e.g. essential or conditionally essential). Also, collection of multiple replicates (2–3 replicates per sample are recommended) is important for increasing statistical certainty through increasing the number of observations per gene. (Biological replicates, i.e. samples from different plates or animals, are more important than technical replicates, i.e. resequencing, for giving a fair picture of stochastic variability of insertion counts, which is critical assessing statistical confidence in observed differences.) Several software packages are available for the analysis of TnSeq data, each based on different theoretical frameworks: ESSENTIALS [17], TnSeq-Explorer [18], TraDIS-Toolkit [19], TnSeqDiff [20], fitness ratios [21], ARTIST [22], and others.

In this chapter, we describe Transit [23], a Python-based software package for analyzing TnSeq data. Transit combines implementations of a range of previously published statistical analysis methods. Transit was originally designed for statistical analysis of Himar1 TnSeq datasets (in which insertions are assumed to be restricted to TA sites), though some of the methods have been adapted for Tn5 data (for which the typical lower saturation causes challenges). Transit has a preprocessor (TPP) for extracting insertion counts from raw sequence (.fastq) files that encodes best practices accumulated from experience in multiple labs over the years. Transit also incorporates tools for quality control (QC analysis) for assessing the quality of datasets. Transit has a graphical user interface (GUI) to make it easy for users to perform most tasks, though this review will focus on running tasks from the command line. The methods are also accessible as a Python library for programmers to call in their own scripts (the source code can be downloaded from GitHub, https://github.com/mad-lab/transit). Further details can be found in the online documentation, https://transit.readthedocs.io/en/latest/.

The analytical tools in Transit are oriented around addressing three types of questions:

1. identifying essential genes in a single (e.g. reference) condition,
2. comparative evaluation of conditional essentiality between two conditions, and
3. analysis of genes showing variability across multiple conditions

Each of these has different use cases (*see* **Note 4**). This chapter describes how to do these analyses using Transit and includes comments about the impact of (and guidelines on) data quality, file formats, and other practical information.

2 Materials

Transit is designed to run on Linux, Macs, and Windows machines. Transit is written in python3 and requires wxPython4, R, and many other packages as dependencies. Transit can be installed in two ways: (1) via "pip3 install," which will download and install the compiled code from the repository on PyPi along with all dependencies (so that it can be run directly from the command line as "transit") or (2) by cloning the source code from GitHub and then manually installing the required packages (see Installation Instructions in the online documentation, https://transit.readthedocs.io/en/latest/). The current version is 3.1.0 as of this writing; for future changes, see the online documentation. Note that BWA must be installed for TPP, and some of the more advanced functions (like ZINB) require R to be installed, along with some specific packages.

In the sections below, we will give examples based on a study of gene requirements for growth of *M. tuberculosis* H37Rv on cholesterol compared to glycerol [24]. In this study, there are 5 TnSeq datasets collection, 2 for glycerol and 3 for cholesterol. For brevity, we will refer to these wig files as G1.wig, G2.wig, C1.wig, C2.wig, and C3.wig.

If installed from GitHub, running Transit would require command sequences like "`python3 $TRANSITDIR/src/transit.py` ...," where $TRANSITDIR is the path where Transit is installed. However, for simplicity, we will simply use the command "`transit`" in the examples below.

Users can get help on most commands (e.g. reminders of arguments and flags) by running Transit without any arguments or with "`--help`."

Most of the analyses in Transit generate tab-separated output files as a convention, which can be opened as spreadsheets in Excel. Lines prefixed with "#" in the output files are comments.

All the input and output files for the examples in this chapter can be accessed online at

http://orca1.tamu.edu/essentiality/transit/examples/index.html.

3 Methods

3.1 Preprocessing (TPP)

Transit has a preprocessing step called TPP (Transit Preprocessor) that maps reads from sequencing a Tn library (files in .fastq or .fasta format) into a genome (.fasta format) and tabulates insertion counts at TA sites. The counts are output in a file format called ".wig" files, which simply have two values on each line—coordinate and insertion counts for each TA site. Wig files have two header lines, the second of which indicates the name of the genome sequence that was used as a reference ("variableStep chrom= H37Rv" indicating H37Rv.fna as the reference sequence, for example). In subsequent analyses, it is critical to use the annotation file corresponding to the genome sequence used in TPP, to ensure consistency of the coordinate system.

TPP uses BWA [25] (which must be installed separately) to map reads into the genome (*see* **Note 5**).

An example of running TPP is as follow:

```
> tpp -bwa /home/bwa/bwa-0.7.12/bwa -ref H37Rv.fna -reads1
G1_R1.fastq
    -reads2 G1_R2.fastq -output G1
```

The inputs are the sequencing data (fastq files for reads 1 and 2) and the reference genome sequence (H37Rv.fna, nucleotide fasta). Multiple intermediate files will be generated for the base filename given by -output. The primary output file would be the wig file, **G1.wig**, which contains the insertion counts at TA sites.

TPP can also map reads to genome sequences containing multiple contigs (or replicons), such as when an organism contains multiple chromosomes and/or plasmids. Separate reference sequences can be provided to TPP as a comma-separated list.

Another important output file from TPP is the .tn_stats file that gets generated. It contains important parameters and diagnostics from the run. As a quick check, it is useful to examine the saturation (fraction of TA sites with insertions, ideally >30%) and NZmean (mean over nonzero sites, ideally >10). The .tn_stats files also report the total reads, the number of reads mapped, and related statistics which can give insight into the degree of attrition and possible reasons for it (for example, whether reads lacked the expected prefix, mapped predominantly to a single site, or matched known sequences related to the transposon vector or primers).

3.1.1 Making Combined Wig Files

When working with a large collection of datasets, it can be cumbersome to have to type multiple wig filenames as inputs to subsequent commands. In such cases, it is convenient to make a "combined_-wig" file, which combines the counts at the same TA sites from multiple wig files on the same line (and appends on the ORF id and gene name from the annotation). To aid in comparing insertion counts between datasets, the counts are normalized using TTR normalization (by default, see below), though other normalization methods (including "nonorm" that preserves raw counts) can be applied using the -n flag. The "#File:" header lines document the datasets that were combined.

```
> transit export combined_wig G1.wig,G2.wig,C1.wig,C2.wig,C3.
wig H37Rv.prot_table glyc_chol_combined_wig.txt    # TTR norm
is implicit

> transit export combined_wig G1.wig,G2.wig,C1.wig,C2.wig,C3.
wig H37Rv.prot_table glyc_chol_combined_wig_raw.txt -n nonorm
```

A recommended practice is to use the combined_wig file for examining and comparing normalized insertion counts in a gene or locus that is indicated to be conditionally essential, to confirm whether the effect is genuine or perhaps influenced by a few outlier insertion counts.

The "**.prot_table**" file contains the coordinates and annotations of genes in the genome (*see* Subheading 3.2.2).

It is often useful to reduce the data in each sample to the mean insertion counts for each gene, which can be done with the "export mean_counts" command:

```
> transit export mean_counts G1.wig,G2.wig,C1.wig,C2.wig,C3.
wig H37Rv.prot_table glyc_chol_gene_means.txt
```

The output file, glyc_chol_gene_means.txt, can be opened as a spreadsheet.

An individual wig file may be normalized itself using the following command:

```
> transit normalize <input.wig> <output.wig> -n [method]
```

where "method" is TTR, betageom, or several others (see documentation). Normalization can also be applied simultaneously to multiple datasets in a combined_wig file by adding the "-c" flag before the filename.

3.1.2 Evaluating Quality of TnSeq Data

To summarize the statistics of multiple samples in a combined_wig file, one may use the "tnseq_stats" command (given either a list of individual wig files or a combined_wig file):

```
> transit tnseq_stats G1.wig G2.wig C1.wig C2.wig C3.wig -o glyc_chol_tnseq_stats.txt
or
> transit tnseq_stats -c glyc_chol_combined_wig_raw.txt -o glyc_chol_tnseq_stats.txt
```

The output file is a tab-separated file that can be opened as a spreadsheet. It contains important information, including the saturation, total counts, mean, NZmean, and max count, for each dataset, along with several statistics on the read-count distribution (skewness, etc.). While the saturation will not change, it is best to run *tnseq_stats* on a combined_wig file to which no normalization has been applied ("-n nonorm"), since otherwise, the mean counts will reflect the results of TTR normalization. If datasets with especially low saturation (e.g. <15%) or low NZmean (e.g. <1) are observed, the researcher might consider excluding them from further analyses.

Users can assess the quality of the TnSeq datasets by generating and examining the **tnseq_stats** table mentioned above. The primary metrics are saturation and mean insertion count (specifically, NZmean). While there are not rigorous criteria for defining "bad" datasets, rules of thumb I use for "good" datasets are density>30% (ideally >50%) and NZmean> 10 (ideally >50). In addition, I look at MaxReadCount and Skewness as indicators. Typically, MaxReadCount will be in the range of a few thousands to tens of thousands. If you see individual sites with counts in the range of 10^5–10^6, it might mean you have some positive selection at a site [e.g. biological (fitness advantage), or an artifact due to things like PCR jackpotting], and this can have the effect of reducing counts and influencing the distribution at all the other sites. If MaxReadCount< 100, that is also probably problematic (either not enough reads or possibly skewing). Also, skewness> 30 often (but not always) signals a problem. The reason why it is not easy to boil all these down to a simple set of criteria is that some of the metrics interact with one another.

A useful tool when evaluating the quality of a collection of TnSeq datasets is to make a correlation plot of the mean insertion counts (averaged at the gene level). While it is difficult to state how much correlation there should be between conditions (or even between replicates of the same condition), the corrplot can often reveal individual samples that stand out as being far less correlated with all the others (which subsequently might be excluded from analyses).

```
> transit corrplot glyc_chol_combined.wig.txt glyc_chol_corr-
plot.png
```

Although saturation and NZmean are the easiest to assess, a more complicated aspect of data quality is the insertion-count distribution (*see* **Note 6**).

3.2 Analyses for Single Conditions

3.2.1 Gumbel Analysis

Analyses of individual conditions might occur when a new strain is being evaluated in a reference condition (rich growth medium, for example), and the goal is to make a preliminary catalog of essential genes. There are two principle methods in Transit: Gumbel analysis, and a Hidden Markov Model (HMM). Gumbel analysis focuses on genes with statistically significant gaps or consecutive sequences of TA sites lacking insertions (empty sites, with counts of 0) [26]. Since most libraries are sub-saturated, empty sites will occur in non-essential regions at random, but long sequences of empty sites are statistically unlikely, and genes containing such gaps are taken as essential. The Gumbel method uses the Extreme Value Distribution to quantify the certainty, and a posterior probability is reported in the output file (with separate thresholds for essential and non-essential genes, and uncertain genes in between). An important advantage of this gap-based approach is that it is tolerant of insertions that sometimes occur at the N- and C-termini of ORFs. A disadvantage is that it works less well with less-saturated libraries (<40%) and can yield many more uncertain calls, especially for shorter genes (with <10 TA sites). Note that the magnitude of insertion counts (and hence normalization) does not matter for the Gumbel analysis.

```
> transit gumbel G1.wig,G2.wig ref.prot_table gumbel_H37Rv_-
glycerol.txt
```

Gumbel can be run on multiple replicates (*see* **Note 7**).

A typical expectation is that around 10–15% of genes will be essential for growth in vitro in most bacterial genomes, as has been observed across many organisms [10], though of course it depends on size of genome and growth condition.

3.2.2 Prot Tables

An important input file to Gumbel, like many other methods in Transit, is a "prot_table," which contains the annotation of the genome (including start and stop coordinates of genes). The prot_-table format is derived from an old format that could be downloaded from GenBank years ago. It is a tab-separated file with the following fields for each gene: function/description, start coord, end coord, strain ("+" or "−"), gene length (in amino acids; unused), 3 unused fields, gene name, and ORF id. More recently, the gff (or gff3) format is being used for genome annotations. Transit has a method to convert a gff file into a prot_table. However, there is greater flexibility in the gff format (especially, varying use of keywords), which is difficult to anticipate, so if the Transit command does not work, users might have to write their own script to convert their gff file to a prot_table.

```
> transit convert gff_to_prot_table <ref.gff> <ref.
prot_table>
```

See notes on including other types of genes in prot_tables (*see* **Note 8**).

3.2.3 Hidden Markov Model

The HMM can also be used to assess TnSeq data in a single condition. It works on a different principle—using a probabilistic model to estimate the state at each TA based on the counts and consistency with adjacent sites [27]. This allows the HMM to smooth over individual outlier values (such as an isolated insertion in any otherwise empty region or empty sites scattered among insertion in a non-essential region) and make a call for a region/gene that integrates information over multiple sites. The important difference from the Gumbel analysis is that the HMM takes into account the magnitudes of insertion counts, which can also carry information about the growth requirement (or fitness effect) of a gene. This allows the HMM to make finer distinctions, utilizing 4 states for individual sites: ES (essential), GD (growth-defect), NE (non-essential), and GA (growth-advantaged). One might see a GD call for a gene whose disruption (by the transposon) impairs growth, so counts are suppressed (compared to the global average, but not all the way to 0), while a gene might be called GA if transposon insertions actually confer a growth advantage, resulting in inflated counts [21]. The gene-level calls are made based on the majority call among the TA sites within each gene. The HMM automatically tunes its internal parameters (e.g. transition probabilities) to the characteristics of the input datasets (saturation and mean insertion counts) and can work over a broad range of saturation levels (as low as 20%, [27]).

```
> transit hmm G1.wig,G2.wig H37Rv.prot_table hmm_H37Rv_gly-
cerol.txt
```

The HMM command has flags for normalization (-n) and also how to handle merging of replicates (-r), though defaults are usually fine. There are actually two output files generated by the HMM, one named on the command line, which will contain the analysis (e.g. state probabilities and call) at each individual TA site, and another, hmm_H37Rv_glycerol_genes.txt ("genes" added as suffix to filename), which contains the call for each gene. There is no "probability" associated with the call for each gene. It is important to keep in mind that the call for a given gene can be influenced by insertions in the adjacent region, which is part of the design of HMMs.

The "-l" flag may be used to apply a LOESS correction to adjust for large-scale variations in insertion counts across the genome (*see* **Note 9**).

Because essential and growth-defect regions are close (both have suppressed counts), the difference in call by the HMM (ES vs GD) can be affected by noise, so we often combine the two categories. Typically, 15–20% of genes in a bacterial genome are called by the HMM as ES or GD, suggesting they are either absolutely required (ES) or contribute to fitness (GD) in the growth condition evaluated.

3.3 Pairwise Comparisons: Resampling

Comparisons between two conditions can be used to identify conditionally essential genes. In addition to binary cases where an essential becomes non-essential or vice versa, we also include genes with *quantitative changes in insertion counts*, reflecting apparent fitness changes for mutants (for example, genes in which the mean insertion count decreases by twofold, though not going all the way to 0). The primary tool in Transit used for pairwise comparisons is "resampling." It is equivalent to a permutation test on the difference of the mean counts between the two conditions for each gene. Of course, because of stochasticity, almost every gene will exhibit some difference in mean insertion counts between any pair of conditions. What matters is whether the difference is statistically significant.

Resampling is a frequentist approach that compares the observed differences of mean insertion counts to a null distribution to determine whether the difference is larger than would be expected by chance (given the same counts but without knowledge of the condition) (*see* **Note 10**).

Normalization is critically important for resampling. If the individual datasets are not properly normalized, it can lead to the appearance of an artificially inflated number of conditional essentials (i.e. false positives, artifacts). The default normalization used in Transit is called TTR (Trimmed Total Read-count) (*see* **Note 11**).

Resampling can be run on 2 sets of wig files (comma-separated) as follows:

```
> transit resampling G1.wig,G2.wig C1.wig,C2.wig,C3.wig H37Rv.
prot_table resampling_glyc_chol.txt -a
```

The full list of options for the resampling command is

```
> transit resampling
usage:
transit resampling <comma-separated .wig control files>
<comma-separated .wig experimental files> <annotation .prot_-
table or GFF3> <output file> [Optional Arguments]
or
transit resampling -c <combined_wig> <samples_metadata>
<ctrl condition name> <exp condition name> <annotation
.prot_table> <output file> [Optional Arguments]
NB: The ctrl and exp condition names should match Condition
names in samples_metadata file.

Optional Arguments.
-s <int>      := Number of samples. Default: -s 10000
-n <string> := Normalization method. (Default: -n TTR)
-h          := Output histogram of the permutations for each
gene. (Default: Turned Off)
-a          := Perform adaptive resampling. (Default: Turned
Off)
-ez          := Exclude rows with zero across conditions.
(Default: Turned Off)
-PC <float> := Pseudocounts used in calculating LFC. (de-
fault: 1)
-l           := Perform LOESS Correction (Default: Turned Off)
-iN <int>   := Ignore TAs occurring within given percent (as
int) of the N terminus. (Def: 0)
-iC <int>   := Ignore TAs occurring within given percent (as
int) of the C terminus. (Def: 0)
--ctrl_lib  := String of letters representing library of
control files in order
```

```
                    e.g. 'AABB'. Default empty. Letters used must
also be used in --exp_lib
                 If non-empty, resampling will limit permutations
to within-libraries.
--exp_lib     :=   String of letters representing library of
experimental files in order
```

where typically the experimental condition represents a treatment and the control condition represents the untreated or reference condition. The samples_metadata used with combined_wig inputs will be explained in the next section.

The input files are implicitly normalized by TTR (though this can be changed using "-n").

TA sites near the termini of ORFs can be trimmed using -iC and -iN. For example, to ignore insertion in the first or last 5% of an ORF, use "-iC 5 -iN 5."

I recommend using the "-a" flag to apply the adaptive version of resampling, which is much faster and generally outputs *p*-values close (within a factor of 2–3) to those obtained by performing all 10,000 samples. It truncates the resampling early for genes when it is clear that they are not going to be significant. Almost all significant genes will still be significant, except for possibly a few marginal differences very close to the 0.05 threshold on the adjusted *p*-value.

If the -h flag is given, histograms of the null distribution from resampling for each gene, along with a demarcation for the observed difference in mean insertion count between the conditions, will be generated in a sub-directory. These images can be examined to determine whether the null distribution for a gene of interest looks appropriately bell-shaped (versus bimodal, for example).

The output file contains the various statistics on the resampling for each gene and a *Padj* column at the end. Users can open the tab-separated file as a spreadsheet and sort by *Padj*. The conditionally essential genes are those with $Padj < 0.05$.

It is difficult to say what the expectations would be for number of conditional essentials, since it depends on the impact of the biological effects of the selection conditions (*see* **Note 12**).

The output also reports the log-fold-change (LFC, base 2) of the mean counts. It can be used to sort the genes to identify those showing the greatest increase or decrease in counts (relatively less or more essential). However, only genes with $Padj < 0.05$ should be considered significant. But these could be uncertain and should be ignored if they have a *Padj* above 0.05. Frequently, genes with the largest magnitude LFC are often smaller genes (with fewer TA sites) or genes with very low insertion counts, which are often not statistically significant. Of course, as with other frequentist statistical tests, genes with *Padj* above 0.05 should not be over-interpreted as evidence that they are *unaffected* by the condition.

Pseudocounts are used in the calculation of LFCs to help reduce noise (*see* **Note 13**).

See notes for analyzing datasets from different libraries using resampling (*see* **Note 14**).

3.3.1 Pathway Analysis

Typically, the significant genes do not all come from the same pathway but often represent a variety of pathways in which genes experience apparent increases or decreases in fitness. In order to get a sense of which pathways might be enriched among the conditionally essential genes, the "pathway_enrichment" command may be used:

```
> transit pathway_enrichment <resampling_file>
<associations> <pathways> <output_file> [-M <FET|GSEA|ONT>]
[-PC <int>]
```

The ⟨associations⟩ file lists mappings of ORF ids to pathway ids in a two-column format. Three example systems of functional categories are *Sanger roles* [28] (123 roles), *COG categories* (Clusters of Orthologous Genes, [29], 20 categories), and *GO terms* (Gene Ontology, http://www.geneontology.org/; there are ~4000 GO terms with at least 1 gene in H37Rv). The corresponding pathway association files for genes in Mtb H37Rv provided in the Transit data directory ($TRANSIT/src/pytransit/data/) are H37Rv_sanger_roles.dat, H37Rv_COG_roles.dat, and H37Rv_GO_terms.txt.

There may be multiple roles/pathway associations for each gene (listed on separate lines) or none (e.g. for hypothetical genes not yet functionally annotated). Note that these files have been expanded to include associations of each gene with all parents of each GO term or role (since these are hierarchical systems). The ⟨pathways⟩ file gives the functional descriptions for each pathway_id/role/GO term (e.g. sanger_roles.dat, COG_roles.dat, and GO_term_names.dat in the data directory). These will be the same for all organisms.

There are three alternative methods used to evaluate significance of pathway enrichment:

- The first is to use *Fisher's exact test*, where the *p*-value based on hypergeometric distribution of observed counts of category members among the hits compared the expected number based on the whole genome [30]. This is the default method ("**-M FET**"). The output file contains a list of significant pathways, sorted by adjusted *p*-value. In addition, the *enrichment* for each gene is reported. Enrichment is defined as the ratio of *observed* pathway members among the significant genes to the *expected* number based on the background proportion of category members in the overall genome. To mitigate the impact small pathways with only a few genes (which might spuriously appear as enriched), pseudocounts of 2 are incorporated in the numerator and denominator. This can be modified with the "-PC" flag. However, this analysis can be of limited use if the total number of significant (conditionally essential) genes is small (<10).

- An alternative approach is *Gene Set Enrichment Analysis* (GSEA; [31]) (using the **"-M GSEA"** flag). GSEA takes into account the ranking of all the genes, without regard to a significance cutoff. First, the genes are sorted by LFC. Alternatively, they can be ranked by the signed log-p-value ($SLPV = sign(LFC) * \log_{10}(pval)$), which effectively ranks the genes by significance from the largest increase in insertion counts to the largest decrease (with insignificant genes falling in the middle of the ranking). Then, the GSEA algorithm calculates a score reflecting the mean rank of a given set of genes and performs a simulation to determine the significance (p-value) by comparing to a null distribution of scores derived from random shuffling of the order. The closer the mean rank of a group of genes is to the top of the entire ranked list, or the closer to the bottom, the more significant. The potential advantage of GSEA over Fisher's exact test is that all the genes in a pathway can contribute to its enrichment, even if only a few (or none) are above the significance cutoff, as long as there is a systematic trend of increased or decreased counts shared by many of them.
- Finally, the *Ontologizer* method [32] has been implemented and is available via the **"-M ONT"** flag. The Ontologizer method acknowledges the hierarchical nature of the Gene Ontology and is designed to take advantage of parent–child relationships among GO terms by computing a conditional version of Fisher's exact test for a node conditioned on the genes in its parents. This can help focus the analysis on nodes in the GO hierarchy showing the most specific enrichment.

Here are some examples of using these various options.

```
# uses Fisher's exact test by default (with PC=2 as pseudo-
counts)
> transit pathway_enrichment resampling_glyc_chol.txt $DATA/
H37Rv_sanger_roles.dat $DATA/sanger_roles.dat pathways_glyc_-
chol_Sanger.txt

# can do this with GO terms too
> transit pathway_enrichment resampling_glyc_chol.txt $DATA/
H37Rv_GO_terms.txt $DATA/GO_term_names.dat pathways_glyc_-
```

```
chol_GO.txt

# with COG categories
> transit pathway_enrichment resampling_glyc_chol.txt $DATA/
H37Rv_COG_roles.dat $DATA/COG_roles.dat pathways_glyc_chol_-
COG.txt

# can also do GSEA method (on any system of functional
categories)
> transit pathway_enrichment resampling_glyc_chol.txt $DATA/
H37Rv_sanger_roles.dat $DATA/sanger_roles.dat pathways_San-
ger_GSEA.txt -M GSEA

# Ontologizer is a specialized method for GO terms
> transit pathway_enrichment resampling_glyc_chol.txt $DATA/
H37Rv_GO_terms.txt $DATA/GO_term_names.dat pathways_Ontologi-
zer.txt -M ONT
```

where $DATA refers to the path to the Transit data directory noted above. In this dataset, 2 COG and 6 Sanger categories related to secondary metabolite/small-molecule metabolism and lipid metabolism are identified as significant. The most significant GO term identified (among 26 significant terms, though many overlap) is `GO:0008202`, "steroid metabolic process" ($Padj = 0.0000089$; 15 out of 75 conditionally essential genes from resampling), which is consistent with the experiment (growth on cholesterol versus glycerol).

3.4 Analyses of Multiple Conditions

Recently, researchers have begun conducting more complex experiments using Tn libraries where a library is assessed across a large number of treatment conditions. For example, one might evaluate responses (conditional essentiality) to treatment with a panel of different antibiotics, possibly at different concentrations. One might assess the library in different stress conditions (heat, cold, hypoxia, SDS, nutrient starvation, iron limitation, etc.) or with media supplemented in various ways (e.g. multiple sources of carbon, nitrogen, sulfur, or phosphorous). Or one might compare survival of a bacterial transposon mutants in animal models with different genetic backgrounds, or for different durations of infection, etc. Analysis of such TnSeq datasets goes well beyond simple pairwise comparisons between conditions and often requires customized statistical analysis to evaluate the effects of the experimental variables. As a first step, Transit has some tools for evaluating variability of insertion counts across conditions. The analysis below focuses on identifying genes that exhibit some statistically significant differences in insertion counts across the panel of conditions. This is a useful starting point to begin to assess effects of the treatments (and similarities among them) based on the subsets of genes that respond.

In this section, we use an example from a study of iron utilization in mycobacteria [33]. The data include 2–3 replicates each of an *M. tuberculosis* H37Rv transposon library grown in in vitro conditions involving several different vehicles for iron delivery, focusing on various forms of heme and mycobactin (6 conditions). The 14 wig files for this example have already been consolidated into a combined wig file (iron_combined_wig4.txt), and a samples metadata file is used to encode which samples belong to which condition.

3.4.1 Genetic Interaction Analysis

A special case of multi-condition analysis that occurs frequently is when TnSeq libraries in a wild-type strain and a mutant strain (e.g. gene knockout) are compared between two conditions (e.g. a stress and a control). Typically, one is looking for genes that are conditionally essentially in the stress condition, but one wants to factor out those genes with a similar response in the wild-type strain and focus on differences unique to the mutant. There are many variations of this scheme which also require a comparison of 4 TnSeq datasets, arranged as 2×2. Transit has a specialized method based on Bayesian analysis [34] to identify significantly interacting genes (which interact with the knockout in the context of the stress). An example of how to run this command is as follows, which takes 4 groups of comma-separated wig files as input:

```
> python3 ../../transit/src/transit.py GI
<wigs_for_strA_cond1> <wigs_for_strA_cond2>
<wigs_for_strB_cond1> <wigs_for_strB_cond2> <annotation .
prot_table or GFF3> <output file>
```

In the output file, significant genes are categorized as "aggravating," "alleviating," or "suppressive" interactions, depending on whether they exhibit increased or decreased insertion counts in the mutant compared to the wild-type in the stress condition.

3.4.2 ANOVA

ANOVA analysis in Transit can be used to identify genes that exhibit significant *variability* of insertion counts across multiple conditions. ANOVA is a traditional statistical method for determining whether there are differences in observations among several groups. This method can be run in Transit using the following command:

```
> transit anova --help
```

```
Usage: python3 transit.py anova <combined wig file>
<samples_metadata file> <annotation .prot_table> <output
file> [Optional Arguments]
Optional Arguments:
-n <string> := Normalization method. Default: -n TTR
--include-conditions <cond1,...> := Comma-sep list of conds to
use for analysis
--ignore-conditions <cond1,...> := Comma-sep list of conds to
ignore
-iN <int> := Ignore TAs within given percentage of N terminus.
Def: -iN 0
-iC <int> := Ignore TAs within given percentage of C terminus.
Def: -iC 0
-PC <int> := pseudocounts to use for calculating LFC. Default:
-PC 5
```

For example,

```
> transit anova iron_combined_wig4.txt iron_samples_metadata.
txt H37Rv.prot_table anova_iron.txt
```

The use of a combined_wig file makes it easy to work with a large number of datasets in this context. The samples metadata file is a spreadsheet (in tab-separated text format) prepared by users that contains information about each sample/dataset/wig file. The headers must contain "Id," "Filename," and "Condition," where Id is a unique name for each sample, Filename is the name of its wig file (incorporated in the combined_wig file), and Condition is a symbolic name used to represent the treatment (typically represented by multiple replicates). The samples metadata file can contain additional information as well (e.g. time points, concentrations, batches, etc.). During the ANOVA analysis, the (TTR-normalized) insertion counts for the TA sites in each gene will be pooled into groups based on the condition labels, before computing the F-statistic and *p*-value. *p*-values are adjusted post hoc by the Benjamini–Hochberg method. Flags `--ignore-conditions` and `--include-conditions` can be used to focus on the analysis on just a subset of desired conditions (provided as a comma-separated list). For example, "`--include-conditions HighFeMBT,Hemin,Hemoglobin`" or "`--ignore-conditions LowFeMBT`."

The output of ANOVA analysis is a tab-separated spreadsheet that can be sorted by adjusted *p*-value. In these data, 181 genes were found to exhibit significant variability in insertion counts among the 6 iron supplementation conditions. If a gene is determined to be significant by ANOVA, it only means that its counts

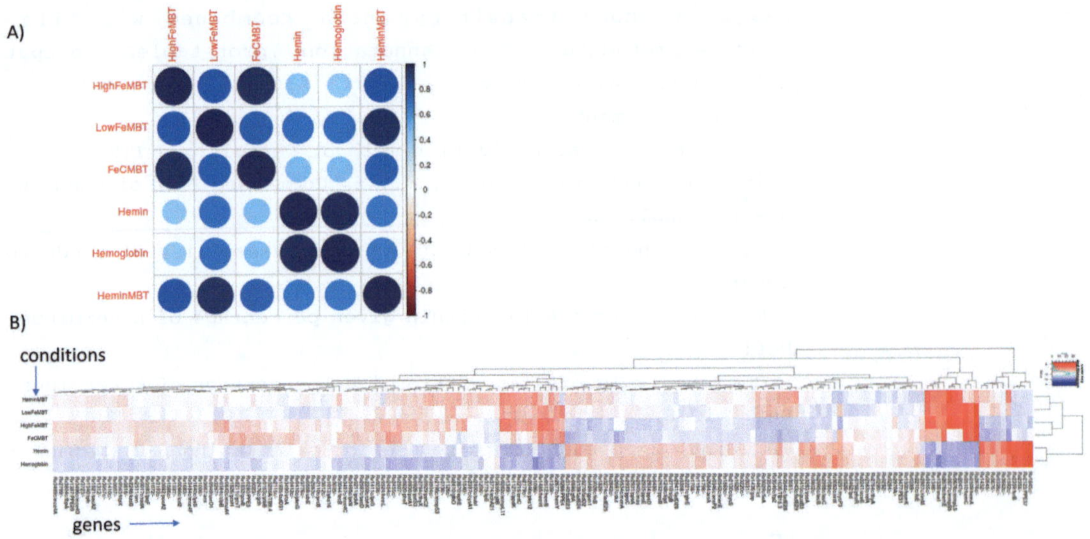

Fig. 1 (**a**) Correlation plot among iron supplementation conditions based on significantly varying genes according to ANOVA. (**b**) Heatmap of same data showing clustering of genes and conditions. Blue means more insertions than average (i.e. less essential), and red means less insertions than average (i.e. more essential)

vary in some condition (at least one) relative to the others, but it does not indicate which one. Like traditional ANOVA analysis, post hoc analyses must be employed to determine which conditions a gene is responding to, such as Tukey's range test (or honestly significant difference), which is based on pairwise comparisons between conditions [35].

To facilitate this post hoc analysis, the ANOVA method in Transit also prints out LFCs for each gene in each condition. The LFCs can be used to look for genes that respond to specific conditions, e.g. by sorting on these columns to look for genes with the most enrichment or depletion in a given condition. Note that *LFCs are computed relative to the mean insertion count for a gene across all conditions.* Thus, there will almost always be some condition (s) with higher counts (representing more fitness for the mutant) and other condition(s) with lower counts (where disruption of gene has less fitness and hence is relatively more essential) than average. Pseudocounts of 5 are used in calculating the LFCs (which helps reduce high-magnitude LFCs for genes with low counts, which are susceptible to noise), though this can be changed with the -PC flag.

The LFC columns in the ANOVA output can be colored as a heatmap in the spreadsheet (using Excel) to make the patterns of variation among the genes more clear. In addition, the ANOVA output file can be used to generate a heatmap (*see* Fig. 1, adapted from [33]) that simultaneously clusters the significant genes and the conditions, which is especially useful for shedding light on the relationships among the conditions apparent in the data:

```
> transit heatmap -anova anova_iron.txt heatmap_iron.png
```

Similarly, the anova file can be used as input the corrplot command (with the "-anova" flag suffixed) to show the similarities among the condition.

```
> transit corrplot anova_iron.txt heatmap_iron.png -anova
```

Importantly, the heatmap and corrplot analyses are based only on the significantly varying genes ($Padj < 0.05$, typically only a few hundreds) in order to enhance the patterns, since otherwise they would be washed out by the rest of the genes in the genome, the majority of which do not exhibit significant variation.

3.4.3 Zero-Inflated Negative Binomial (ZINB)

An alternative method for identifying genes exhibiting variability of insertion counts across conditions is Zero-Inflated Negative Binomial (ZINB) regression, which has recently been added to Transit [36]. One of the limitations of ANOVA analysis is that it assumes the data are Normally distributed. However, transposon insertion counts clearly violate this assumption (for several reasons). The ZINB method is designed to be a better model for insertion count data by: (a) representing the nonzero counts using a Negative Binomial distribution and (b) representing the saturation (TA sites with counts of 0) using a zero component in a mixture model. Thus, when analyzing the counts in a gene across multiple conditions, ZINB can detect variations in either the magnitudes in the insertion counts or the local level of saturation. This enables ZINB to be a little more sensitive than ANOVA, and empirical studies suggest that ZINB can detect about 30–50% more significant genes.

The usage of the ZINB command in Transit is as follows:

```
> transit zing --help
Usage: transit zinb <combined wig file> <samples_metadata
file> <annotation .prot_table> <output file> [Optional Argu-
ments]
Optional Arguments:
 -n <string>        :=  Normalization method. Default: -n TTR
 --condition        :=  column name in samples_metadata to use.
Default: "Condition"
 --ignore-conditions <cond1,...> :=  Comma separated list of
```

```
conditions to ignore
 --include-conditions <cond1,...> := Comma separated list of
conditions to include
 -iN <float>     := Ignore TAs within given percentage of the
N terminus. Def: 5
 -iC <float>     := Ignore TAs within given percentage of the
C terminus. Def: 5
 -PC <N>          := Pseudocounts used in calculating LFCs in
output file. Default: -PC 5
 --covars ... :=  Comma-sep. list of variables (columns in
metadata) to use as covariates
 --interactions ... :=  Comma-sep. list of variables to test
for interactions
 --gene <ORF id or gene name> := Run method for just one gene
and print model output.
```

For example,

```
> transit zinb iron_combined_wig4.txt iron_samples_metadata.
txt H37Rv.prot_table zinb_iron.txt
```

If there are just two conditions provided, then using ZINB analysis to identify genes with variable insertion counts is equivalent to detecting conditional essentiality. Hence, ZINB can be viewed as an alternative to resampling in this limiting case, and anecdotal testing suggests that there is a great deal of overlap. ZINB can sometimes identify additional genes as conditionally essential where there is a difference in either local saturation or magnitudes of counts alone, even if the overall mean count between conditions is not significantly different and hence not detected by resampling (*see* [36] for a more thorough comparison of the differences in significant hits between ZINB and resampling on example datasets).

In the output file for ZINB, multiple columns of information are provided on the mean counts in each condition, LFCs (relative to the mean across all conditions), nonzero mean, and saturation. For convenience, the—gene flag can be used to run the analysis just on a single gene, and the Transit will print out the condition-dependent and condition-independent models, likelihoods, etc., as in this example:

```
> transit zinb iron_combined_wig4.txt iron_samples_metadata.
txt H37Rv.prot_table output.txt --gene glpK
```

The significance (*p*-value) of each gene is determined by a likelihood ratio test (LRT). First, a condition-dependent model is generated by fitting independent parameters for each condition (such as mean and dispersion of insertion counts in the Negative Binomial and saturation as the mixing coefficient with the zero component). Then, a condition-independent ZINB model is generated by fitting a common set of parameters based on the counts pooled across all conditions. Finally, a likelihood ratio test is performed to determine whether the increase in likelihood of the condition-dependent model is justified, given the increased number of parameters. A *p*-value is derived for each gene using a chi-square distribution, and then the *p*-values are adjusted post hoc by the Benjamini–Hochberg procedure to control the overall FDR.

One of the advantages of the ZINB model is that it is implemented in a generalized linear model (GLM) framework, which allows incorporation of experimental variables (attributes of the conditions) as covariates. One application of this idea is to factor out the effect of an attribute that is known to affect the main condition in a way that is not of interest. For example, suppose we are interested in identifying genes that exhibit variable responses to different antibiotics. Furthermore, suppose the samples were each cultured for varying amounts of time (e.g. 0, 1, or 2 weeks). It is natural to expect that there will be variation in insertion counts between samples at different time-points, even if they are treated with the same drug. To evaluate the effect of the drug, it is desirable to subtract out any systematic effects on the insertion counts due to time (independent of drug). If a "Time" column is included in the samples metadata file, and a column with the header "Drug" encodes the drug treatment for each sample, then this may be achieved by using the `--covars` flag, as in this hypothetical example:

```
> transit zinb combined_wig.txt samples_metadata.txt H37Rv.
prot_table ZINB_cond_drug_covar_time.txt --condition Drug --
covars Time
```

If there are multiple covariates, they can be specified using a comma-separated list (i.e. corresponding to separate columns in the metadata file). The `--covars` flag could also potentially be used to correct for batch effects (where the variation in insertion counts in some samples appears to be determined by the batch of the experiment or data collection).

A similar approach can be used to test for interactions of variables with the main condition (*see* **Note 15**)

In the resulting output file, significant genes are those that exhibit some variability among the drug treatments that is dependent on carbon source.

As with ANOVA, the ZINB output file can be used to make a heatmap showing the clustering of the significant genes and the conditions:

```
> transit heatmap -zinb zinb_iron.txt zinb_iron_heatmap.png
```

4 Summary

Transit is designed to be a platform for statistical analysis of TnSeq data, with a focus on analysis of Himar1 transposon libraries (where insertions are restricted to TA dinucleotides). Some of the analytical methods can be applied to other transposons, like Tn5, though currently, they do not work as robustly. Although Transit has a graphical user interface (GUI), some of the more recent tools that have been added can only be invoked at the command line, which has been the focus of this chapter. The Transit Preprocessor (TPP) provides a way of processing raw sequencing data files and reducing the raw data to TA site insertion counts in the form of .wig files. The "tnseq_stats" command provides important summary statistics on datasets which is useful for diagnostics (i.e. identifying poor quality datasets that might need to be excluded or recollected). The analytical tools can be divided into 3 major tasks. First, individual datasets from an organism in a single (e.g. reference) condition can be used to identify essential genes using methods such as the Gumbel analysis (gaps) or a Hidden Markov Model. Second, pairs of conditions can be compared to identify conditionally essential genes using resampling (a permutation test on mean counts). The recently added ability to apply resampling on datasets mapped to different genome sequences has proven useful for studying differences in gene essentiality between libraries made from different strains, such as clinical isolates. Several methods for pathway enrichment analysis have also been added to gain additional insight from functional similarities among conditionally essential genes.

However, more recent developments in Transit have focused on supplying tools for analyzing larger collections of datasets from experiments involving multiple conditions. To keep things manageable when working with large collections of datasets, many of the tools in Transit have been extended to use "combined_wig" files and accompanying metadata files that encode relevant information about the different conditions. A starting point for analyzing such complex experiments is to identify genes exhibiting statistically significant variability across the conditions, using ANOVA or ZINB analysis, and then to begin to cluster and assess genes based on the similarity of their patterns of count variations, utilizing correlation plots and heatmaps. ZINB can be used to perform more sophisticated analyses through the exploitation of variables relating the different conditions as covariates and/or interactions (including capturing the trend or dependence of insertion counts on quantitative variables such as time or concentration). Genetic interaction (GI) analysis can be used to evaluate experiments where two different experimental variables are evaluated, producing a $2 \times 2 = 4$-way comparison of conditions, and test for significant interactions (e.g. suppressive, alleviating, or aggravating), which is especially useful for identifying genes associated with phenotypic changes in a knockout strain compared to a wild-type strain.

Transit continues to evolve and improve, especially through feedback and suggestions from users (send email to ioerger@cs.tamu.edu). In the future, we hope to add new statistical methods to support analysis of more complex experiments, improve integration with the GUI, and also extend and improve the analyses to TnSeq libraries made with other transposons, especially Tn5.

5 Notes

1. For example, TnSeq has been used to study genes required for metabolizing/utilizing specific nutrients such as cholesterol [24] or iron [33], tolerance of stress conditions [37, 38], genetic interactions with knocked-out genes in libraries made from null mutants [39–41], and virulence in animal models [4, 5, 9, 42, 43].
2. For Himar1, there is an apparent non-permissiveness of TA sites with a G at flanking positions ±2 bp, with G or C at ±3 [44], and also a general preference for insertions in more bendable regions of the DNA [45]. Similarly, only a generalized sequence preference has been identified for Tn5, but it is insufficient to make insertion locations predictable [46].
3. Some studies have observed large-scale chromosomal biases, where mean insertion counts exhibit a trend based on the position of the chromosome, such as a gradation based on distance between the origin and the termination of replication. Adjustments for such biases can be made through methods such as the LOESS correction [47], which equalizes the smoothed mean across the whole genome.

4. A common case of single-condition analysis might be assessing essential genes and pathways in a new bacterial species. A common case of pairwise analysis might be comparing essentiality in a stress condition, such as starvation, iron limitation, low pH, hypoxia, antibiotic exposure, or growth in an animal model, compared to a reference condition, such as growth on rich medium. However, recently, more complex TnSeq experiments are being conducted involving multiple experimental variables/treatments, such as varying antibiotic concentrations, varying durations (the number of days or weeks in culture or in vivo), comparison of survival in different animal breeds/species/genotypes, or supplementation with various nutrients, making tools for task multi-condition analysis necessary to explore/characterize patterns of response over larger sets of experimental conditions.
5. Since the reads represent junctions between the transposon and chromosome, the first step is identifying reads (in read 1) with a prefix matching the terminus of the transposon and the stripping this prefix off to map the genomic suffix. Many protocols introduce random shifts in the location of the prefix sequence within the reads, to reduce sequencing problems when all reads begin with the same nucleotides. By default, TPP searches for a prefix corresponding to Himar1, but other transposons can be accommodated by specifying the search sequence using the "-prefix" flag. Illumina sequencers often provide pairs of reads, and while read 2 is not absolutely necessary for mapping reads (i.e. optional), Sassetti et al. have shown how to embed random nucleotide barcodes in read 2 that can be used by TPP to reduce raw read counts at each TA site to counts of unique DNA templates, which helps reduce noise due to PCR jackpotting effects [15]. Because of the lengths of the prefix in read 1 and barcodes (and surrounding constant regions and genomic regions in read 2), it is recommended to use a read length of at least 75 bp (i.e. 75×75 bp paired-end sequencing). Various flags can be used to adjust the methods and parameters for mapping reads, such as the number of mismatches allowed or constraints on location of prefix.
6. Generally speaking, the number of TA sites with low counts is most abundant, and sites with high counts are far less frequent. While it is not theoretically guaranteed, well-behaved datasets typically conform closer to a geometric distribution (though possibly with some excess dispersion, which could be modeled as a negative binomial). In contrast, lower quality datasets often exhibit a significant skew away from this distribution. Skewed datasets are often more dominated by excessively high counts at a few sites, while the counts at the vast majority of remaining sites are quite low by comparison. This can cause problems with

normalization; even though TTR is designed to be robust to a few outliers, it is only linear transformation and cannot correct for skew in the data. This can lead to an inflation of apparently differentially essential genes (artifacts) when skewed datasets are compared to other conditions.

The Transit GUI provides plots of the insertion-count distribution, including a QQ plot (quantile–quantile plot) against an ideal geometric, to visually assess the degree of skew. The closer to a diagonal, the better. Aside from "skewness" as a metric itself, the tnseq_stats table also reports the Pickand's tail index [48]. Anecdotal experience suggests that dataset with $PTI > 1.0$ (and maybe even $PTI > 0.5$) is potentially problematic.

The causes of skew are (currently) not well understood, though, anecdotally, it often seems to be associated with loss of diversity (e.g. in animal-passeged samples) over "over-selection" of libraries. For example, culturing too long under a stress condition can amplify the effects of even small fitness differences.

What can be done with skewed datasets? One solution is to apply the beta-geometric correction (BGC) [16]. This is a nonlinear normalization procedure that adjusts the insertion-count distribution to look more like a conventional geometric distribution. This can be accomplished using the "-n betageom" flag to the "normalize" command. BGC normalization dramatically squashes down the sites with the highest counts, but if it is applied uniformly to all the samples in a comparison, it often greatly reduces the number of (apparently) significant hits, hopefully getting rid of false positives and retaining only a few true positives where the trend of insertions truly supports a change in essentiality (fitness difference).

```
>transit normalize G1.wig G1_BGC.wig -n betageom
```

7. If multiple replicates are available, they can be provided to Gumbel via a comma-separated list, as shown above, and the data will be merged. The output file will contain values for each gene, such as the number of TA sites, the number of sites with insertion, the longest run of sites without insertion, and finally a posterior probability ("zbar") and call (ES=essential, NE=non-ess, U=uncertain, S=too short to analyze). This command also has flags for explicitly trimming (ignoring) TA sites in termini of genes, etc.
8. Note that one can add other types of genes to a prot_table, such as tRNAs, rRNAs, and ncRNAs. All that is required it to include lines with the expected format, including start and end coordinates and strand. Furthermore, the same approach can be used to represent other types of loci, such as operons (spanning multiple genes) or intergenic regions, to be analyzed for

essentiality. However, note that most intergenic regions and ncRNAs are relatively small compared to typical protein-coding regions, and hence they often have only a few (if any) TA sites, making essentiality analysis highly uncertain (except for the most highly saturated libraries [44])

9. Because the HMM takes the magnitudes of insertions into account, it can be affected by large-scale variations in insertion counts across the genome. In the GUI, a plot of the smoothed mean insertion count can be generated. If there are noticeable differences, the local deviations from the global mean can be subtracted out using the LOESS correction [17] ("-l" flag for the hmm command). While this is typically not necessary for Tn libraries in *M. tuberculosis*, chromosomal biases can be more severe in faster-replicating species, like *Vibrio cholera* [3] and *E. coli* [49].

10. First, the normalized insertion counts are pooled over all TA sites and replicates for both conditions. Then, a null distribution is created by repeatedly drawing two random samples (of the same size as the original number of observations) from the pooled counts and calculating the difference of means 10,000 times. From this, a two-tailed p-value is derived for the observed difference, and the p-values are adjusted post hoc for multiple testing via the Benjamini–Hochberg correction [50] to limit the overall false discovery rate (FDR) to 5%.

    This comparative approach to identifying conditionally essential genes is preferred over simply combining the results of individual analyses of each condition (e.g. Gumbel) and selecting genes that are called essential in one condition but not the other, because the results of that approach could be influenced by which genes fall just above or below the significance cutoff. Resampling is a direct comparison that looks at differences in insertion counts. The resampling procedure is designed to be (somewhat) robust to noise (e.g. outliers, high variance among counts) and is appropriately less sensitive for smaller genes and more sensitive with more replicates.

11. For each dataset, the total insertions over all TA sites are tabulated, after removing the top and bottom 1% of sites (to reduce the influence of outliers). The sites are divided by this total and then scaled up so that the mean insertion count over the whole genome is 100. While the calculation is simple, it puts counts from different datasets on a comparable basis. Furthermore, it maintains a balance so that counts from less-saturated datasets are inflated proportionally, so that the mean insertion for most genes (averaged over multiple TA sites) stays about the same on average. This objective is important for the assumptions of resampling (see explanation in [23]). A way to verify that TTR normalization is doing the right thing is to

make a scatter plot of the mean insertion counts for each gene between any two datasets; the data should scatter along the X= Y diagonal. Individual points can deviate from this line due to experimental noise, sampling error for small genes with few TA sites, or biological differences (if the datasets represent distinct experimental conditions). But if there is a systematic deviation away from the X= Y line in this scatter plot (due to improper normalization), it would likely produce an excess of conditional essentials detected by resampling.

12. In some cases, the treatment (e.g. a high-stress condition) might affect the essentiality (and fitness) of hundreds of genes. In other cases, biologically weak selection criteria might affect only a few (or perhaps no) genes. If the experimenter feels that there are too few hits (e.g. compared to their expectations, possibly lacking genes known to be in affected pathways), then it might help to collect additional replicates, which can increase the sensitivity of detection. If, subjectively, too many significant genes are output, it might be a sign of problems with skewing of the data. The user can follow the QC guidelines above to possibly identify lower quality datasets to try excluding, or they might consider applying BGC normalization ("-n betageom") to the datasets.

13. To dampen the appearance of high-magnitude LFCs from genes with low insertion counts (which are more susceptible to noise), one can increase the pseudocounts using "-PC." The LFC is calculated using the following formula, which uses PC=1 by default, to avoid the result being undefined for genes with means of 0 in either condition:

$$LFC = \log_2((\text{mean in expt condition} + PC)/(\text{mean in ctrl condition} + PC))$$

For example, a gene with a mean count of 1 in the first condition and 4 in the second would appear to have a fold-change of 4 and LFC of 2.0 (with PC= 0). However, this is typically in the range of noise for TnSeq experiments. Considering that TTR normalization adjusts the mean insertion count of each sample to around 100, raising the pseudocounts to a level of PC= 5 would reduce this to $\log_2(9/6) = 0.58$, masking out differences below this level.

By loading the resampling output file into the GUI (bottom panel), a *volcano plot* can be generated (via "Choose Action"), showing a scatter plot of the *p*-values versus the LFC for each gene. Genes that become more essential (in the experimental condition over control) are on the left (negative LFC), and significant genes are above the dashed red line (*p*-value threshold adjusted for 5% FDR).

14. If the replicates in each condition represent samples from multiple libraries, the sensitivity of resampling can sometimes be increased by performing separate random draws from the counts for each library and computing the difference of the means between conditions summed separately for each library. This can help reduce the variance in the difference in the mean insertion count between conditions by effectively subtracting out variability due to differences in abundance at each TA site between libraries. The transposon library each sample in the two conditions can be specified symbolically using a string of characters. For example "`--exp_lib AABB --ctrl_lib AB`" would indicate that the first 2 experimental samples came from library A and the second two samples from library B; there can be different numbers of samples in the control condition, but they should use the same codes (in this case, one sample each from library A and library B).
15. For example, suppose we are interested in genes that respond differentially to a panel of drugs. Furthermore, suppose TnSeq data were collected for cultures grown on media containing one of several carbon sources, e.g. glycerol, glucose, or cholesterol. In order to test whether carbon source interacts with drug, ZINB will fit a model based on the cross product of all combinations of the two variables and compare it (using an LRT) to a condition-independent model (where the counts are pooled for the main condition). Assuming there is a column with the header "CarbonSource" in the samples metadata, the interaction may be tested as in the following hypothetical example:

```
>transit zinb combined_wig.txt samples_metadata.txt H37Rv.prot_table ZINB_cond_drug_interac_carbon.txt--condition Drug--interactions CarbonSource
```

Acknowledgements

This work has been supported by NIH grants U19 AI107774 and P01 AI143575.

References

1. van Opijnen T, Camilli A (2013) Transposon insertion sequencing: a new tool for systems-level analysis of microorganisms. Nat Rev Microbiol 11(7):435–442
2. Barquist L, Boinett CJ, Cain AK (2013) Approaches to querying bacterial genomes with transposon-insertion sequencing. RNA Biol 10(7):1161–1169
3. Chao M, Abel S, Davis B, Waldor M (2016) The design and analysis of transposon insertion sequencing experiments. Nat Rev Microbiol 14 (2):119–128
4. Langridge GC, Phan M, Turner D, Perkins T, Parts L, Haase J, Charles I, Maskell D, Peters S, Dougan G, et al. (2009) Simultaneous assay of every salmonella typhi gene using one million transposon mutants. Genome Res 19

(12):2308–2316. http://www.ncbi.nlm.nih.gov/pubmed/19826075

5. Gawronski JD, Wong SM, Giannoukos G, Ward DV, Akerley BJ (2009) Tracking insertion mutants within libraries by deep sequencing and a genome-wide screen for Haemophilus genes required in the lung. Proc Natl Acad Sci USA 106(38):16422–16427
6. Goodman AL, McNulty NP, Zhao Y, Leip D, Mitra RD, Lozupone CA, Knight R, Gordon JI (2009) Identifying genetic determinants needed to establish a human gut symbiont in its habitat. Cell Host Microbe 6(3):279–289
7. Wetmore K, Price M, Waters R, Lamson J, He J, Hoover C, Blow M, Bristow J, Butland G, Arkin A, Deutschbauer A (2015) Rapid quantification of mutant fitness in diverse bacteria by sequencing randomly bar-coded transposons. mBio 6(3):e00306-15
8. Jensen P, Zhu Z, van Opijnen T (2017) Antibiotics disrupt coordination between transcriptional and phenotypic stress responses in pathogenic bacteria. Cell Rep 20 (7):1705–1716
9. Zhang YJ, Reddy MC, Ioerger TR, Rothchild AC, Dartois V, Schuster BM, Trauner A, Wallis D, Galaviz S, Huttenhower C, Sacchettini JC, Behar SM, J RE (2013) Tryptophan biosynthesis protects mycobacteria from CD4 T-cell-mediated killing. Cell 155 (6):1296–1308
10. Luo H, Lin Y, Gao F, Zhang CT, Zhang R (2014) DEG 10, an update of the Database of Essential Genes that includes both protein-coding genes and non-coding genomic elements. Nucleic Acids Research 42: D574–D580
11. Rubin EJ, Akerley BJ, Novik VN, Lampe DJ, Husson RN, Mekalanos JJ (1999) In vivo transposition of mariner-based elements in enteric bacteria and mycobacteria. PNAS 96 (4):1645–1650
12. Sassetti CM, Boyd DH, Rubin EJ (2001) Comprehensive identification of conditionally essential genes in mycobacteria. PNAS 98 (22):12712–12717. https://doi.org/10.1073/pnas.231275498. http://www.pnas.org/content/98/22/12712.abstract. http://www.pnas.org/content/98/22/12712.full.pdf+html
13. Reznikoff WS (2003) Tn5 as a model for understanding DNA transposition. Mol Microbiol 43(5):1199–1206
14. Lampe DJ, Churchill ME, Robertson HM (1996) A purified mariner transposase is sufficient to mediate transposition in vitro. Eur Mol Biol Organ J 15(19):5470–5479
15. Long J, DeJesus M, Ward D, Baker R, Ioerger T, Sassetti C (2015) Identifying essential genes in *Mycobacterium tuberculosis* by global phenotypic profiling. In: Lu LJ (ed) Methods in molecular biology: gene essentiality, vol 1279. Springer, Berlin, p 79–95
16. DeJesus MA, Ioerger TR (2016) Normalization of transposon-mutant library sequencing datasets to improve identification of conditionally essential genes. J Bioinform Comput Biol 14(3):1642004
17. Zomer A, Burghout P, Bootsma HJ, Hermans PW, van Hijum SA (2012) ESSENTIALS: software for rapid analysis of high throughput transposon insertion sequencing data. PLoS ONE 7(8):e43012
18. Solaimanpour S, Sarmiento F, Mrazek J (2015) Tn-seq explorer: a tool for analysis of high-throughput sequencing data of transposon mutant libraries. PLoS ONE 10(5):e0126070
19. Barquist L, Mayho M, Cummins C, Cain AK, Boinett CJ, Page AJ, Langridge GC, Quail MA, Keane JA, Parkhill J (2016) The TraDIS toolkit: sequencing and analysis for dense transposon mutant libraries. Bioinformatics 32(7):1109
20. Zhao L, Anderson MT, Wu W, T Mobley HL, Bachman MA (2017) TnseqDiff: identification of conditionally essential genes in transposon sequencing studies. BMC Bioinformatics 18 (1):326
21. van Opijnen T, Bodi KL, Camilli A (2009) Tn-seq: high-throughput parallel sequencing for fitness and genetic interaction studies in microorganisms. Nat Methods 6(10):767–772
22. Pritchard JR, Chao MC, Abel S, Davis BM, Baranowski C, Zhang YJ, Rubin EJ, Waldor MK (2014) ARTIST: high-resolution genome-wide assessment of fitness using transposon-insertion sequencing. PLoS Genet 10(11):e1004782
23. DeJesus MA, Ambadipudi C, Baker R, Sassetti C, Ioerger TR (2015) TRANSIT–a software tool for Himar1 TnSeq analysis. PLoS Comput Biol 11(10):e1004401
24. Griffin JE, Gawronski JD, DeJesus MA, Ioerger TR, Akerley BJ, Sassetti CM (2011) High-resolution phenotypic profiling defines genes essential for mycobacterial growth and cholesterol catabolism. PLoS Pathog 7(9):e1002251. https://doi.org/10.1371/journal.ppat.1002251
25. Li H, Durbin R (2009) Fast and accurate short read alignment with Burrows–Wheeler transform. Bioinformatics 25(14):1754–1760
26. DeJesus MA, Zhang YJ, Sassetti CM, Rubin EJ, Sacchettini JC, Ioerger TR (2013) Bayesian

analysis of gene essentiality based on sequencing of transposon insertion libraries. Bioinformatics 29(6):695–703

27. DeJesus MA, Ioerger TR (2013) A Hidden Markov Model for identifying essential and growth-defect regions in bacterial genomes from transposon insertion sequencing data. BMC Bioinformatics 14:303
28. Cole ST, Brosch R, Parkhill J (1998) Deciphering the biology of mycobacterium tuberculosis from the complete genome sequence. Nature 393(6685):537–544. http://dx.doi.org/10.1038/31159
29. Galperin MY, Makarova KS, Wolf YI, Koonin EV (2015) Expanded microbial genome coverage and improved protein family annotation in the COG database. Nucl Acids Res 43: D261–D269
30. Irizarry RA, Wang C, Zhou Y, Speed TP (2009) Gene set enrichment analysis made simple. Stat Methods Med Res 18(6):565–575
31. Subramanian A, Tamayo P, Mootha VK, Mukherjee S, Ebert BL, Gillette MA, Paulovich A, Pomeroy SL, Golub TR, Lander ES, Mesirov JP (2005) Gene set enrichment analysis: a knowledge-based approach for interpreting genome-wide expression profiles. PNAS 102(43):15545–15550
32. Grossmann S, Bauer S, Robinson PN, Vingron M (2007) Improved detection of overrepresentation of Gene-Ontology annotations with parent-child analysis. Bioinformatics 23 (22):3024–3031
33. Zhang L, Hendrickson RC, Meikle V, Lefkowitz EJ, Ioerger TR, Niederweis M (2020) Comprehensize analysis of iron utilization by Mycobacterium tuberculosis. PLoS Pathog 16 (3):e1008337
34. DeJesus MA, Nambi S, Smith CM, Baker RE, Sassetti CM, Ioerger TR (2017) Statistical analysis of genetic interactions in Tn-Seq data. Nucl Acids Res 45(11):e93
35. Tukey J (1949) Comparing individual means in the analysis of variance. Biometrics 5 (2):99–114
36. Subramaniyam S, DeJesus MA, Zaveri A, Smith CM, Baker RE, Ehrt S, Schnappinger D, Sassetti CM, Ioerger TR (2019) Statistical analysis of variability in tnseq data across conditions using zero-inflated negative binomial regression. BMC Bioinf 20(1):603
37. Xu W, DeJesus MA, Rücker N, Engelhart CA, Wright MG, Healy C, Lin K, Wang R, Park SW, Ioerger TR, Schnappinger D, Ehrt S (2017) Chemical genetic interaction profiling reveals determinants of intrinsic antibiotic resistance in Mycobacterium tuberculosis. Antimicrob Agents Chemother 61(22):e01334–17
38. Matern WM, Rifat D, Bader JS, Karakousis PC (2018) Gene enrichment analysis reveals major regulators of Mycobacterium tuberculosis gene expression in two models of antibiotic tolerance. Front Microbiol 9:610
39. Kieser KJ, Boutte CC, Kester JC, Baer CE, Barczak AK, Meniche X, Chao MC, Rego EH, Sassetti CM, Fortune SM, Rubin EJ (2015) Phosphorylation of the peptidoglycan synthase PonA1 governs the rate of polar elongation in mycobacteria. PLoS Pathog 11(6): e1005010
40. Nambi S, Long JE, Mishra BB, Baker R, Murphy KC, Olive AJ, Nguyen HP, Shaffer SA, Sassetti CM (2015) The oxidative stress network of mycobacterium tuberculosis reveals coordination between radical detoxification systems. Cell Host Microbe 17(6):829–837
41. Baranowski C, Welsh MA, Sham LT, Eskandarian HA, Lim HC, Kieser KJ, Wagner JC, McKinney JD, Fantner GE, Ioerger TR, Walker S, Bernhardt TG, Rubin EJ, Rego EH (2018) Maturing Mycobacterium smegmatis peptidoglycan requires non-canonical crosslinks to maintain shape. Elife 7:e37516
42. Fu Y, Waldor M, Mekalanos J (2013) Tn-seq analysis of vibrio cholerae intestinal colonization reveals a role for t6ss-mediated antibacterial activity in the host. Cell Host Microbe 14 (6):652–663
43. Dragset MS, Ioerger TR, Loevenich M, Haug M, Sivakumar N, Marstad A, Cardona PJ, Klinkenberg G, Rubin EJ, Steigedal M, Flo TH (2019) Global assessment of Mycobacterium avium subsp. hominissuis genetic requirement for growth and virulence. mSystems 4: e00402-19
44. DeJesus MA, Gerrick ER, Xu W, Park SW, Long JE, Boutte CC, Rubin EJ, Schnappinger D, Ehrt S, Fortune SM, Sassetti CM, Ioerger TR (2017) Comprehensive essentiality analysis of the mycobacterium tuberculosis genome via saturating transposon mutagenesis. MBio 8(1):e02133-16
45. Lampe DJ, Grant TE, M RH (1998) Factors affecting transposition of the Himar1 mariner transposon in vitro. Genetics 149(1):179–187
46. Ason B, Reznikoff WS (2004) DNA sequence bias during Tn5 transposition. J Mol Biol 335:1213–1225
47. Chao MC, Pritchard JR, Zhang YJ, Rubin EJ, Livny J, Davis BM, Waldor MK (2013) High-resolution definition of the Vibrio cholerae essential gene set with hidden Markov model-based analyses of transposon-insertion

sequencing data. Nucleic Acids Res 41 (19):9033–9048

48. Pickands J (1975) Statistical inference using extreme order statistics. Ann Stat 3:119–131

49. Warr AR, Hubbard TP, Munera D, Blondel CJ, zur Wiesch PA, Abel S, Wang X, Davis BM, Waldor MK (2019) Transposon-insertion sequencing screens unveil requirements for EHEC growth and intestinal colonization. PLoS Pathog 15(8):e1007652

50. Benjamini Y, Yekutieli D (2005) False discovery rate controlling confidence intervals for selected parameters. J Am Stat Assoc 100 (469):71–81

Chapter 23

Geptop 2.0: Accurately Select Essential Genes from the List of Protein-Coding Genes in Prokaryotic Genomes

Qing-Feng Wen, Wen Wei, and Feng-Biao Guo

Abstract

Computational tool composites alternative way to identify essential genes and it is low-cost and time-efficient. Based on experimental essentiality sets deposited in the databases DEG and OGEE as reference, we developed an automatically computational tool named Geptop to select essential genes from the set of protein-coding genes in a prokaryotic genome, which utilizes the strategy of reciprocally best hit for homology search and evolutionary distance for weight assigning. The latest version of Geptop is 2.0 (http://guolab.whu.edu.cn/geptop), which can predict gene essentiality with the mean AUC 0f 0.84 in prokaryotes and is more stable. The chapter is to briefly introduce the tool and tell how to use it.

Key words Essential genes, Geptop 2.0, Prediction tools, Input and output

1 Introduction

Essential genes are critical for the survival of organisms, and play an important role in synthetic biology, especially the construction of minimal genome [1, 2]. It is meaningful to study essential genes, for example, essential genes of human pathogens might be used as antibiotic targets [3]. However, most of experimental methods to identify essential genes are expensive and time-consuming, thus computational identification of essential genes represents an important alternative approach. In 2013, our group developed Geptop, which is a prediction tool based on orthology and phylogeny, and both online service and standalone package are provided [4, 5]. Since its first release, Geptop has been used and commented positively by scientists [6–11]. For example, Nigatu et al recommended geptop "due to the high accuracy and the availability of a Web server" and "it is the most used computational tool" in identification of essential genes [12].

Geptop utilizes the information of essential genes in Database of Essential Genes (DEG) [13], and relies on orthologous search and evolutionary distance to quantize gene essentiality. Comparing

Ren Zhang (ed.), *Essential Genes and Genomes: Methods and Protocols*, Methods in Molecular Biology, vol. 2377,
https://doi.org/10.1007/978-1-0716-1720-5_23,

with other prediction models, Geptop is competitive especially in AUC index [5, 11]. Besides, Geptop can finish tasks in a short time and can be capable of completing prediction of far distance species.

2 Materials

2.1 Reference Genomes and Cross-Species Validation

The species whose essential genes have been genome-scale identified by experiments can be used as reference genomes. All these informations are got from DEG (http://www.essentialgene.org/). Totally 37 species are used as reference genomes (Table 1), and the minimal evolution distance is calculated between one species and other 36 species via the composition vector method [14]. Besides, the average minimal distance among 37 species is 0.303, and the average cross-species validation AUC is 0.84 which represents excellent prediction. Excellent performance in prediction of species with distant relationships suggests that Geptop can perform stably and can provide reliable prediction in species with far evolutionary distance.

2.2 Introduction of Web Page

The web page of Geptop is shown in Fig. 1. Here, users can find information about the first version and citation. Guideline and standalone package are also available on this page, and essential genes predicted in 968 genomes can be found on this website. The most important is the online service frame where users can fill in their email and upload their whole genome (proteomes), whose essential genes need to be predicted.

3 Methods

Geptop is quite easy to use, even for people who do not have any programming knowledge, online service can satisfy their needs. Otherwise, users can download standalone version of Geptop and run Geptop on their own computer. We will introduce both ways in the following part.

3.1 Input and Output of Geptop

The whole sequence genome can be downloaded on *ftp://ftp.ncbi.nlm.nih.gov/genomes/genbank/bacteria/* or *ftp://ftp.ncbi.nlm.nih.gov/genomes/archive/old_genbank/Bacteria/*. Note that the input must be whole-genomic (proteomic) sequence of fasta format (Fig. 2). And the name of each gene or protein must be unique in whole genomes.

No matter whether users run Geptop on our online service or their own computer, the output format remains the same as shown in Fig. 3, which exhibits the classification, essentiality score, and protein information.

Table 1
The information of reference species

Species	Phylum	Minimal distance	AUC
Acinetobacter baylyi ADP1	g-proteobacteria	0.397	0.897
Bacillus subtilis 168	Firmicutes	0.487	0.962
Caulobacter crescentus NA1000	a-proteobacteria	0.471	0.932
Escherichia coli MG1655	Proteobacteria	0.326	0.944
Francisella novicida U112	g-proteobacteria	0.494	0.858
Mycobacterium tuberculosis H37Rv	Actinobacteria	0.497	0.759
Mycoplasma genitalium G37	Firmicutes	0.498	0.713
Mycoplasma pulmonis UAB CTIP	Firmicutes	0.496	0.861
Pseudomonas aeruginosa UCBPP-PA14	g-proteobacteria	0.065	0.814
Salmonella enterica serovar Typhi Ty2	g-proteobacteria	0.110	0.961
Salmonella enterica serovar typhimurium LT2	g-proteobacteria	0.018	0.710
Staphylococcus aureus N315	Firmicutes	0.074	0.829
Staphylococcus aureus NCTC 8325	Firmicutes	0.074	0.885
Streptococcus pneumoniae TIGR4	Firmicutes	0.145	0.821
Streptococcus pneumoniae R6	Firmicutes	0.145	0.721
Streptococcus sanguinis SK36	Firmicutes	0.416	0.971
Helicobacter pylori 26695	e-proteobacteria	0.488	0.613
Porphyromonas gingivalis ATCC 33277	Bacteroidetes	0.483	0.835
Bacteroides thetaiotaomicron VPI-5482	Bacteroidetes	0.423	0.796
Burkholderia thailandensis E264	b-proteobacteria	0.239	0.926
Salmonella enterica subsp. *enterica serovar typhimurium* str. 14028S	g-proteobacteria	0.024	0.654
Sphingomonas wittichii RW1	a-proteobacteria	0.489	0.823
Shewanella oneidensis MR-1	g-proteobacteria	0.483	0.912
Pseudomonas aeruginosa PAO1	g-proteobacteria	0.065	0.945
Campylobacter jejuni subsp. *jejuni* NCTC 11168 = ATCC 700819	e-proteobacteria	0.093	0.703
Salmonella enterica serovar typhimurium SL1344	g-proteobacteria	0.018	0.886
Bacteroides fragilis 638R	Bacteroidetes	0.423	0.735
Burkholderia pseudomallei K96243	b-proteobacteria	0.239	0.736
Streptococcus pyogenes MGAS5448	Firmicutes	0.082	0.943

(continued)

Table 1 (continued)

Species	Phylum	Minimal distance	AUC
Streptococcus pyogenes NZ131	Firmicutes	0.082	0.923
Synechococcus elongatus PCC 7942	Cyanophyta	0.497	0.772
Rhodopseudomonas palustris CGA009	a-proteobacteria	0.486	0.903
Streptococcus agalactiae A909	Firmicutes	0.419	0.905
Acinetobacter baumannii ATCC 17978	g-proteobacteria	0.397	0.838
Agrobacterium f	a-proteobacteria	0.486	0.905
Brevundimonas subvibrioides ATCC 15264	a-proteobacteria	0.471	0.944
Campylobacter jejuni subsp. *jejuni* 81-176	e-proteobacteria	0.093	0.749

3.2 Usage of Online Service

In the bottom of Fig. 1, there is a gray frame in which user can submit their genome sequence. A default value of essentiality score cutoff is set as 0.24, which can be changed by users according to their needs, and the higher essentiality score cutoff is set, the fewer number of genes will be identified as essential. The result can be shown to users both on website or sent to their email. Note that gene sequence and protein sequence are both available on Geptop online service.

3.3 Usage of Standalone Version

Before running standalone script of Geptop, the following requirement should be met. Both Linux and Windows environment are available, and Python should be at least version 2 (http://www.python.org/). In addition, Biopython library (http://biopython.org/) and Standalone NCBI BLAST (ftp://ftp.ncbi.nih.gov/blast/executable) need to be installed first. We use command line and optional parameters to run Geptop. And the usage is described below.

```
Usage: python geptop2.py -i protein.faa [Optional parameters: -p path of BLAST executable -s essentiality score cutoff -o output file].
```

4 Additional Noted Information for users of Geptop

1. The uploaded sequence must be whole genome or proteome, and each gene should have a unique name. Fasta format is needed strictly. Both protein sequence and gene sequence of a

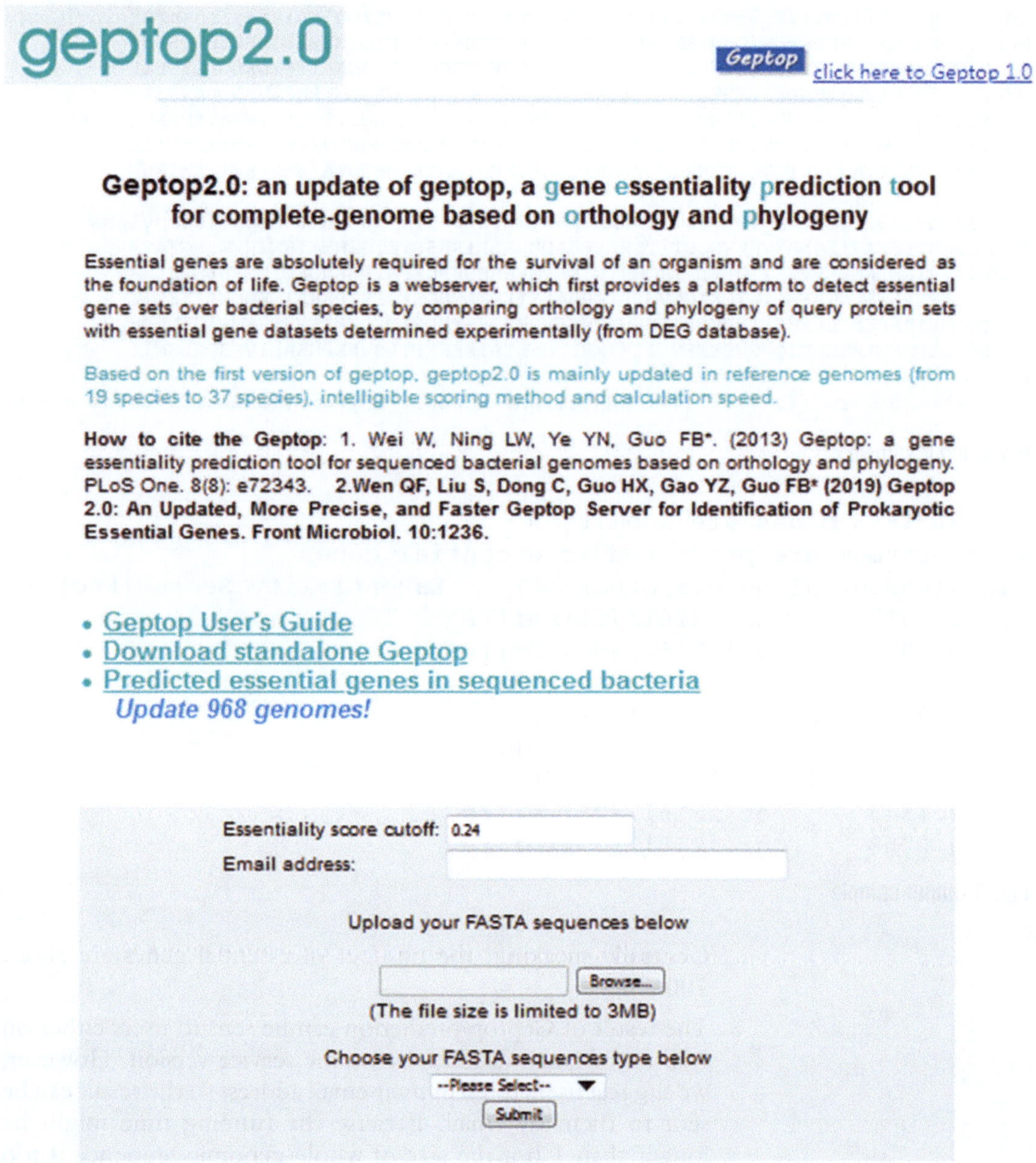

Fig. 1 The web page of Geptop

species are available; however, when standalone version runs on users' own computer, only protein sequence is acceptable unless users add a switch script that switches gene sequence to protein sequence before running Geptop.

2. We suggest our default essentiality score cutoff; however, users can change it as they want especially when the number of predicted essential genes are doubtful such as 1000 or more.

```
>gi|121612751|ref|YP_999720.1| chromosomal replication initiation protein [Campy
MNPSQILENLKKELNENEYENYLSNLKFNEKQSKADLLVFNAPNELMAKFIQTKYGKKIAHFYEVQSGNK
AIINIQAQSAKQSNKSTKIDIAHIKAQSTILNPSFTFDSFVVGDSNKYAYGACKAITHKDKLGKLYNPIF
VYGPTGLGKTHLLQAVGNASLEMGKKVIYATSENFINDFTSNLKNGSLDKFHEKYRNCDVLLIDDVQFLG
KTDKIQEEFFFIFNEIKNNDGQIIMTSDNPPNMLKGITERLKSRFAHGIIADITPPQLDTKIAIIRKKCE
FNDINLSNDIINYIATSLGDNIREIEGIIISLNAYATILGQEITLELAKSVMKDHIKEKKENITIDDILS
LVCKEFNIKPSDVKSNKKTQNIVTARRIVIYLARALTALTMPQLANYFEMKDHTAISHNVKKITEMIEND
ASLKAKIEELKNKILVKSQS
>gi|121612492|ref|YP_999721.1| DNA polymerase III subunit beta [Campylobacter je
MKLSINKNTLESAVILCNAYVEKKDSSTITSHLFFHADEDKLLIKASDYEIGINYKIKKIRVESSGFATA
NAKSIADVIKSLNNEEVVLETIDNFLFVRQKNTKYKLPMFNHEDFPNFPNTEGKNQFDIDSSDLSRSLKK
ILPSIDTNNPKYSLNGAFLDIKTDKINFVGTDTKRLAIYTLEKANNQEFSFSIPKKAIMEMQKLFYEKIE
IFYDQNMLIAKNENFEFFTKLINDKFPDYEKVIPKTFKQELSFSTEDFIDSLKKISVVTEKMRLHFNKDK
IIFEGISLDNMEAKTELEIQTGVSEEFNLTIKIKHLLDFLTSIEEEKFTLSVNEPNSAFIVKSQGLSMII
MPMIL
>gi|121612557|ref|YP 999722.1| DNA gyrase subunit B [Campylobacter jejuni subsp.
```

Fig. 2 Input sample

```
#Total 1653 genes are submitted
#501 of them are predicted as essential genes
Class(essential gene:1,others:0)    Essentiality Score  Protein
1    0.7517      gi|121612751|ref|YP_
1    0.6987      gi|121612492|ref|YP_
1    0.8323      gi|121612557|ref|YP_
0    0.0    gi|121612266|ref|YP_
0    0.0537      gi|121613203|ref|YP_
0    0.0    gi|121612732|ref|YP_
0    0.1343      gi|121613706|ref|YP_
0    0.0268      gi|121612537|ref|YP_
```

Fig. 3 Output sample

Generally speaking, the number of essential genes are about 200–400.

3. The result of Geptop prediction can be sent to users either on website or email if users use online service version. However, we suggest users to fill in their email address so the result can be sent to them by email. Because the running time might be longer than 1 h if the size of whole-genomic sequence is too large, and in this situation, getting result through email will be highly welcomed.

5 Conclusion

As a gene essentiality prediction tool, Geptop performs well for its user-friendly webpage and easy usage. The workflow of predicting essential genes is summarized in Fig. 4. We hope that this tool can help more scientists do wonderful research concerning essential genes.

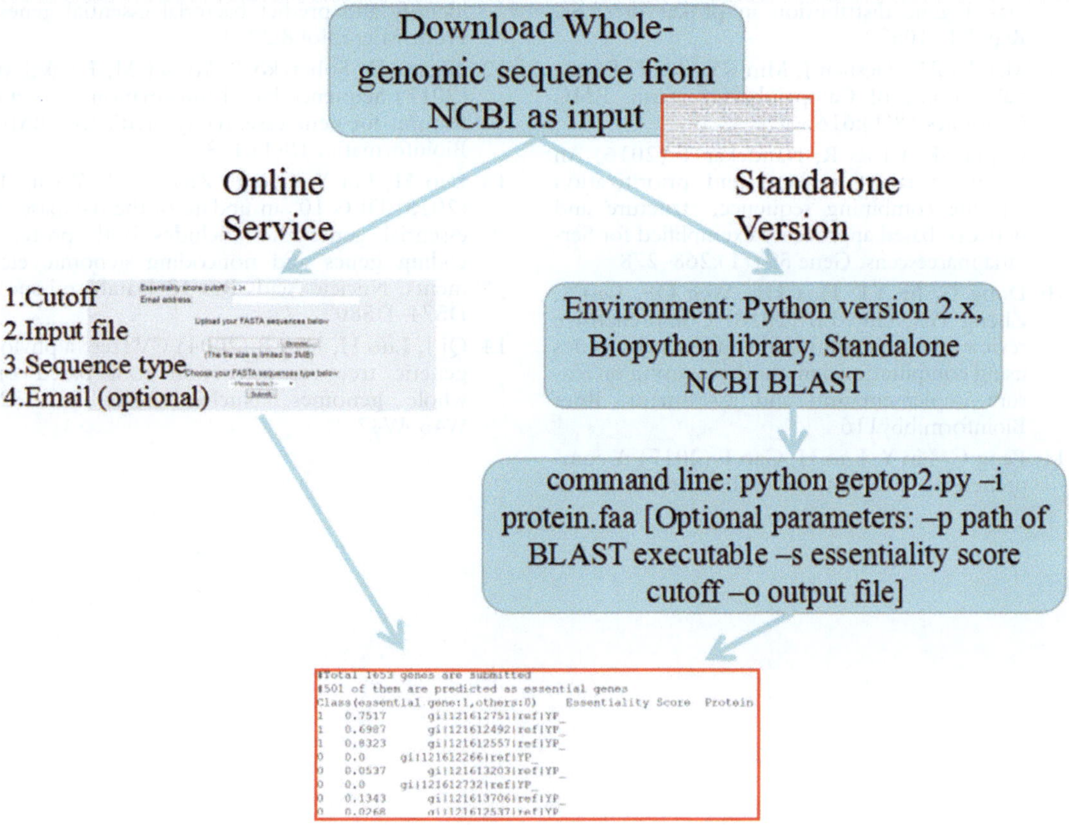

Fig. 4 The workflow of predicting essential genes

Acknowledgments

We thank the book editor for his advice and encouragement. This work was supported by national natural science foundation of China [31871335].

References

1. Juhas M, Eberl L, Church GM (2012) Essential genes as antimicrobial targets and cornerstones of synthetic biology. Trends Biotechnol 30(11):601–607
2. Glass JI, Assad-Garcia N, Alperovich N, Yooseph S, Lewis MR, Maruf M, Hutchison CA, Smith HO, Venter JC (2006) Essential genes of a minimal bacterium. Proc Natl Acad Sci U S A 103(2):425–430
3. Peters JM, Colavin A, Shi H, Czarny TL, Larson MH, Wong S, Hawkins JS, Lu CHS, Koo B-M, Marta E (2016) A comprehensive, CRISPR-based functional analysis of essential genes in bacteria. Cell 165(6):1493–1506
4. Wei W, Ning LW, Ye YN, Guo FB (2013) Geptop: a gene essentiality prediction tool for sequenced bacterial genomes based on orthology and phylogeny. PLoS One 8(8):e72343
5. Wen QF, Liu S, Dong C, Guo HX, Gao YZ, Guo FB (2019) Geptop 2.0: an updated, more precise, and faster Geptop server for identification of prokaryotic essential genes. Front Microbiol 10:1236
6. Rancati G, Moffat J, Typas A, Pavelka N (2017) Emerging and evolving concepts in gene essentiality. Nat Rev Genet 19(1):34–49
7. Gao N, Lu G, Lercher MJ, Chen W-H (2017) Selection for energy efficiency drives strand-

biased gene distribution in prokaryotes. Sci Rep 7(1):10572

8. Mandal RK, Tieshan J, Min KY (2017) Essential genome of Campylobacter jejuni. BMC Genomics 18(1):616
9. Gupta SK, Gross R, Dandekar T (2016) An antibiotic target ranking and prioritization pipeline combining sequence, structure and network-based approaches exemplified for Serratia marcescens. Gene 591(1):268–278
10. Dong C, Jin YT, Hua HL, Wen QF, Luo S, Zheng WX, Guo FB (2018) Comprehensive review of the identification of essential genes using computational methods: focusing on feature implementation and assessment. Brief Bioinform:bby116
11. Peng C, Lin Y, Luo H, Gao F (2017) A comprehensive overview of online resources to identify and predict bacterial essential genes. Front Microbiol 8:2331
12. Nigatu D, Sobetzko P, Yousef M, Henkel W (2017) Sequence-based information-theoretic features for gene essentiality prediction. BMC Bioinformatics 18(1):473
13. Luo H, Lin Y, Gao F, Zhang CT, Zhang R (2013) DEG 10, an update of the database of essential genes that includes both protein-coding genes and noncoding genomic elements. Nucleic Acids Res 42(Database issue): D574–D580
14. Qi J, Luo H, Hao B (2004) CVTree: a phylogenetic tree reconstruction tool based on whole genomes. Nucleic Acids Res 32: W45–W47

Index

Ren Zhang (ed.), *Essential Genes and Genomes: Methods and Protocols*, Methods in Molecular Biology, vol. 2377, https://doi.org/10.1007/978-1-0716-1720-5,

MIX
Papier aus verantwortungsvollen Quellen
Paper from responsible sources
FSC® C105338

If you have any concerns about our products,
you can contact us on
ProductSafety@springernature.com

In case Publisher is established outside the EU,
the EU authorized representative is:
Springer Nature Customer Service Center GmbH
Europaplatz 3, 69115 Heidelberg, Germany

Printed by Libri Plureos GmbH
in Hamburg, Germany